# Nonlinear Oligopolies

Gian Italo Bischi • Carl Chiarella
Michael Kopel • Ferenc Szidarovszky

# Nonlinear Oligopolies

## Stability and Bifurcations

Springer

Professor Gian-Italo Bischi
Dipartimento di Economia
e Metodi Quantitativi (DEMQ)
Università di Urbino "Carlo Bo"
via Saffi n.42, I-61029 Urbino
Italy
gian.bischi@uniurb.it

Professor Carl Chiarella
School of Finance and Economics
University of Technology, Sydney
P.O. Box 123
Broadway, NSW 2007
Australia
carl.chiarella@uts.edu.au

Professor Michael Kopel
University of Graz
Institute of Organization
and Economics of Institutions
University of Graz
Universitätsstraße 15/E4, A-8010 Graz
Austria
michael.kopel@uni-graz.at

Professor Ferenc Szidarovszky
Systems and Industrial Engineering Department
University of Arizona
Tucson, AZ 85721-0020
USA
szidar@sie.arizona.edu

ISBN 978-3-642-02105-3 e-ISBN 978-3-642-02106-0
DOI 10.1007/978-3-642-02106-0
Springer Heidelberg Dordrecht London New York

Library of Congress Control Number: 2009933992



*Cover design:* WMXDesign GmbH, Heidelberg, Germany

Printed on acid-free paper

Springer is part of Springer Science+Business Media (www.springer.com)

# Preface

Oligopoly theory is one of the most intensively studied areas of mathematical economics. On the basis of the pioneering works of Cournot (1838), many researchers have developed and extensively examined the different variants of oligopoly models. Initially, the existence and uniqueness of the equilibrium of the different types of oligopolies was the main concern, and later the dynamic extensions of these models became the focus. The classical result of Theocharis (1960) asserts that under discrete time scales and static expectations, the equilibrium of a single-product oligopoly without product differentiation and with linear price and cost functions is asymptotically stable if and only if it is a duopoly. In the continuous time case, asymptotic stability is guaranteed for any number of firms. In these cases the resulting dynamical systems are also linear, where local and global asymptotic stability are equivalent to each other. The classical book of Okuguchi (1976) gives a comprehensive summary of the earlier results and developments. The multiproduct extensions have been discussed in Okuguchi and Szidarovszky (1999); however, nonlinear features were barely touched upon in these contributions.

With the development of the critical curve method by Gumowski and Mira (1980) (see also Mira et al. (1996)) for discrete time systems and the introduction of continuously distributed information lags by Invernizzi and Medio (1991) in continuous time systems, increasing attention has been given to the global dynamics of nonlinear oligopolies. The authors of this book have devoted a great deal of research effort to this area. Their cooperation has resulted in several joint conference presentations and a large number of journal publications. The development of the theory of nonlinear dynamic oligopolies has now reached a stage where the authors feel it has become necessary and worthwhile to collect and summarize the most important results in a book form.

This book may be regarded as a continuation of the work of Okuguchi and Szidarovszky (1999) and is focused mainly on the nonlinearity of oligopoly models. It consists of six chapters and a sequence of appendices. Chapter 1 introduces and discusses the classical Cournot model with a large variety of demand and cost functions. With these examples we try to illustrate a large collection of different types of best response functions, as well as show the existence of unique and multiple equilibria. Dynamic processes are introduced in the second part of the chapter, where we discuss static and adaptive expectations, partial adjustment towards the best response, and gradient adjustments. An introduction to the analysis of global dynamics is given through specific examples of duopolies and symmetric and semi-symmetric oligopolies. Chapter 2 is devoted to concave oligopolies. The existence and uniqueness of the equilibrium is proved in general, and conditions are given for the local asymptotic stability of the equilibrium. In the discrete time case, it is required that the speeds of adjustment be sufficiently small for all firms, and in the continuous time case local stability is always guaranteed regardless of the values of the speeds of adjustments. Global dynamics are investigated in the cases of symmetric and semi-symmetric firms. The relation between gradient adjustment and partial adjustment towards the best response is then briefly outlined. The global dynamics of continuous time models are illustrated by assuming continuously distributed time lags, and we show that time lags may destroy stability. At the critical values of model parameters at which stability is lost, a Hopf bifurcation occurs giving rise to the possibility of the birth of limit cycles. General oligopolies are discussed in Chapter 3. Oligopolies with isoelastic price functions are first considered, conditions are given for the local asymptotic stability of the equilibrium in both the discrete and continuous time cases, and global dynamics are examined in the case of discrete time models. We assume next that the cost function of each firm depends on the output of the rest of the industry in addition to its own output level. A special case of this model results in parabolic best response functions, and we show the complexity of the global dynamics that can occur in such cases. Modified and extended oligopoly models are introduced and examined in Chapter 4, including market share attraction games, labor-managed oligopolies, models with intertemporal demand attraction, and the effects of production adjustment costs. The last section of this chapter is devoted to the case of partially cooperating firms in which the payoff function of each firm includes its own profit and a share of the profits of its competitors. Local and global stability analyses are carried out for these models. Chapter 5 considers three issues related to the firms' uncertain knowledge of the demand function. In the first part it is assumed that the firms misspecify the price function, and the dynamics of the model depend on the way the firms estimate the price function and also on the adjustment process the firms select. The steady states of these resulting dynamical systems usually differ from the full-information equilibria, and so "subjective" equilibria occur. In the second part of this chapter, special adaptive learning processes are introduced, where the firms adaptively learn (update) the unknown parameter of the price function. For all of the models, both the local and global dynamics are studied. In the third part, it is assumed that the price function is estimated by the firms with random errors. Each firm faces a multiobjective problem by maximizing the expected profit and minimizing the variance of the profit. If the weighting method is used to transform the problem to a single-objective optimization problem, then the resulting model is equivalent to that of oligopolies with misspecified price functions. Finally, Chapter 6 gives a brief overview of the very large and complex field covered in this book, as well as an outline of future research directions.

Five appendices are included in the book. Appendix A presents the fundamentals of Lyapunov stability theory, while Appendix B presents conditions for local and global asymptotic stability by using the linearization procedure. Appendix C introduces the main concepts of noninvertible maps and critical curves, which serve as the theoretical basis for the analysis of the global dynamics of discrete time models. Appendix D introduces the mathematical tools needed to examine continuous time models with continuously distributed time lags. Appendix E demonstrates a special determinantal identity that is very helpful in computing the characteristic polynomials of matrices with a particular structure. Finally, Appendix F gives sufficient and necessary conditions for the asymptotic stability of two-dimensional systems based on their quadratic characteristic polynomials.

As can be seen from the foregoing description, the authors have tried to give a comprehensive review of the different model variants and the mathematical methodology used in analyzing nonlinear oligopolies. A large collection of references are cited and listed in the bibliography; however, the authors have not attempted to give a complete collection of all the important works in this area. The interested reader should consult Kopel (2009), which gives an up-to-date survey of oligopoly dynamics and cites all of the important references.

The authors sincerely hope that this book will help graduate students, researchers in mathematical economics, economists, and applied mathematicians to understand the central issues and major methodologies of this fascinating and exciting field. Hopefully, the book will inspire them to become interested in initiating, or continuing, their own research agenda in this area.

The authors have benefitted from discussions with many of their research collaborators who are too numerous to list here, but special thanks are particularly due to Laura Gardini and Iryna Sushko. The authors also thank Stephanie Ji-Won Ough for the tremendous job she has done in turning the various drafts into an excellent manuscript and for coping with a great deal of deadline pressure. Finally, acknowledgement should be made to the institutions of the authors: the University of Urbino; the University of Technology, Sydney; The University of Graz; the University of Arizona – for providing financial support for this project.

Urbino — Gian Italo Bischi
Sydney — Carl Chiarella
Graz — Michael Kopel
Tucson — Ferenc Szidarovszky
May, 2009

# Contents

# Notation

| | |
|---|---|
| $A, B$ | Coefficients of specific forms of the price (inverse demand) function |
| $C_k$ | Cost function of firm $k$ |
| $C_2$ | Cycle of period two, or 2-cycle |
| $C_4$ | Cycle of period four, or 4-cycle |
| $CS$ | Critical set of a continuous map |
| $CS_{-1}$ | Set of rank-1 preimages of $CS$ |
| $CS_k$ | Set of rank-$k$ images of $CS$ |
| $Det$ | Determinant of a matrix |
| $E, E_k, E^*$ | Equilibrium points |
| $\boldsymbol{H}, \bar{\boldsymbol{H}}$ | Special Jacobian matrices |
| $\boldsymbol{I}$ | Identity matrix |
| $\boldsymbol{J}$ | Jacobian matrix |
| $K_k$ | Output adjustment cost function |
| $L$ | Identical capacity limit of the firms |
| $L_k$ | Capacity limit of firm $k$ |
| $LC$ | Critical curve |
| $LC_{-1}$ | Rank-1 preimage of LC |
| $LC_k$ | Set of rank-$k$ images of $LC$ |
| $N$ | Number of firms |
| $O$ | Origin |
| $O_{-1}^{(k)}$ | The $k$th Rank-1 preimage of the origin |
| $\boldsymbol{P}$ | Transformation matrix |
| $Q$ | Output of the entire industry |
| $\bar{Q}$ | Equilibrium output of the entire industry |
| $Q_k$ | Aggregated output of all firms except firm $k$ |
| $Q_k^E$ | Expectation of $Q_k$ |
| $Q_k^{E,prior}$ | Expectation of $Q_k$ at the beginning of the time period |
| $Q_k^{E,post}$ | Expectation of $Q_k$ at the end of the time period |
| $\tilde{Q}_k$ | Belief about $Q_k$ by firm $k$ |
| $R_k$ | Best response of firm $k$ |

| | |
|---|---|
| $\widetilde{R}_k$ | Best response of firm $k$ as a function of $Q$ (in Chapters 2-4) |
| $\widetilde{R}_k$ | Believed best response of firm $k$ (in Chapter 5) |
| $S(t)$ | Cumulative effect of past consumption |
| $T$ | Map |
| $T^{(k)}$ | Restriction of the map $T$ to region $\mathbb{D}^{(k)}$ |
| $T^{-n}(x)$ | Rank-$n$ preimages of $x$ |
| $Tr$ | Trace of a matrix |
| $T^t$ | $t^{\text{th}}$ iteration of map $T$ |
| $U(A)$ | Neighborhood of $A$ |
| $V$ | Lyapunov function |
| $W$ | Competitive wage rate |
| $Z_k$ | Region of points having $k$ distinct rank-1 preimages |
| $\mathcal{B}(A)$ | Basin of attractor $A$ |
| $\mathcal{B}(E)$ | Basin of equilibrium $E$ |
| $\mathcal{B}_0(A)$ | Immediate basin of $A$ |
| $\mathbb{D}$ | Phase space |
| $\mathbb{D}^{(k)}$ | Segments of the phase space |
| $\mathbb{R}^2$ | 2-dimensional vector space |
| $\mathbb{R}^2_+$ | Non-negative quadrant of $\mathbb{R}^2$ |
| $a$ | Common speed of adjustment of the firms |
| $a_k$ | Constant speed of adjustment of firm $k$, or $\alpha'_k(0)$ in dynamic models |
| $c_k, d_k, e_k$ | Cost function coefficients |
| $f$ | Price (or inverse demand) function |
| $\widetilde{f}_k$ | Believed price function by firm $k$ |
| $h_k$ | Number of labor units in firm $k$ |
| $p$ | Price |
| $\bar{p}$ | Equilibrium price |
| $\widetilde{p}_k$ | Price believed by firm $k$ |
| $r_k$ | Derivative of the best response of firm $k$ at the equilibrium |
| $r_{kx}, r_{kQ}$ | Partial derivatives of the best response of firm $k$ at the equilibrium |
| $w(t-s, T, m)$ | Weighting function |
| $x_k$ | Output of firm $k$ |
| $\bar{x}_k$ | Equilibrium output of firm $k$ |
| $x_k^L$ | Limiting quantity |
| $x_k^M$ | Monopoly quantity |
| $\widetilde{x}_l$ | Output of firm $l$ believed by other firms |
| $z_k^*$ | Solution of best response equations |
| $\Delta p_k$ | Discrepancy between actual and believed prices by firm $k$ |
| $\alpha_k$ | Sign-preserving adjustment function for firm $k$ |
| $\beta_S$ | Coefficient of $S(t)$ in discrete time models |
| $\gamma_S$ | Coefficient of $S(t)$ in continuous time models |

| | |
|---|---|
| $\gamma_{kl}$ | Cooperation level of firm $k$ towards firm $l$ |
| $\varepsilon_k$ | Learning parameter in learning models |
| $\varphi_k$ | Payoff (profit) function of firm $k$ |
| $\bar{\varphi}_k$ | Payoff (profit) of firm $k$ at the equilibrium |
| $\widetilde{\varphi}_k$ | Believed payoff function of firm $k$ |
| $\lambda$ | Eigenvalue (or multiplier) |
| $\lambda(C)$ | Eigenvalue (or multiplier) of cycle $C$ |
| $\tau(\boldsymbol{x}_0)$ | Trajectory starting at initial point $\boldsymbol{x}_0$ |
| $\psi_k$ | Payoff function of partially cooperating firm $k$ |
| $\partial B$ | Boundary of region $B$ |
| $\|.\|$ | Norm of a vector or a matrix |

# Chapter 1
# The Classical Cournot Model

In this chapter we will introduce the classical Cournot model, which is also known as the single-product quantity setting oligopoly model without product differentiation. In the first section of the chapter the Cournot model will be discussed as an $N$-firm static game and the best responses of the firms and the equilibria will be determined in a series of examples, many of which will be built upon in developing the ideas in subsequent chapters. Section 1.2 introduces the dynamic adjustment processes via which we shall assume that firms adjust output over time. We will in particular discuss expectation formation processes and adaptive adjustments and gradient adjustments. The final section will illustrate by simple examples the complexity of the dynamics that can arise in these models due to certain nonlinear features to be described below. The fundamental techniques for the global analysis of the dynamics of such models will be explained in Sect. 1.3.

## 1.1 Introduction

The basic model can be described as follows. Consider an industry of $N$ firms producing a homogeneous product. Let $k = 1, 2, \dots, N$ denote the firms and let $x_k$ be the output quantity of firm $k$. We assume that the inverse demand (or price) function depends on the total output level of the industry, so the market price may be written $p = f\left(\sum_{k=1}^{N} x_k\right)$. The particular form of the function $f$ can be derived from microeconomic principles (see for example, Vives (1999)), and several function types are discussed in the literature.

An important example of an inverse demand function which is linear is obtained by assuming that the utility function of a typical consumer is quadratic,

$$U(q) = aq - \frac{1}{2}bq^2, \qquad (a, b > 0),$$

where $q$ is the quantity of the good purchased by the consumer. If we denote the market price of the good by $p$, then for a sufficiently large income the consumer

G.I. Bischi et al., *Nonlinear Oligopolies*, DOI 10.1007/978-3-642-02106-0_1, 

solves the optimization problem

$$\max(U(q) - pq).$$

Assuming an interior optimum, the first order condition implies that

$$0 = U'(q) - p = a - bq - p,$$

so that the individual demand at the price $p$ is therefore

$$q(p) = \frac{a}{b} - \frac{1}{b}p.$$

Consider now $n$ heterogenous consumers with quadratic utility and preference parameters $a_i$ and $b_i$. From the previous description we know that for any fixed price consumer $i$ will buy the amount $q_i = (a_i - p)/b_i$, so the total demand becomes

$$D = \sum_{i=1}^{n} q_i = \sum_{i=1}^{n} \frac{a_i}{b_i} - \sum_{i=1}^{n} \frac{1}{b_i} p,$$

and hence the relationship between total demand and market price is linear. Notice that if price increases, demand decreases and that there is a maximum price, usually referred to as the *reservation price*, above which demand reduces to zero. If we denote by $Q = \sum_{k=1}^{N} x_k$ the quantity supplied by the $N$ firms in the industry and we assume that at the price $p$ the market clears, that is $D = Q$, then it also follows that the relation between industry output and price is linear. Hence, by inverting this relationship we finally obtain

$$p = f(Q) = A - BQ,$$

where

$$A = \sum_{i=1}^{n} \frac{a_i}{b_i} \Big/ \sum_{i=1}^{n} \frac{1}{b_i}, \quad B = 1 \Big/ \sum_{i=1}^{n} \frac{1}{b_i}.$$

Obviously, this representation is only valid for $Q \leq A/B$, that is as long as the industry output is below the *market saturation* point. Otherwise, we have $p = 0$.

In the case of a general inverse demand function the profit of firm $k$ $(1 \leq k \leq N)$ is the difference between its revenue and its cost and so is given by

$$\varphi_k(x_1, \ldots, x_N) = x_k f\left(\sum_{l=1}^{N} x_l\right) - C_k(x_1, \ldots, x_N), \tag{1.1}$$

where $C_k$ is the cost function of firm $k$.[1] Our formulation takes into account the fact that the cost of each firm depends not only on its own output but also on the outputs

[1] In the game theory context the profit functions are usually called the payoff functions, and the firms are called the players. We will occasionally make use of these terms throughout this book.

of the competitors. The firms have to compete in the secondary market to ensure capital, manpower, energy, material, etc. for their production processes. The technological and intellectual spillover between companies is another cost externality which adds to the interdependence of the firms. In the literature on oligopoly theory the interdependence of the firms through their cost functions is either ignored by assuming that the cost of firm $k$ is $C_k(x_k)$, or it is assumed that the cost of firm $k$ depends on its own production level $x_k$ and also on the total production level of the rest of the industry, which we will denote by $Q_k = \sum_{l \neq k} x_l$ so that the cost function of firm $k$ may be written more generally as $C_k(x_k, Q_k)$. In the rest of the book we will consider various cases where cost externalities arise. Note that under this assumption the profit of any firm $k$ just depends on its own output and the output of the rest of the industry, it does not depend on the individual output level of any competitor. For this reason it is convenient to rewrite the profit function of firm $k$ as

$$\varphi_k(x_1, \ldots, x_N) = x_k f(x_k + Q_k) - C_k(x_k, Q_k). \tag{1.2}$$

Taken together, the above set-up yields a static $N$-person game, where the players are the firms, the strategy set of firm $k$ is the interval $[0, L_k]$, where $L_k$ is the capacity limit of firm $k$ and its payoff function is given by (1.2). If we assume that all firms are rational in the sense that they want to maximize their own profits, then we can derive the firms' best responses. That is, if firm $k$ knows the total production $Q_k$ of the rest of the industry, then it will select a production level $x_k$ that maximizes its profit (1.2). For each value of $Q_k$ let $R_k(Q_k)$ denote the set of all optimal solutions, that is

$$R_k(Q_k) = \left\{ x_k \mid x_k = \arg \max_{0 \leq x_k \leq L_k} \{x_k f(x_k + Q_k) - C_k(x_k, Q_k)\} \right\}, \tag{1.3}$$

which is called the *best response* or *best reply mapping* of firm $k$. In the general case this is a point-to-set mapping, and in this case it is usually called the *best reply correspondence*. In the case of a unique optimal solution, $R_k(Q_k)$ is called the *best reply* or *reaction function* of firm $k$. The *Nash equilibrium* of the game is a simultaneous production vector $(\bar{x}_1, \ldots, \bar{x}_N)$ which is a best response for each firm, under the assumption that all others maintain their corresponding equilibrium production levels. This concept can be mathematically expressed for all $k$ as,

$$\bar{x}_k \in R_k(\bar{Q}_k) \text{ with } \bar{Q}_k = \sum_{l \neq k} \bar{x}_l. \tag{1.4}$$

At the equilibrium all firms simultaneously select their best responses to the corresponding equilibrium choices of the competitors. In other words, no firm has any interest to deviate unilaterally from its equilibrium level.

In the following examples we will show that best responses might have a large variety of forms, and also, that oligopolies may have no equilibrium at all. Furthermore, in the case of existence there may be multiple equilibria, and the number of equilibria may be finite or infinite. In the case of multiple equilibria, the problem of equilibrium selection arises. In such situations, the non-negativity of the profits and the dynamic evolution of the oligopoly game, determined by the adjustment processes and the degree of bounded rationality of the players, can be used to determine which equilibria are realistic and which are not. We will return to this problem in later chapters.

*Example 1.1.* Consider the case of a linear oligopoly where the price function has the form $f(Q) = \max\{0, A - BQ\}$ with $Q = \sum_{k=1}^{N} x_k$ and $C_k(x_k) = d_k + c_k x_k$ $(1 \leq k \leq N)$ with $A$, $B$, $c_k$, $d_k$ being all positive. Note that the max operation ensures that the price is zero for total output above the market saturation point $A/B$. In this case $\varphi_k$ is strictly concave in $x_k$ with derivative

$$\frac{\partial \varphi_k}{\partial x_k} = \begin{cases} A - BQ_k - 2Bx_k - c_k & \text{if} \quad Q_k + x_k < \frac{A}{B}, \\ -c_k & \text{if} \quad Q_k + x_k > \frac{A}{B}, \end{cases}$$

and this derivative does not exist if $Q_k + x_k = A/B$.

If for any firm $k$ it is the case that $A - c_k \leq 0$, then $\partial \varphi_k / \partial x_k$ is always negative, so the best response of this firm is always zero, and hence entry for this firm is blocked. Hence such firms do not participate in production, and therefore we can ignore them in all further discussions. If for firm $k$, the capacity limit $L_k$ is sufficiently large, then with $A > c_k$, its monopoly quantity is $x_k^M = (A - c_k)/(2B)$, which can be obtained from the first order condition with $Q_k = 0$.

In order to determine the best response of the firms, consider firm $k$ and assume that the total production level $Q_k$ of the rest of the industry is fixed. Notice first that the best response of this firm cannot exceed $A/B - Q_k$, that is, the total industry output cannot be larger than the market saturation point. In contrast, assume that $x_k > A/B - Q_k$, then the price is zero, and by decreasing the value of $x_k$ by a small amount, the price will be still zero and the cost decreases. So the payoff of this firm would increase contradicting the assumption that $x_k$ is the firm's best response. Therefore with fixed values of $Q_k$ the best response of firm $k$ is selected in the interval $[0, \bar{L}_k]$ with $\bar{L}_k = \min\{L_k, A/B - Q_k\}$. If the capacity limits of the firms are sufficiently small, that is, when $\sum_{k=1}^{N} L_k \leq A/B$, then the zero segment of the price function cannot occur, so $\bar{L}_k = L_k$ for all $k$ and $Q_k$. For the sake of simplicity in the following discussion we will assume that this is the case. Since $\varphi_k$ is strictly concave in $x_k$, the best response of firm $k$ is unique and is given as

$$R_k(Q_k) = \begin{cases} 0 & \text{if} \quad \frac{\partial \varphi_k}{\partial x_k} \Big|_{x_k = 0} \leq 0, \\ L_k & \text{if} \quad \frac{\partial \varphi_k}{\partial x_k} \Big|_{x_k = L_k} \geq 0, \\ z_k^* & \text{otherwise,} \end{cases}$$

where $z_k^*$ is the solution of

$$\frac{\partial \varphi_k}{\partial x_k} = 0,$$

implying in the present case that

$$z_k^* = -\frac{1}{2}Q_k + \frac{A - c_k}{2B}. \tag{1.5}$$

Straightforward calculations reveal that

$$R_k(Q_k) = \begin{cases} 0 & \text{if} \quad Q_k \geq (A - c_k)/B, \\ L_k & \text{if} \quad Q_k \leq (A - c_k - 2BL_k)/B, \\ -\frac{1}{2}Q_k + (A - c_k)/(2B) & \text{otherwise.} \end{cases} \tag{1.6}$$

In the case of two firms, when $Q_1 = x_2$ and $Q_2 = x_1$, we can illustrate graphically the existence of a unique equilibrium. Figure 1.1 shows the best response functions of the two firms in the situation where $L_1 < x_1^M$ and $L_2 < x_2^M$. If $L_1 \geq x_1^M$, then the vertical segment of $R_1(x_2)$ disappears and we simply have $R_1(0) = x_1^M$. A similar situation occurs when $L_2 \geq x_2^M$. The best replies intersect at a unique point, which is the Nash equilibrium. It can also be proved that with an arbitrary value of $N$, the oligopoly always has a unique equilibrium (see for example Sect. 2.1, and Okuguchi and Szidarovszky (1999)). If the market saturation point and the capacity limits are sufficiently large, then we can even compute the unique equilibrium.

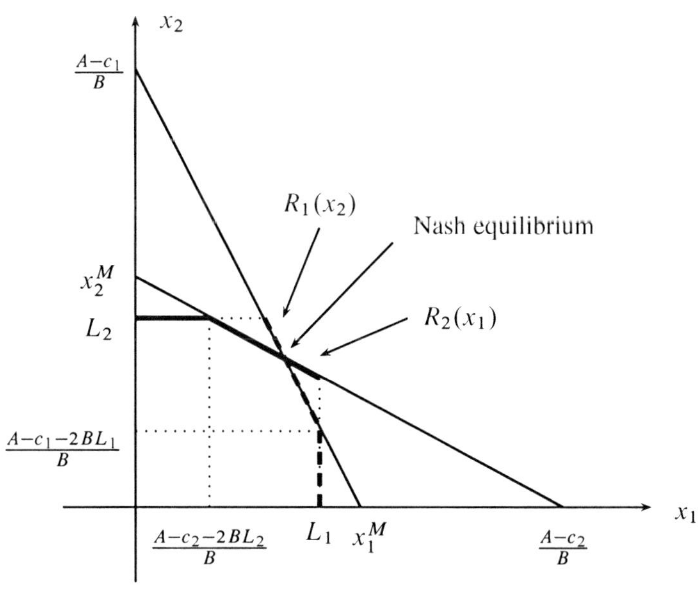


**Fig. 1.1** Example 1.1; the Cournot model in the case of duopoly ($N = 2$) with linear price and cost functions. The figure shows the reaction functions $R_1(x_2)$ (*dashed line*), $R_2(x_1)$ (*solid line*) and the unique equilibrium

Assume that all equilibrium outputs are positive, the other case can be examined similarly. The first order conditions imply that

$$\begin{aligned}\frac{\partial\varphi_k}{\partial x_k} &= \frac{\partial}{\partial x_k}[x_k(A - Bx_k - BQ_k) - (d_k + c_k x_k)] \\ &= A - 2Bx_k - BQ_k - c_k \\ &= A - Bx_k - BQ - c_k = 0,\end{aligned} \tag{1.7}$$

where $Q$ is the total output of the industry. So

$$x_k = \frac{A - BQ - c_k}{B} = \frac{A - c_k}{B} - Q. \tag{1.8}$$

By summing this last equation over all firms we obtain for $Q$ the single equation

$$Q = \frac{NA - \sum_{i=1}^N c_i}{B} - NQ, \tag{1.9}$$

implying that at the equilibrium

$$\bar{Q} = \frac{NA - \sum_{i=1}^N c_i}{(N+1)B}. \tag{1.10}$$

Notice that $\bar{Q} < A/B$, so the price is always positive. From (1.8) and (1.10) we can compute the equilibrium output levels of the firms as

$$\bar{x}_k = \frac{A - c_k}{B} - \frac{NA - \sum_{i=1}^N c_i}{(N+1)B} = \frac{A - (N+1)c_k + \sum_{i=1}^N c_i}{(N+1)B}. \tag{1.11}$$

The output levels in (1.11) can be an equilibrium only if they are all non-negative and below the corresponding capacity limits. The equilibrium price is then

$$\bar{p} = A - B\bar{Q} = \frac{A + \sum_{i=1}^N c_i}{N+1}.$$

At the equilibrium, the profit of firm $k$ is given by

$$\begin{aligned}\bar{\varphi}_k = \bar{x}_k\bar{p} - (d_k + c_k\bar{x}_k) &= \bar{x}_k\left(\frac{A + \sum_{i=1}^N c_i}{N+1} - c_k\right) - d_k \\ &= \frac{1}{(N+1)^2 B}\left(A - (N+1)c_k + \sum_{i=1}^N c_i\right)^2 - d_k.\end{aligned}$$

Notice that with zero fixed cost the equilibrium profit of firm $k$ is non-negative, and if $\bar{x}_k > 0$ and $d_k$ is sufficiently small, then $\bar{\varphi}_k$ is necessarily positive. If capacity limits are present and this "unconditional" equilibrium becomes infeasible, then the "conditional" equilibrium can still be computed, but cannot be represented by simple equations. Okuguchi and Szidarovszky (1999) discuss algorithms to compute such equilibria. ▼

If nonlinearity (which was in the form of capacity constraints in the above example) is introduced into the models, then usually numerical methods are required to compute the equilibrium in the general case. Analytical methods are available in only very special cases, for example by assuming symmetric or semi-symmetric firms. If all firms have identical capacity limits and cost functions, and their initial outputs are also the same, then the oligopoly is called *symmetric*. If $(N-1)$ firms are identical in this sense and one firm is different, then we have a *semi-symmetric* case. We will frequently make use of such special cases in later chapters.

*Example 1.2.* Assume again a linear price function $f(Q) = \max\{0, A - BQ\}$ but quadratic cost functions $C_k(x_k) = c_k x_k + e_k x_k^2$. The profit of firm $k$ now has the form

$$\varphi_k(x_1, \ldots, x_N) = \begin{cases} x_k(A - Bx_k - BQ_k) - (c_k x_k + e_k x_k^2) & \text{if} \quad x_k + Q_k \leq \frac{A}{B}, \\ -(c_k x_k + e_k x_k^2) & \text{otherwise.} \end{cases}$$

For the sake of simplicity we assume again that $\sum_{k=1}^{N} L_k \leq A/B$, that is, the zero segment of the price function cannot occur.

(i) Assume first that for all $k$, $0 < e_k$. Then the cost function is convex, so that marginal costs are increasing in $x_k$, and the profit is concave in $x_k$. Since

$$\frac{\partial \varphi_k}{\partial x_k} = A - 2Bx_k - BQ_k - c_k - 2e_k x_k,$$

the best response is unique and has the form

$$R_k(Q_k) = \begin{cases} 0 & \text{if} \quad A - BQ_k - c_k \leq 0, \\ L_k & \text{if} \quad A - 2BL_k - BQ_k - c_k - 2e_k L_k \geq 0, \\ (A - BQ_k - c_k)/(2(B + e_k)) & \text{otherwise,} \end{cases}$$

which is piece-wise linear, similar to the case of the previous example where both demand and cost were linear. Notice that if $A \leq c_k$, then $R_k(Q_k) = 0$ regardless of the value of $Q_k$, so we assume that $A > c_k$ for all firms. In the case of duopoly the $x_1$ intercept of $R_1(x_2)$ is the monopoly output $x_1^M$ of firm 1, and the $x_2$ intercept of $R_2(x_1)$ is the monopoly output $x_2^M$ of firm 2. It can be proved (see Chap. 2) that there is always a unique Nash equilibrium in this case.

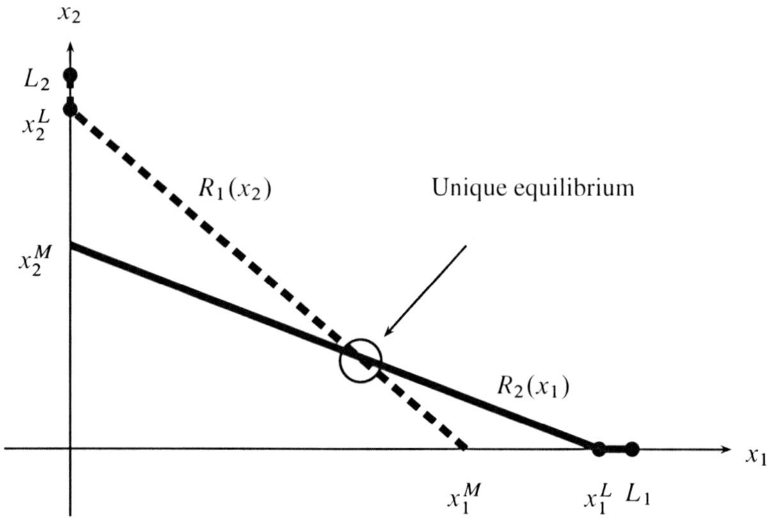


**Fig. 1.2** Example 1.2; the Cournot model with linear price function and quadratic cost function in the case of duopoly ($N = 2$). The reaction functions $R_1(x_2)$, $R_2(x_1)$ and the unique equilibrium. The figure illustrates case (ii) when $B^2 < 4(B + e_1)(B + e_2)$ and $x_k^L > x_k^M$, $k = 1, 2$

(ii) Assume next that for all $k$, $-B < e_k < 0$, then the cost function is concave, however $\varphi_k$ remains concave in $x_k$, so the best response remains the same as above. However, this case raises the possibility of multiple equilibria. Consider a duopoly ($N = 2$). Figure 1.2 depicts the reaction functions in the case where

$$B^2 < 4(B + e_1)(B + e_2),$$

that is when marginal costs are decreasing but not too strongly.[2] Furthermore, the "limit quantities" $x_k^L = (A - c_k)/B$, that is the corresponding quantity levels which guarantee that the other firm is kept out of the market, are larger than the monopoly quantities $x_k^M = (A - c_k)/(2(B + e_k))$. Under these conditions there is still a unique interior equilibrium given by

$$\begin{aligned} E &= (\bar{x}_1, \bar{x}_2) \\ &= \left( \frac{2(B + e_2)(A - c_1) - B(A - c_2)}{4(B + e_1)(B + e_2) - B^2}, \right. \\ &\qquad \left. \frac{2(B + e_1)(A - c_2) - B(A - c_1)}{4(B + e_1)(B + e_2) - B^2} \right) \end{aligned} \tag{1.12}$$

[2] This interpretation is based on the fact that the condition is satisfied if $-e_k$ ($k = 1, 2$) does not get too close to $B$.

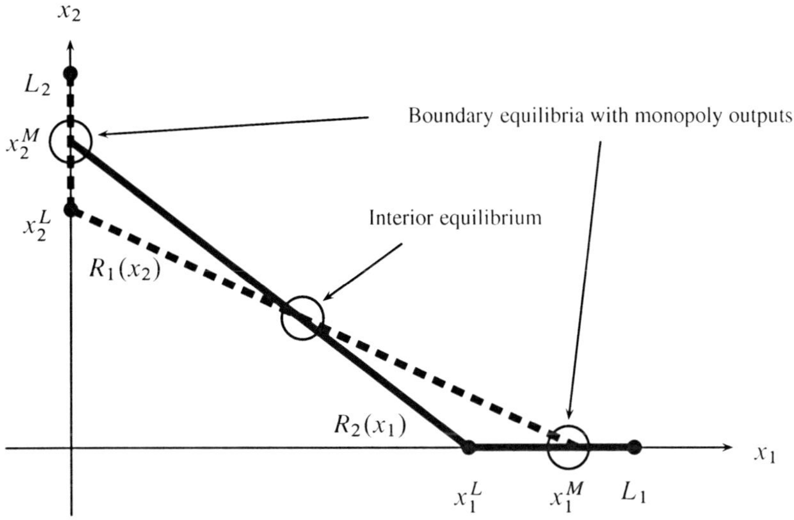


**Fig. 1.3** Example 1.2; the Cournot model with linear price function and quadratic cost function in the case of duopoly ($N = 2$). The figure shows case (ii) when $B^2 > 4(B + e_1)(B + e_2)$ and $x_k^L < x_k^M$, $k = 1, 2$. Three equilibria occur in this case

and the equilibrium profits are

$$\bar{\varphi}_k = (B + e_k)(\bar{x}_k)^2, \qquad k = 1, 2.$$

If in contrast

$$B^2 > 4(B + e_1)(B + e_2),$$

so that marginal costs are decreasing strongly, then the uniqueness of the equilibrium is no longer guaranteed. For example, Fig. 1.3 shows a case where

$$x_k^L = \frac{A - c_k}{B} < \frac{A - c_k}{2(B + e_k)} = x_k^M,$$

so that there is an interior equilibrium and there are also two boundary equilibria given by

$$E_1 = \left(\frac{A - c_1}{2(B + e_1)}, 0\right) \quad \text{and} \quad E_2 = \left(0, \frac{A - c_2}{2(B + e_2)}\right),$$

where we assume again that $A > c_k$ for both firms. Observe in addition, that $E_k$ includes the monopoly output for firm $k$ ($k = 1, 2$). At the boundary equilibrium $E_k$, the profit of firm $k$ is

$$(A - c_k)^2/(4(B + e_k)) > 0.$$

In the borderline case, when

$$B^2 = 4(B + e_1)(B + e_2),$$

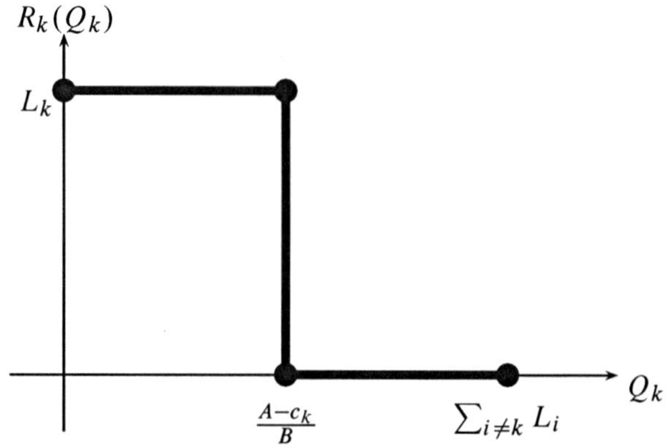


**Fig. 1.4** Example 1.2; the Cournot model with linear price function and quadratic cost function in the case of duopoly ($N = 2$). The figure shows the reaction function of a typical firm in case (iii) when $e_k = -B$. The number of equilibria may be $1, 3$ or infinite

the two straight lines either coincide or are parallel. Therefore there are either infinitely many equilibria, or a unique boundary equilibrium.

(iii) In the case where $e_k = -B$ for all $k$, the profit function assumes the linear form

$$\varphi_k = x_k(A - BQ_k - c_k),$$

therefore

$$R_k(Q_k) = \begin{cases} 0 & \text{if } A - BQ_k - c_k < 0, \\ L_k & \text{if } A - BQ_k - c_k > 0, \\ \text{arbitrary } x_k & \text{if } A - BQ_k - c_k = 0. \end{cases}$$

We can assume again that $c_k < A$, otherwise $R_k(Q_k) = 0$ for all $Q_k$. This best response function is illustrated in Fig. 1.4 in the case when

$$\frac{A - c_k}{B} < \sum_{i \neq k} L_i.$$

In the case when the last inequality becomes an equality, the vertical segment moves to $Q_k = \sum_{i \neq k} L_i$. If however the above relation is violated with strict inequality, then $R_k(Q_k) = L_k$ for all $Q_k$. Depending on the values of $(A - c_k)/B$ and $L_k$, in the duopoly case the number of equilibria can be 1, 3 or infinite; Fig. 1.5 shows a case where three equilibria exist.

(iv) Assume finally that for all $k$, $e_k < -B$. In this case $\varphi_k$ is convex in $x_k$, so the best response is located at an endpoint of the feasible interval $[0, L_k]$ and is of the form

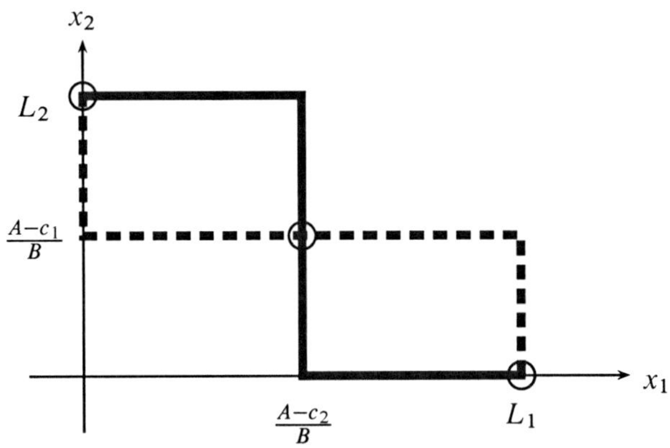


**Fig. 1.5** Example 1.2; the Cournot model with linear price function and quadratic cost function in the case of duopoly ($N = 2$). The figure shows case (iii) when $e_k = -B$, and there exist three equilibria

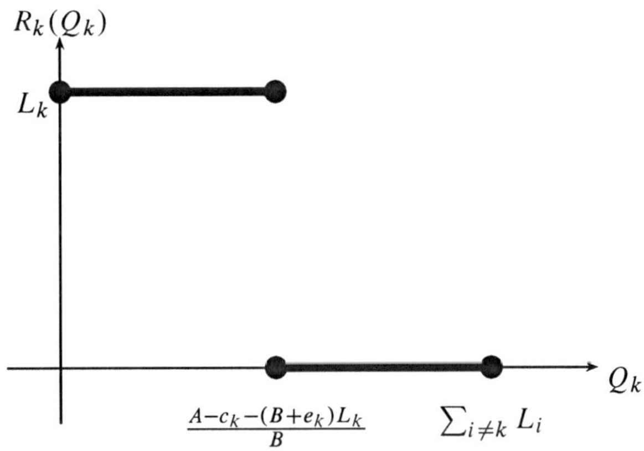


**Fig. 1.6** Example 1.2; the Cournot model with linear price function and quadratic cost function. The figure shows case (iv) when $e_k < -B$. The best response of the typical firm is determined by the fact that the profit function is linear in this case

$$R_k(Q_k) = \begin{cases} L_k & \text{if} \quad L_k(A - BL_k - BQ_k) - (c_k L_k + e_k L_k^2) > 0, \\ 0 & \text{if} \quad L_k(A - BL_k - BQ_k) - (c_k L_k + e_k L_k^2) < 0, \\ \{0; L_k\} & \text{if} \quad L_k(A - BL_k - BQ_k) - (c_k L_k + e_k L_k^2) = 0. \end{cases}$$

This function is illustrated in Fig. 1.6 in the case when

$$0 < (A - c_k - (B + e_k)L_k)/B < \sum_{i\neq k} L_i.$$

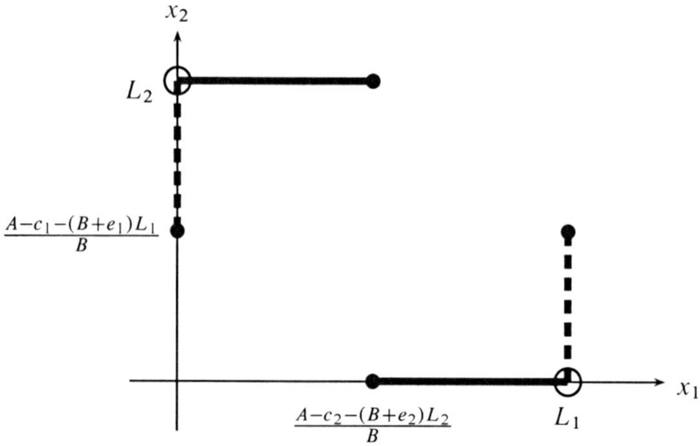


**Fig. 1.7** Example 1.2; the Cournot model with linear price function and quadratic cost function in the case of duopoly ($N = 2$). The figure shows case (iv) when $e_k < -B$ and the existence of two equilibria with convex profit functions

In the duopoly case ($N = 2$) the Nash equilibrium is at the intersection of the two best response functions. The number of equilibria can be 1, 2 or 3 depending on the relative order of magnitude of the values $(A - c_k - (B + e_k)L_k)/B$ and $L_l$ $(l \neq k)$. In Fig. 1.7 we show the case of two equilibria $(L_1, 0)$ and $(0, L_2)$.

Notice that in all cases at $x_k = 0$ the profit of firm $k$ is zero, therefore at the best response it has to be non-negative. Hence, at any equilibrium the profit of each firm is also non-negative. ▼

*Example 1.3.* Consider again the duopoly in which $N = 2$, furthermore take $L_1 = L_2 = 1.5$, $C_1(x_1) = 0.5x_1$, $C_2(x_2) = 0.5x_2$ and assume that the price function is given by

$$f(Q) = \begin{cases} 1.75 - 0.5Q & \text{if} \quad 0 \le Q \le 1.5, \\ 2.5 - Q & \text{if} \quad 1.5 \le Q \le 2.5, \\ 0 & \text{if} \quad Q \ge 2.5. \end{cases} \tag{1.13}$$

Notice that the cost functions are linear but that the price function is piece-wise linear. Because of the kink in the price function the profit functions are not differentiable at $Q = 1.5$. By calculating and comparing the left and right hand derivatives of the profit function, it is easy to show that there are infinitely many equilibria and they form the set

$$\bar{X} = \{(\bar{x}_1, \bar{x}_2) | 0.5 \le \bar{x}_1 \le 1, \quad 0.5 \le \bar{x}_2 \le 1, \quad \bar{x}_1 + \bar{x}_2 = 1.5\}.$$

Notice that the total output of the two firms is unique, satisfying $x_1 + x_2 = 1.5$, but this total output can be divided between the two firms in infinitely many different ways. At any equilibrium, $\bar{Q} = 1.5$, so the equilibrium price is $f(\bar{Q}) = 1$, and therefore the profit of firm $k$ is always positive, being given by

$$\varphi_k(\bar{x}_1, \bar{x}_2) = \bar{x}_k \cdot 1 - 0.5\bar{x}_k = 0.5\bar{x}_k. \qquad \blacktriangledown$$

*Example 1.4.* In this example we assume linear cost functions, $C_k(x_k) = c_k x_k$ with some positive constant $c_k$, and a quadratic price function where

$$f(Q) = \begin{cases} A - Q^2 & \text{if} \quad 0 \leq Q \leq \sqrt{A}, \\ 0 & \text{if} \quad Q > \sqrt{A}. \end{cases}$$

It is also assumed that $A > c_k$ for all $k$. Notice that at the best response of firm $k$ it is the case that $Q_k + x_k \leq \sqrt{A}$, otherwise the value of $x_k$ can be decreased by a small amount, when the price is still zero and the cost would decrease. Therefore at the best response of all firms the total output has to be less than or equal to $\sqrt{A}$. For the sake of simplicity assume that $\sum_{k=1}^{N} L_k \leq \sqrt{A}$, the other case can be discussed in a similar way. By assuming an interior optimum, the first order condition implies that

$$\frac{\partial}{\partial x_k}[x_k(A - (x_k + Q_k)^2) - c_k x_k] = A - 3x_k^2 - 4x_k Q_k - Q_k^2 - c_k = 0.$$

If $c_k \geq A$, then $\varphi_k$ is strictly decreasing in $Q_k$, so the best response of firm $k$ is always zero. Therefore we may assume that $c_k < A$ for all $k$. The solution of the above quadratic equation is

$$z_k^* = \frac{1}{3}\left(\sqrt{Q_k^2 + 3(A - c_k)} - 2Q_k\right).$$

Since the payoff function of firm $k$ is strictly concave in $x_k$, the best response assumes the form

$$R_k(Q_k) = \begin{cases} 0 & \text{if} \quad z_k^* < 0, \\ L_k & \text{if} \quad z_k^* > L_k, \\ z_k^* & \text{otherwise.} \end{cases}$$

This function is illustrated in Fig. 1.8. Simple differentiation shows that $z_k^*$ is strictly decreasing and convex in $Q_k$. It can be proved that there is always a unique equilibrium. Since at $x_k = 0$ the profit of firm $k$ is zero, the profits at the best responses and therefore the equilibrium profits must be non-negative for all firms. In the case of an interior equilibrium the equilibrium quantities can be derived in closed-form. The first order condition may be rewritten as

$$A - Q^2 + x_k(-2Q) - c_k = 0,$$

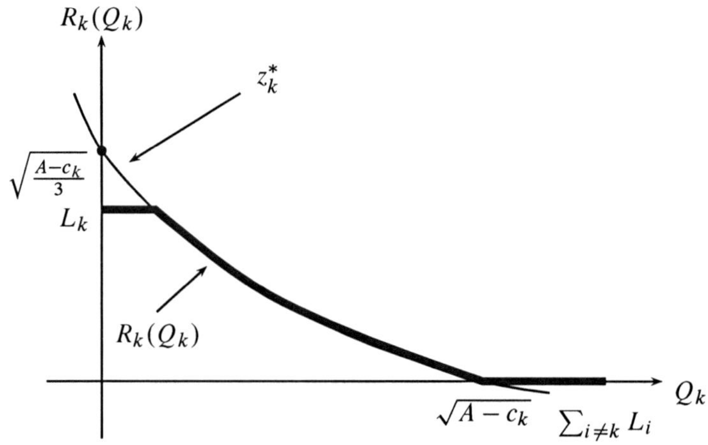


**Fig. 1.8** Example 1.4; the best response function (*thick line*) of a typical firm $k$ with a linear cost function and quadratic price function

implying that at the interior equilibrium

$$\bar{x}_k = \frac{A - \bar{Q}^2 - c_k}{2\bar{Q}}.$$

Summation over all $N$ firms yields

$$\bar{Q} = \frac{NA - N\bar{Q}^2 - \sum_{l=1}^N c_l}{2\bar{Q}},$$

and therefore

$$\bar{Q}^2 = \frac{NA - \sum_{l=1}^N c_l}{N+2}.$$

The individual quantities in equilibrium are then obtained as

$$\begin{aligned} \bar{x}_k &= \frac{1}{2\sqrt{(NA - \sum_{l=1}^N c_l)/(N+2)}} \left( A - \frac{NA - \sum_{l=1}^N c_l}{N+2} - c_k \right) \\ &= \frac{2A + \sum_{l=1}^N c_l - (N+2)c_k}{2\sqrt{(N+2)(NA - \sum_{l=1}^N c_l)}}. \end{aligned} \tag{1.14}$$

For positivity of all equilibrium quantities, additional conditions are required, namely that

$$c_k < \frac{2A + \sum_{l\neq k} c_l}{N+1} \quad \text{for all } k.$$

Obviously, if firm $k$'s unit costs $c_k$ are too high (for a given number of firms $N$), production might not be feasible (so that firm $k$ offers $x_k = 0$). Furthermore, for increasing $N$ (and given unit costs) some (high-cost) firms might drop out of the market. The equilibrium price is given by

$$\bar{p} = \frac{2A + \sum_{l=1}^{N} c_l}{N+2} > 0$$

and the equilibrium profit of firm $k$ is

$$\bar{\varphi}_k = \frac{(2A + \sum_{l=1}^{N} c_l - (N+2)c_k)^2}{2(N+2)\sqrt{(N+2)(NA - \sum_{l=1}^{N} c_l)}}.$$

▼

*Example 1.5.* Assume again linear cost functions, $C_k(x_k) = d_k + c_k x_k$, but isoelastic (hyperbolic) price function, $f(Q) = A/Q$. The form of the profit of firm $k$ depends on whether $Q_k$ is positive or zero. If $Q_k > 0$, then

$$\varphi_k(x_1, \ldots, x_N) = \frac{Ax_k}{x_k + Q_k} - (d_k + c_k x_k),$$

and if $Q_k = 0$, then

$$\varphi_k(x_1, \ldots, x_N) = \begin{cases} A - (d_k + c_k x_k) & \text{if} \quad x_k > 0, \\ -d_k & \text{if} \quad x_k = 0, \end{cases}$$

where we assume that firm $k$ cannot exit the market, so with zero production level it must face fixed costs. Notice that if $Q_k = 0$, then with any $x_k > 0$, the revenue of firm $k$ is always $A$. In this case firm $k$ has no best response and its interest is to select a very small output level, since the supremum of its profit occurs at $x_k = 0$. Assume next that $Q_k > 0$. In maximizing $\varphi_k$, the first order condition is

$$\frac{AQ_k}{(x_k + Q_k)^2} - c_k = 0.$$

Since $\varphi_k$ is strictly concave in $Q_k$, the best response of firm $k$ is

$$R_k(Q_k) = \begin{cases} 0 & \text{if} \quad \sqrt{\frac{AQ_k}{c_k}} - Q_k \leq 0, \\ L_k & \text{if} \quad \sqrt{\frac{AQ_k}{c_k}} - Q_k \geq L_k, \\ \sqrt{AQ_k/c_k} - Q_k & \text{otherwise}. \end{cases}$$

This function is illustrated in Fig. 1.9. We note that the best response is first increasing and then decreasing. This is in contrast to the examples considered previously, where the best responses were decreasing everywhere. Some authors consider

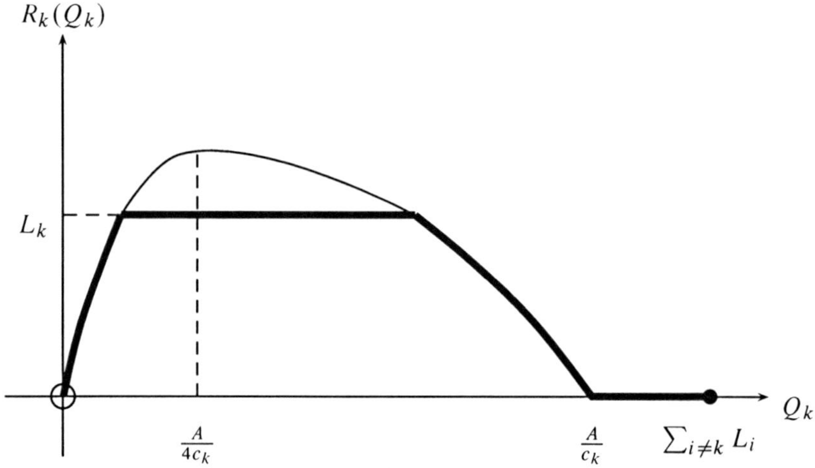


**Fig. 1.9** Example 1.5; the best response function (*thick line*) of a typical firm $k$ with a linear cost function and hyperbolic price function

$\bar{x}_1 = \cdots = \bar{x}_N = 0$ as a trivial equilibrium in a limiting sense.[3] In a non-trivial equilibrium, when $\bar{Q} > 0$, still some equilibrium outputs might be zero, when the marginal costs, $c_k$, for some firms are very large. By assuming that the value of $L_k$ is sufficiently large for all firms, the positive equilibrium can be computed as follows. Since for all $k$,

$$x_k = \sqrt{\frac{A(Q - x_k)}{c_k}} - (Q - x_k),$$

we have

$$c_k Q^2 = A(Q - x_k),$$

implying that

$$x_k = \frac{AQ - c_k Q^2}{A}.$$

Summing this equation over all $N$ firms, we obtain

$$Q = \frac{NAQ - Q^2 \sum_{k=1}^{N} c_k}{A}.$$

So the total output of all firms is

$$\bar{Q} = \frac{(N-1)A}{\sum_{k=1}^{N} c_k},$$

[3] See Agliari et al. (2005, 2006), Agliari (2006) and Matsumoto and Serizawa (2007).

and by substituting it into the above expression for $x_k$, the equilibrium output of firm $k$ is given by

$$\bar{x}_k = \frac{(N-1)A}{\sum_{l=1}^N c_l} - \frac{A(N-1)^2 c_k}{\left(\sum_{l=1}^N c_l\right)^2},$$

and the equilibrium profit of firm $k$ is given by

$$\bar{\varphi}_k = \frac{A\bar{x}_k}{\bar{Q}} - c_k\bar{x}_k - d_k = \left(\frac{\sum_{l=1}^N c_l}{N-1} - c_k\right)\bar{x}_k - d_k = A\left(1 - \frac{(N-1)c_k}{\sum_{l=1}^N c_l}\right)^2 - d_k.$$

In order to guarantee that all equilibrium outputs of the firms are positive, we have to assume that

$$c_k < \frac{\sum_{l \neq k} c_l}{N-2},$$

that is, the marginal costs cannot be too high. ▼

The examples above considered the case in which the cost function of a firm depends only on its own output. We will next present two particular examples including cost externalities, with linear price and cost functions, where the fixed costs are equal to zero and the marginal cost of each firm depends on the output of the rest of the industry.

*Example 1.6.* In the case of $N$ firms assume a linear price function $f(Q) = A - BQ$, and furthermore assume that the marginal cost of each firm is a function of the output of the rest of the industry, $M_k(Q_k)$. If zero fixed cost is assumed, then the cost function of firm $k$ is given as (see Howroyd and Russell (1984), Russell et al. (1986) and Furth (2009))

$$C_k(x_k, Q_k) = x_k M_k(Q_k),$$

so the profit of firm $k$ is

$$x_k(A - Bx_k - BQ_k) - x_k M_k(Q_k),$$

by assuming that $x_k + Q_k \leq A/B$. Notice that this function is strictly concave in $x_k$, so in the case of sufficiently small capacity limits there is a unique best response function given by

$$R_k(Q_k) = \begin{cases} 0 & \text{if} \quad A - BQ_k - M_k(Q_k) \leq 0, \\ L_k & \text{if} \quad A - 2BL_k - BQ_k - M_k(Q_k) \geq 0, \\ z_k^* & \text{otherwise}, \end{cases}$$

where $z_k^*$ is the solution of the equation

$$A - 2Bz_k - BQ_k - M_k(Q_k) = 0$$

inside the interval $(0, L_k)$. That is,

$$z_k^* = \frac{A - BQ_k - M_k(Q_k)}{2B}.$$

We mention here that for an arbitrary value of $Q_k$, the profit of each firm $k$ is zero with $x_k = 0$, so the payoff at the best response also must be non-negative. Hence at any equilibrium the firms have non-negative profit values.

If $M_k(Q_k)$ is a linear function, then $z_k^*$ is also linear in $Q_k$, so $R_k(Q_k)$ is a piece-wise linear function similar to Example 1.1. If we assume that $M_k(Q_k)$ is a quadratic function, then $z_k^*$ is also quadratic in $Q_k$. Thus if we write

$$M_k(Q_k) = \alpha_k + \beta_k Q_k + \gamma_k Q_k^2,$$

then

$$z_k^* = \frac{(A - \alpha_k) + (-B - \beta_k)Q_k - \gamma_k Q_k^2}{2B}.$$

Let $\mu_k > 1$ be a given constant and select

$$\alpha_k = A, \quad \beta_k = -B(1 + 2\mu_k) \quad \text{and} \quad \gamma_k = 2B\mu_k,$$

then we have the relatively simple form

$$z_k^* = \mu_k Q_k(1 - Q_k).$$ ▼

*Example 1.7.* Consider again the oligopoly of the previous example with the only difference being that the marginal cost of each firm $k$ is a hyperbola of the form

$$M_k(Q_k) = \frac{c_k}{1 + \gamma_k Q_k}.$$

In this case $R_k(Q_k)$ has the same structure as in the previous example with

$$z_k^* = \frac{A - BQ_k - M_k(Q_k)}{2B} = \frac{1}{2B}\left(A - BQ_k - \frac{c_k}{1 + \gamma_k Q_k}\right).$$

In Chap. 3 we will give a detailed analysis of this example. ▼

In our last example we show an oligopoly for which no equilibrium exists.

*Example 1.8.* Consider the case of two firms, $N = 2$, with capacity limits $L_1 = L_2 = 0.5$, linear price function $f(Q) = 1 - Q$ with $Q = \sum_{k=1}^{2} x_k$, and discontinuous cost functions

$$C_k(x_k) = \begin{cases} 10 & \text{if} \quad x_k = 0, \\ 10x_k + 5 & \text{if} \quad 0 < x_k \le \frac{1}{2}. \end{cases} \tag{1.15}$$

The higher costs at zero reflect exit barriers, which do not occur when the firms start producing. We will show that this oligopoly has no equilibrium. On the contrary, assume that $(\bar{x}_1, \bar{x}_2)$ is an equilibrium. Assume first that $x_1 > 0$, then

$$\varphi_1(x_1, \bar{x}_2) = x_1(1 - x_1 - \bar{x}_2) - (10x_1 + 5) = -x_1^2 - (9x_1 + x_1\bar{x}_2 + 5)$$

with derivative

$$\frac{\partial \varphi_1}{\partial x_1}(x_1, \bar{x}_2) = -2x_1 - 9 - \bar{x}_2 < 0.$$

Therefore $\varphi_1$ is strictly decreasing in $x_1$. Assume next that $x_1 = 0$. Then $\varphi_1(0, \bar{x}_2) = -10$ with $\lim_{x_1 \to 0^+} \varphi_1(x_1, \bar{x}_2) = 0 \cdot f(Q) - 5 = -5 > \varphi_1(0, \bar{x}_2)$ showing that at $\bar{x}_2$, firm 1 has no best response. Hence no equilibrium exists. ▼

## 1.2 Dynamic Adjustment Processes

In this section dynamic adjustment processes in the Cournot model will be introduced. If all firms simultaneously select the corresponding output levels of an equilibrium, then none of the firms can change unilaterally its output level and increase profit. So without coordination and cooperation between the firms, the output level of all firms will remain steady at the equilibrium levels. If the selected output levels do not form an equilibrium, then at least one firm is able to increase its profit by changing its output level unilaterally. Since the firms are rational, all firms will do the same. Since the firms change their output levels simultaneously, they cannot reach their best response levels, because the competitors simultaneously move away from their previously assumed output levels at the same time. In this way the firms usually would not reach an equilibrium, so output changes are again undertaken, and a dynamic process develops. The model of the resulting process depends on the assumed nature of the time scales and on the way the firms adjust output levels, which in turn depends on their expectation formation.

In the *discrete time* case let $t = 0, 1, 2 \cdots$ denote the time periods, then here we shall assume that in each time period each firm changes its output level to the best response based on its latest belief of the total production level of the rest of the industry. This process can be written as

$$x_k(t+1) = R_k\left(Q_k^E(t+1)\right), \tag{1.16}$$

where $Q_k^E(t+1)$ is the total output of the rest of the industry expected by firm $k$ for the next time period $t+1$. We emphasize here the fact that expectation is not meant in its probabilistic sense, rather it is a deterministic predicted value. The most simple expectation scheme is the one in which the firms use the latest available information,

$$Q_k^E(t+1) = \sum_{l \neq k} x_l(t), \tag{1.17}$$

which is sometimes called the *static*, or *naive*, or *Cournot expectation*.

The firms are also able to develop certain learning procedures based on earlier data. The most popular such learning scheme is obtained when the firms adjust their expectations *adaptively* according to

$$Q_k^E(t+1) = Q_k^E(t) + a_k \left( \sum_{l \neq k} x_l(t) - Q_k^E(t) \right), \tag{1.18}$$

with $a_k$ being a positive constant known as the *speed of adjustment* of firm $k$. It is usually assumed that $0 < a_k \leq 1$ for all $k$. The interpretation of this dynamic learning scheme is that, if firm $k$ underestimated (overestimated) the output of the rest of the industry in the previous time period, then in the next time period this firm wants to increase (decrease) its estimate. This increase (decrease) is represented by the second term, and the coefficient $a_k$ determines the speed (or rate) of adjustment. If the expectation of a firm were correct in the previous time period, then there would be no need to change the expectation, in this case the second term would be zero. Notice that the special case of $a_k = 1$ reduces to the static or Cournot expectation.

Mathematically, the dynamic process (1.16), together with naive expectations (1.17) form the $N$-dimensional dynamical system

$$x_k(t+1) = R_k \left( \sum_{l \neq k} x_l(t) \right) \qquad (k = 1, 2, \ldots, N), \tag{1.19}$$

to which we will refer as *best response dynamics with naive expectations*.

Under the adaptive expectations scheme (1.18), the dynamic process (1.16) becomes the $2N$-dimensional dynamical system

$$x_k(t+1) = R_k \left( a_k \sum_{l \neq k} x_l(t) + (1 - a_k) Q_k^E(t) \right), \tag{1.20}$$

$$Q_k^E(t+1) = a_k \sum_{l \neq k} x_l(t) + (1 - a_k) Q_k^E(t), \tag{1.21}$$

for $k = 1, 2, \ldots, N$. We will refer to this process as the *best response dynamics with adaptive expectations*.

In the latter formulation we have formally $2N$ state variables, however it is easy to show that the best response dynamics with adaptive expectations are actually driven by the $N$ expectation variables and the production outputs can be computed directly from them. In fact, for all $k$, (1.18) can be written as

$$Q_k^E(t+1) = a_k \sum_{l \neq k} x_l(t) + (1-a_k) Q_k^E(t) = a_k \sum_{l \neq k} R_l\left(Q_l^E(t)\right) + (1-a_k) Q_k^E(t). \tag{1.22}$$

The dynamic process now reduces to an $N$-dimensional dynamical system in the expected variables $Q_1^E(t), \ldots, Q_N^E(t)$, and at each time period $t$ the output of firm $k$ is given as

$$x_k(t) = R_k\left(Q_k^E(t)\right),$$

which is a static mapping from beliefs to realizations in the sense that both sides of the mapping are computed at the same time $t$.

In most industries any increase of the output level of any firm requires time, new hirings, purchase of new machinery, or sometimes even the opening up of a new plant. Therefore output changes are made gradually. For example, in the case of the dynamic process (1.19) instead of selecting the best response directly, the new output level of firm $k$ is selected somewhere in between the current level and the best response to ensure that the output level change occurs in the right direction. This concept of *partial adjustment towards the best response with naive expectations* can be described by the modified $N$-dimensional dynamical system

$$x_k(t+1) = a_k R_k\left(\sum_{l \neq k} x_l(t)\right) + (1-a_k)x_k(t), \tag{1.23}$$

for some $a_k \in (0, 1]$. In the case of $a_k = 0$ the output level would never change, therefore this value is excluded. Notice that in the case of $a_k = 1$, the partial adjustment towards the best response with naive expectations (1.23) reduces to best response dynamics with naive expectations (1.19).

In the special case of two firms ($N = 2$) both dynamical systems (1.22) and (1.23) have the common form

$$\begin{aligned} y_1(t+1) &= a_1 R_1(y_2(t)) + (1-a_1)y_1(t), \\ y_2(t+1) &= a_2 R_2(y_1(t)) + (1-a_2)y_2(t) \end{aligned}$$

with $y_1 = x_1$ and $y_2 = x_2$ in (1.23), and $y_1 = Q_2^E$, $y_2 = Q_1^E$ and $a_1$ and $a_2$ being interchanged in (1.22). If $N > 2$, then systems (1.22) and (1.23) are equivalent if $R_k\left(\sum_{l \neq k} y_l(t)\right) = \sum_{l \neq k} R_l(y_l(t))$ holds for all $k$. In the symmetric case (when $R_k \equiv R$), this condition holds if $R(Q_k) = rQ_k$ with some constant $r$.

It is important to realize that dynamic adjustment processes of the kind considered above are defined on the action space $\Pi_{k=1}^N [0, L_k]$ and incorporate only the firms' quantity decision. In order to obtain economically feasible trajectories, we need to keep in mind the fact that prices (and profits) have to be non-negative in the long run, though it is possible (as we shall indeed find) that over some periods negative profits may occur. In some of the models we study it will be possible to ensure non-negative prices simply by selecting suitable parameter values. For example, for the $N$-firm oligopoly model with linear inverse demand function a sufficient condition for non-negative prices is $\sum_{k=1}^{N} L_k \leq A/B$ (see Example 1.1) and for the model with quadratic price function and linear costs, we can simply select $\sum_{k=1}^{N} L_k \leq \sqrt{A}$ (see Example 1.4).

If the time scales are *continuous*, then output changes are made continuously, without direct jumps to the best response levels. It is always assumed that in each time period the output level moves in a direction towards the best response. This concept is modeled by an $N$-dimensional system of ordinary differential equations of the form

$$\dot{x}_k(t) = a_k \left( R_k(\sum_{l \neq k} x_l(t)) - x_k(t) \right) \qquad (k = 1, 2, \ldots, N). \tag{1.24}$$

Here $a_k > 0$ is a given constant and also called the *speed of adjustment* of firm $k$. This is the continuous time counterpart of the discrete system (1.23), which is also called the partial adjustment dynamics.

*Example 1.9.* Consider again the case of linear oligopolies with linear inverse demand and linear cost functions, which was discussed earlier in Example 1.1. By ignoring the non-negativity condition of the outputs and assuming that $L_k = \infty$ for all $k$, the best reply of firm $k$ is given as (see (1.6))

$$R_k(Q_k) = -\frac{1}{2}Q_k + \frac{A - c_k}{2B}.$$

Since for all $k$, $R_k(Q_k)$ is linear with identical derivative, the dynamical systems (1.22) and (1.23) have the same coefficient matrix, so the asymptotic behavior of the discrete dynamics with adaptive expectations and with adaptive adjustments are equivalent. The dynamical system (1.23) for partial adjustment towards the best response can be written as

$$x_k(t+1) = a_k \left( -\frac{1}{2} \sum_{l \neq k} x_l(t) + \frac{A - c_k}{2B} \right) + (1 - a_k) x_k(t), \tag{1.25}$$

which is a linear system with coefficient matrix

$$\begin{pmatrix} 1 - a_1 & -\frac{a_1}{2} & \ldots & -\frac{a_1}{2} \\ -\frac{a_2}{2} & 1 - a_2 & \ldots & -\frac{a_2}{2} \\ \vdots & \vdots & \ddots & \vdots \\ -\frac{a_N}{2} & -\frac{a_N}{2} & \ldots & 1 - a_N \end{pmatrix}.$$

In Chap. 2 (Theorem 2.1) we will see that the eigenvalues of this matrix lie inside the unit circle if and only if $a_k < 4$ for all $k$, and

$$\sum_{k=1}^{N} \frac{a_k}{4 - a_k} < 1.$$

In the case of linear systems local and global asymptotic stability are the same, so the equilibrium is globally asymptotically stable if and only if the above conditions are satisfied.

In the case of continuous time scales the dynamical system for partial adjustment (1.24) can be written as

$$\dot{x}_k(t) = a_k \left( -\frac{1}{2} \sum_{l \neq k} x_l(t) + \frac{A - c_k}{2B} - x_k(t) \right), \tag{1.26}$$

which is again a linear system with coefficient matrix

$$\begin{pmatrix} -a_1 & -\frac{a_1}{2} & \dots & -\frac{a_1}{2} \\ -\frac{a_2}{2} & -a_2 & \dots & -\frac{a_2}{2} \\ \vdots & \vdots & & \vdots \\ -\frac{a_N}{2} & -\frac{a_N}{2} & \dots & -a_N \end{pmatrix}.$$

In Chap. 2 (Theorem 2.2) we will see that all eigenvalues of this matrix always have negative real parts so the equilibrium is locally asymptotically stable. The linearity of the system implies that the Nash equilibrium is also globally asymptotically stable. ▼

Introducing the non-negativity conditions and the capacity limits into the model makes the best reply functions nonlinear. Nonlinearity can also occur by assuming nonlinear cost or price functions. Then the corresponding dynamical systems become nonlinear, and local asymptotic stability does not imply global asymptotic stability. This observation points to the need to perform detailed global analysis of the dynamical behavior. The next section will present the foundation of the relevant methodology.

In models (1.20)–(1.21), for the best response dynamics with adaptive expectations, and (1.23) and (1.24) for the dynamics of partial adjustment towards the best response with naive expectations, we have used simple linear adjustment rules. However these can be easily extended to the nonlinear case by introducing sign-preserving adjustment functions. A real-variable, real-valued function $\alpha : \mathbb{R} \to \mathbb{R}$ is called sign-preserving, if $\alpha(x)$ has the same sign as $x$, that is,

$$\alpha(x) \begin{cases} > 0 & \text{if } x > 0, \\ = 0 & \text{if } x = 0, \\ < 0 & \text{if } x < 0. \end{cases} \tag{1.27}$$

Assume now that for all $k$, $\alpha_k$ is a sign-preserving function, then the dynamical system (1.20)–(1.21) for the best response with adaptive expectations can be extended to

$$x_k(t+1) = R_k\left(Q_k^E(t) + \alpha_k(\sum_{l \neq k} x_l(t) - Q_k^E(t))\right), \tag{1.28}$$

$$Q_k^E(t+1) = Q_k^E(t) + \alpha_k\left(\sum_{l \neq k} x_l(t) - Q_k^E(t)\right). \tag{1.29}$$

Similarly the discrete time dynamical system (1.23) for the dynamics of partial adjustment towards the best response with naive expectations becomes

$$x_k(t+1) = x_k(t) + \alpha_k\left(R_k(\sum_{l \neq k} x_l(t)) - x_k(t)\right), \tag{1.30}$$

whilst the continuous time dynamical system (1.24) for the same process becomes

$$\dot{x}_k(t) = \alpha_k\left(R_k(\sum_{l \neq k} x_l(t)) - x_k(t)\right). \tag{1.31}$$

Another important class of adjustment processes that has been investigated in the literature on dynamic oligopolies by many authors is that of the gradient adjustment process. This adjustment process is based on the observation that if for firm $k$ at a certain time period, $\partial\varphi_k/\partial x_k$ is positive, then it is in firm $k$'s interest to increase the output level, if $\partial\varphi_k/\partial x_k$ is negative, then the firm wants to decrease it, and if $\partial\varphi_k/\partial x_k = 0$, then firm $k$ believes that it is already at its maximum level, so it wants to maintain the same output level. This idea can be mathematically realized in the gradient adjustment processes

$$x_k(t+1) = x_k(t) + \alpha_k\left(\frac{\partial\varphi_k(x_1(t),\ldots,x_N(t))}{\partial x_k}\right) \quad (1 \leq k \leq N), \tag{1.32}$$

in discrete time and

$$\dot{x}_k(t) = \alpha_k\left(\frac{\partial\varphi_k(x_1(t),\ldots,x_N(t))}{\partial x_k}\right) \quad (1 \leq k \leq N), \tag{1.33}$$

in continuous time, where $\alpha_k$ is a sign-preserving function. Notice that dynamic processes based on best response functions require the solution of optimization problems in order to determine the best responses. In contrast gradient adjustment processes do not need the computation of best responses, rather they need only local information about the profit functions. Therefore the uniqueness of best responses is not an issue with gradient adjustment processes. Observe, however, that in the case of gradient adjustment, we need to check whether the obtained quantity is non-negative and also whether it is below the capacity limit.

Clearly the steady states of the dynamic processes (1.28)–(1.29), for the generalised best response with adaptive expectations, and (1.30)–(1.31) for the generalised partial adjustment towards the best response with naive expectations, are the *Nash equilibria*. However only interior equilibria can be the steady states of the gradient adjustment processes (1.32)–(1.33). Therefore boundary equilibria can be obtained as the limits of the trajectories as $t \to \infty$ only in special cases. The foregoing reasoning is based on the fact that a point is a steady state of best response based adjustment if and only if the output levels equal the best responses for all firms, that is, when they are at an equilibrium. However in the case of gradient adjustment a point is a steady state if and only if all partial derivatives are zero, which is not the case if the equilibrium lies on the boundary. Therefore even in the case of asymptotic stability the trajectory does not need to converge to the equilibrium, since the solutions of the first order conditions may lie outside the feasible region, so they are not necessarily steady states. This behavior may be regarded as a drawback of gradient adjustment processes.

*Example 1.10.* In the case of linear oligopoly, discussed in Example 1.9, we can calculate

$$\begin{aligned}\frac{\partial \varphi_k}{\partial x_k} &= \frac{\partial}{\partial x_k}\left\{x_k\left(A - Bx_k - B\sum_{l\neq k} x_l\right) - (c_k x_k + d_k)\right\}\\ &= A - 2Bx_k - B\sum_{l\neq k} x_l - c_k,\end{aligned}$$

so the gradient adjustment dynamical system (1.32) in discrete time with linear sign-preserving functions ($\alpha_k(x) = a_k x$ with $a_k > 0$) can be written as

$$\begin{aligned}x_k(t+1) &= x_k(t) + a_k\left(-2Bx_k(t) - B\sum_{l\neq k} x_l(t) + A - c_k\right)\\ &= 2Ba_k\left(-\frac{1}{2}\sum_{l\neq k} x_l(t) + \frac{A - c_k}{2B}\right) + (1 - 2Ba_k)x_k(t),\end{aligned}$$

which is the same as the dynamical system (1.25) for partial adjustment towards the best response, with $a_k$ replaced by $2Ba_k$. The continuous time system (1.33) with linear sign-preserving functions now assumes the form

$$\begin{aligned}\dot{x}_k(t) &= a_k\Bigg(-B\sum_{l\neq k}x_l(t) - 2Bx_k(t) + A - c_k\Bigg)\\ &= 2Ba_k\Bigg(-\frac{1}{2}\sum_{l\neq k}x_l(t) - x_k(t) + \frac{A-c_k}{2B}\Bigg),\end{aligned}$$

which is the same as system (1.26) with $a_k$ replaced by $2Ba_k$. ▼

The dynamical behavior of these adjustment process systems largely depends on the type and the parameters of the adjustment schemes as well as on the analytical properties of the best response functions, which in turn depend on the shapes of the price and cost functions.

There has been some criticism of the modeling of boundedly rational firms in dynamic oligopoly models using the previously discussed adjustment processes (see for example, Friedman (1977, 1982)). The essence of the criticism is that the firms ignore the fact that their current actions will have an impact on the future actions of the competitors (that is the limit of the adjustment process itself may not be an equilibrium of the repeated game). Therefore, it has been suggested that it would be more reasonable to assume that firms operating in markets over many time periods would seek to maximize a discounted stream of profits over a finite or infinite time horizon taking the strategic behavior of their competitors into account. Beside the fact that such an approach necessarily assumes a high degree of information and rationality on the part of the firms, one justification for the interest in models of the type studied in this book is given by more recent results demonstrating that myopic play is (approximately) optimal if the discount factor is very small (see Dana and Montrucchio (1986, 1987)). Moreover, non-equilibrium adjustment processes like the adjustment processes presented above can be shown to implicitly rely on a combination of "lock-in" and impatience, and this may serve as a further explanation for the players' myopia (see Fudenberg and Levine (1998), and Tirole (1988)). In any case, in this book we follow the argument that the kind of adjustment processes introduced above can "... be interpreted as a crude way of expressing the bounded rationality of agents" (Vives (1999), p. 49). Readers interested in dynamic games where players are more rational and forward-looking might want to consult the book by Dockner et al. (2000) who present a variety of models and summarize many interesting results. In this book we will mainly concentrate on best response based dynamic processes.

## 1.3 An Introduction to the Analysis of Global Dynamics

The purpose of this section is to introduce the main concepts and tools for the analysis of the global properties of a discrete time dynamical system. In order to do so we will use the example of a simple Cournot oligopoly with linear inverse demand and quadratic costs. This example has already been introduced in Sect. 1.1 (see Example 1.2), where we denoted the linear price function as $p = f(Q) = A - BQ$ and the quadratic production cost functions as $C_k(x_k) = c_k x_k + e_k x_k^2$. In order to avoid trivial best responses we assume again that $A > c_k$ for $k = 1, 2$.

### 1.3.1 A Cournot Duopoly Game

We first consider a duopoly game ($N = 2$), where the firms use partial adjustment towards the best response. The reaction functions in this case become

$$R_1(x_2) = \begin{cases} 0 & \text{if } z_1^* < 0, \\ L_1 & \text{if } z_1^* > L_1, \\ z_1^* & \text{otherwise,} \end{cases} \tag{1.34}$$

and

$$R_2(x_1) = \begin{cases} 0 & \text{if } z_2^* < 0, \\ L_2 & \text{if } z_2^* > L_2, \\ z_2^* & \text{otherwise,} \end{cases} \tag{1.35}$$

where $z_k^* = \frac{A - c_k - BQ_k}{2(B+e_k)}$ $(k = 1, 2)$ with $Q_1 = x_2$ and $Q_2 = x_1$. If the duopolists partially adjust their quantities towards the best replies (based on naive expectations) and if the speeds of adjustment are constant, the dynamical system is generated by the iteration of the map $T_a : [0, L_1] \times [0, L_2] \to [0, L_1] \times [0, L_2]$, where

$$T_a : \begin{cases} x_1(t+1) = (1 - a_1)x_1(t) + a_1 R_1(x_2(t)) \\ x_2(t+1) = (1 - a_2)x_2(t) + a_2 R_2(x_1(t)) \end{cases}, \tag{1.36}$$

with $0 < a_k \leqslant 1$. Recall from Sect. 1.2 that the best reply dynamics with naive expectations is obtained as a special case with $a_k = 1$ for $k = 1, 2$. We have also shown in Sect. 1.2 that in a duopoly partial adjustment towards the best response and the best reply dynamics with adaptive expectations are equivalent. Hence, the results obtained in this section also describe what happens if best reply dynamics with adaptive expectations are considered. Using (1.36) together with the steady state conditions $x_k(t + 1) = x_k(t)$, $k = 1, 2$, leads to the equations $x_1 = R_1(x_2)$, $x_2 = R_2(x_1)$, which shows that the steady states of this dynamical system coincide with the Cournot–Nash equilibria of the underlying game and that they are located at the intersections of the reaction curves. Clearly, the steady states do not depend on the adjustment speeds $a_k$. As demonstrated in Sect. 1.1, the number of equilibria depends on the marginal costs. If marginal costs are increasing or even decreasing but not too strongly such that $B + e_k > 0$ and

$$B^2 < 4(B + e_1)(B + e_2), \tag{1.37}$$

then for $x_k^L > x_k^M$ $(k = 1, 2)$ we have a unique interior equilibrium. The quantities at this interior equilibrium are given by

$$E = (\bar{x}_1, \bar{x}_2)$$
$$= \left( \frac{2(B+e_2)(A-c_1) - B(A-c_2)}{4(B+e_1)(B+e_2) - B^2}, \frac{2(B+e_1)(A-c_2) - B(A-c_1)}{4(B+e_1)(B+e_2) - B^2} \right).$$

On the other hand, if $-B < e_k < 0$, $x_k^L < x_k^M$ $(k = 1, 2)$ as before, but

$$B^2 > 4(B+e_1)(B+e_2), \tag{1.38}$$

then a situation of multiple equilibria might be obtained. This is the situation depicted in Fig. 1.3, where in addition to the interior equilibrium there also appear two boundary equilibria. The two coexisting boundary equilibria are given by

$$E_1 = (x_1^M, 0); \quad E_2 = (0, x_2^M),$$

where

$$x_1^M = \frac{A-c_1}{2(B+e_1)}; \quad x_2^M = \frac{A-c_2}{2(B+e_2)},$$

are the monopoly quantities.

Let us first try to give conditions for the global asymptotic stability of an equilibrium, which would also imply its uniqueness. We recall that an equilibrium is globally asymptotically stable if any trajectory starting from an initial condition in the strategy space converges to the equilibrium as $t \to \infty$. In the case of the model (1.36) the strategy space is given by the trapping region $\mathbb{D} = [0, L_1] \times [0, L_2]$. However the map (1.36), whose iteration gives the time evolution of the duopoly game, is not differentiable in the whole strategy space $\mathbb{D}$ because the reaction functions are piecewise differentiable functions defined by

$$R_k(Q_k) = \begin{cases} 0 & \text{if} \quad Q_k \geq \frac{A-c_k}{B}, \\ L_k & \text{if} \quad Q_k \leq \frac{A-c_k-2(B+e_k)L_k}{B}, \\ (A - c_k - BQ_k)/\big(2(B+e_k)\big) & \text{otherwise.} \end{cases}$$

Accordingly, the phase space $\mathbb{D}$ can be subdivided into nine regions defined by the break points of the reaction functions (see Fig. 1.10), such that the map $T_a$ is differentiable (indeed linear in this case) inside each of them, it is defined differently in each region and it is not differentiable on the boundaries between the regions. Depending on the possible combination of the reaction functions the different components of the map are given by

$$T_a|_{\mathbb{D}^{(1)}} : \begin{cases} x_1(t+1) = (1-a_1)x_1(t) + a_1(A - c_1 - Bx_2)/\big(2(B+e_1)\big), \\ x_2(t+1) = (1-a_2)x_2(t) + a_2(A - c_2 - Bx_1)/\big(2(B+e_2)\big), \end{cases}$$

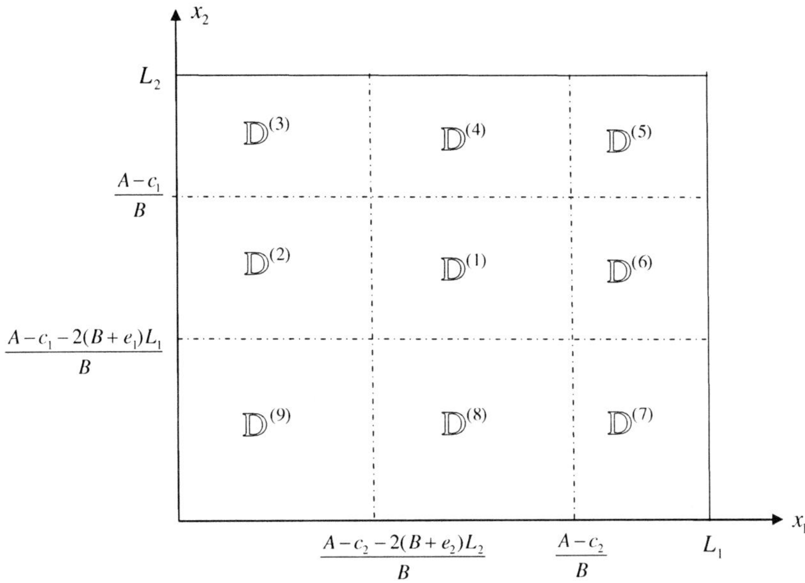


**Fig. 1.10** Phase space regions for the Cournot duopoly game where firms use partial adjustment towards the best response

$$T_a|_{\mathbb{D}^{(2)}} : \begin{cases} x_1(t+1) = (1-a_1)x_1(t) + a_1(A - c_1 - Bx_2)/\big(2(B+e_1)\big), \\ x_2(t+1) = (1-a_2)x_2(t) + a_2 \cdot L_2, \end{cases}$$

$$T_a|_{\mathbb{D}^{(3)}} : \begin{cases} x_1(t+1) = (1-a_1)x_1(t) + a_1 \cdot 0, \\ x_2(t+1) = (1-a_2)x_2(t) + a_2 \cdot L_2, \end{cases}$$

$$T_a|_{\mathbb{D}^{(4)}} : \begin{cases} x_1(t+1) = (1-a_1)x_1(t) + a_1 \cdot 0, \\ x_2(t+1) = (1-a_2)x_2(t) + a_2(A - c_2 - Bx_1)/(2(B+e_2)), \end{cases}$$

$$T_a|_{\mathbb{D}^{(5)}} : \begin{cases} x_1(t+1) = (1-a_1)x_1(t) + a_1 \cdot 0, \\ x_2(t+1) = (1-a_2)x_2(t) + a_2 \cdot 0, \end{cases}$$

$$T_a|_{\mathbb{D}^{(6)}} : \begin{cases} x_1(t+1) = (1-a_1)x_1(t) + a_1(A - c_1 - Bx_2)/\big(2(B+e_1)\big), \\ x_2(t+1) = (1-a_2)x_2(t) + a_2 \cdot 0, \end{cases}$$

$$T_a|_{\mathbb{D}^{(7)}} : \begin{cases} x_1(t+1) = (1-a_1)x_1(t) + a_2 \cdot L_1, \\ x_2(t+1) = (1-a_2)x_2(t) + a_2 \cdot 0, \end{cases}$$

$$T_a|_{\mathbb{D}^{(8)}} : \begin{cases} x_1(t+1) = (1-a_1)x_1(t) + a_1 \cdot L_1, \\ x_2(t+1) = (1-a_2)x_2(t) + a_2(A - c_2 - Bx_1)/\big(2(B+e_2)\big), \end{cases}$$

$$T_a|_{\mathbb{D}^{(9)}} : \begin{cases} x_1(t+1) = (1-a_1)x_1(t) + a_1 \cdot L_1, \\ x_2(t+1) = (1-a_2)x_2(t) + a_2 \cdot L_2. \end{cases}$$

The derivative of the best response function of firm $k$ is either zero or $-B/(2(B+e_k))$, or does not exist in the cases when $Q_k = (A - c_k)/B$ and $Q_k = (A - c_k - 2(B + e_k)L_k)/B$.

Hence we need to consider the four different Jacobian matrices given by

$$J^{(1)} = \begin{pmatrix} 1-a_1 & -\frac{a_1 B}{2(B+e_1)} \\ -\frac{a_2 B}{2(B+e_2)} & 1-a_2 \end{pmatrix}; \quad J^{(2)} = J^{(6)} = \begin{pmatrix} 1-a_1 & -\frac{a_1 B}{2(B+e_1)} \\ 0 & 1-a_2 \end{pmatrix};$$

$$J^{(4)} = J^{(8)} = \begin{pmatrix} 1-a_1 & 0 \\ -\frac{a_2 B}{2(B+e_2)} & 1-a_2 \end{pmatrix};$$

$$J^{(3)} = J^{(5)} = J^{(7)} = J^{(9)} = \begin{pmatrix} 1-a_1 & 0 \\ 0 & 1-a_2 \end{pmatrix}.$$

Select a diagonal matrix $P = \begin{pmatrix} x & 0 \\ 0 & 1 \end{pmatrix}$ with $x > 0$, then the row norms of these Jacobians generated by the matrix $P$ are bounded by the row norm of the matrix

$$\begin{pmatrix} x & 0 \\ 0 & 1 \end{pmatrix} \begin{pmatrix} 1-a_1 & \frac{a_1 B}{2(B+e_1)} \\ \frac{a_2 B}{2(B+e_2)} & 1-a_2 \end{pmatrix} \begin{pmatrix} \frac{1}{x} & 0 \\ 0 & 1 \end{pmatrix} = \begin{pmatrix} 1-a_1 & \frac{a_1 Bx}{2(B+e_1)} \\ \frac{a_2 B}{2(B+e_2)x} & 1-a_2 \end{pmatrix}, \tag{1.39}$$

which is below one if and only if

$$1 - a_1 + \frac{a_1 Bx}{2(B+e_1)} < 1,$$

and

$$1 - a_2 + \frac{a_2 B}{2(B+e_2)x} < 1.$$

Since we assume that $0 < a_k \leq 1$ $(k = 1, 2)$, these relations can be rewritten as

$$\frac{B}{2(B+e_2)} < x < \frac{2(B+e_1)}{B},$$

and a feasible $x$ exists if and only if $B^2 < 4(B+e_1)(B+e_2)$.

Hence under this condition the equilibrium is unique and is globally asymptotically stable regardless of whether it is interior or not. (See Appendix B, Theorem B.3 for the relevant theoretical background.)

Next we will examine the local asymptotic stability of an interior steady state $E$. Let us consider the Jacobian matrix evaluated at the steady state,

$$J = \begin{pmatrix} 1-a_1 & -a_1 \frac{B}{2(B+e_1)} \\ -a_2 \frac{B}{2(B+e_2)} & 1-a_2 \end{pmatrix}.$$

The characteristic equation of this Jacobian is given by $\lambda^2 + p\lambda + q = 0$, where $p = -2 + a_1 + a_2$ and $q = (1 - a_1)(1 - a_2) - a_1 a_2 B^2 / (4(B + e_1)(B + e_2))$. The necessary and sufficient conditions for the eigenvalues to be located inside the unit circle, which are conditions for the local asymptotic stability of the interior Nash equilibrium $E$, are given by the inequalities (see Appendix F, Lemma F.1)

$$1 + p + q > 0 \ , \ 1 - p + q > 0 \ , \quad q < 1. \tag{1.40}$$

These inequalities, respectively, reduce to

$$\frac{B^2}{4(B + e_1)(B + e_2)} < 1,$$
$$\frac{B^2}{4(B + e_1)(B + e_2)} < 1 + 2\frac{2 - a_1 - a_2}{a_1 a_2},$$
$$\frac{B^2}{4(B + e_1)(B + e_2)} > 1 - \frac{a_1 + a_2}{a_1 a_2}.$$

Observe that the first stability condition coincides with condition (1.37) under which this is the only equilibrium and so is globally asymptotically stable. The other conditions do not affect the stability properties, because the second condition is implied by the first one (since $0 < a_k \le 1$) and the last condition is always satisfied (since the left hand side is positive, whereas the right hand side is negative). If $B^2 > 4(B + e_1)(B + e_2)$, then the interior equilibrium is unstable. This is the situation in case (ii) of Example 1.2, where we might have three equilibria with an unstable interior equilibrium.

Consider now the case shown in Fig. 1.3 and the monopoly equilibrium $(0, x_2^M)$. In the neighborhood of this equilibrium $x_2^L < x_2 < L_2$, so $R_1(x_2) = 0$. Furthermore $x_1 = 0$ or a small positive value. Notice that the segments where $R_1(x_2) = L_1$, or $R_2(x_1) = L_2$ are empty, which implies that the sets $\mathbb{D}^{(k)}$ for $k = 3, 2, 9, 8, 7$ are also empty. Therefore any point in a small neighborhood of the equilibrium $(0, x_2^M)$ is in the region $\mathbb{D}^{(4)}$ where the Jacobian matrix is

$$\begin{pmatrix} 1 - a_1 & 0 \\ -\frac{a_2 B}{2(B+e_2)} & 1 - a_2 \end{pmatrix}. \tag{1.41}$$

Let

$$\boldsymbol{P} = \begin{pmatrix} x & 0 \\ 0 & 1 \end{pmatrix} \tag{1.42}$$

be a diagonal matrix with $x > 0$. Then the row norm generated by this matrix is bounded by the row norm of the matrix

$$\begin{pmatrix} x & 0 \\ 0 & 1 \end{pmatrix} \begin{pmatrix} 1 - a_1 & 0 \\ \frac{a_2 B}{2(B+e_2)} & 1 - a_2 \end{pmatrix} \begin{pmatrix} \frac{1}{x} & 0 \\ 0 & 1 \end{pmatrix} \tag{1.43}$$

which is below one if

$$1 - a_2 + \frac{a_2 B}{2(B + e_2)x} < 1,$$

since $0 < a_k \leq 1$ for $k = 1, 2$. This relation can be rewritten as

$$x > \frac{B}{2(B + e_2)},$$

so a feasible positive $x$ exists. From the local stability result of Appendix B we conclude that the monopoly equilibrium $(0, x_2^M)$ is locally asymptotically stable. The stability of the other monopoly equilibrium $(x_1^M, 0)$ can be proved similarly.

This provides a first conclusion with regard to the equilibrium selection problem, because even if we obtain three Nash equilibria, from an evolutionary perspective a stability argument suggests that the interior equilibrium will not be selected. It remains an open question, however, as to which one of the two monopoly equilibria is more likely to be observed in the long run. The situation is even more intricate, since in addition to the two asymptotically stable boundary equilibria, in the strategy space another attracting set might coexist. This can be demonstrated by considering the best reply dynamics obtained for $a_k = 1, k = 1, 2$. In the case when $x_2^M > (A - c_1)/B$ and $x_1^M > (A - c_2)/B$ we have $(R_1(0), R_2(0)) = (x_1^M, x_2^M)$ and $\left(R_1(x_2^M), R_2(x_1^M)\right) = (0, 0)$. Therefore, under best reply dynamics the periodic cycle $C_2 = \left\{(0, 0); \left(x_1^M, x_2^M\right)\right\}$ coexists with the two stable monopoly equilibria. It is also easy to see that $C_2$ is stable, so it may even occur that an adjustment process fails to converge towards any Nash equilibrium in the long run. In such a situation, where several attractors coexist, the question of which attractor will be reached in the long run crucially depends on the initial conditions and the observed outcome becomes path dependent. Each of these long run outcomes has its own basin of attraction (see Appendix C for definitions of these concepts from the qualitative theory of dynamical systems) and any external random factor (a so-called "historical accident") that causes a displacement of some of the initial outputs may cause the trajectory to move across a basin boundary and, consequently, it will converge to a different attractor.

We can shed some light on this issue by using a mixture of analytical, geometrical and numerical methods, an approach which is typically used in the study of the global dynamical properties of nonlinear systems of dimension greater than one (see for example Mira et al. (1996), Brock and Hommes (1997) and Puu (2003)).

To get a better feeling for the global dynamics of our duopoly game where firms use partial adjustment towards the best response, we numerically compute the basins of attraction for the coexisting attractors. Let the reservation price be $A = 450$ and the slope of the linear inverse demand function be $B = 30$. For the sake of simplicity, we consider identical firms with cost parameters $c_1 = c_2 = c = 275$ and $e_1 = e_2 = e = -17$, so that production costs are increasing, but marginal costs are decreasing. (Similar values were chosen by Cox and Walker (1998) in an experimental setup). In order to guarantee non-negative prices, we select $L_1 = L_2 = 7.5$, which ensures that $L_1 + L_2 \leq A/B$. For these parameter values condition (1.38) is fulfilled and the interior equilibrium is unstable. In addition, $A - c_k - 2(B_k + e_k)L_k < 0$ implying that the output space $[0, L_1] \times [0, L_2]$ is divided into only four regions rather than the nine shown in Fig. 1.10).

- In Fig. 1.11a the basins of attraction of $E_1$, $E_2$, and the coexisting 2-cycle $C_2$ are shown for the best reply dynamics, namely for $a_1 = a_2 = 1$. The basin of attraction of $E_1$ is represented by the light-grey region, the basin of $E_2$ by the dark-grey region, and the basin of the cycle $C_2$ by the white region. The peculiar rectangular-shaped structure of the basins is related to the particular structure of the best reply process, $x_1(t+1) = R_1(x_2(t)), x_2(t+1) = R_2(x_1(t))$, where next period's output of firm $i$ only depends on the current output of the other firm. This implies that the eigenvectors associated with the unstable equilibrium $E$ (that belongs to the basin boundaries) are parallel to the coordinate axes. Moreover, the map which generates the dynamics transforms vertical lines into horizontal lines and vice versa. Hence, the invariant sets associated with the unstable node $E$, that form the boundaries of the basins, are formed by vertical and horizontal lines (on this point see also Bischi et al. (2000b)).
- If the speeds of adjustment are smaller than 1, important differences can be observed in the global dynamics. For example, Fig. 1.11b has been obtained with $a_1 = 0.97, a_2 = 0.98$, leaving all the other parameters unchanged. Now the stable 2-cycle has both periodic points characterized by positive coordinates, namely $C_2 = \{(0.19, 0.13); (6.39, 6.38)\}$, and the structure of the basins is different, in particular the basin of the cycle $C_2$ is smaller. The rectangular shape of the basins is lost since in the case of partial adjustment the eigenvectors associated with $E$ are no longer parallel to the coordinate axes.
- If the speeds of adjustment are even further decreased, the basin of the cycle $C_2$ shrinks; see Fig. 1.11c obtained with $a_1 = 0.93, a_2 = 0.95$. The periodic points of $C_2$ approach the boundary of its basin and after a contact with such a boundary, the cycle $C_2$ becomes unstable. As a consequence, the whole strategy space is shared by the basins of the two asymptotically stable boundary Nash equilibria $E_1$ and $E_2$, as depicted in Fig. 1.11d obtained with $a_1 = 0.9, a_2 = 0.92$.

Our analysis suggests the following insights. First, the basins of the Nash equilibria $E_1$ and $E_2$ are always simply connected. We emphasize this fact since later on we will encounter examples where the basins will not have such a simple structure. Second, whereas the local asymptotic stability of the boundary Nash equilibria does not depend on the adjustment speeds, the shape of the basins changes significantly when adjustment speeds become smaller. If the players' speeds of adjustment are lower, then the size of the basins of the equilibria is larger. As far as local asymptotic stability is concerned, it is well-known in the literature that decreasing the speeds of adjustments usually stabilizes the system (see for instance Fisher (1961), McManus and Quandt (1961) and some results to be presented in Chap. 2). Here, however, we emphasize that (in the present example) this also holds for the global dynamics. Finally, since the firm with the smaller adjustment speed has the larger basin, this firm is more likely to achieve the role of the monopolist, if initial production quantities are selected randomly from a close to uniform distribution.

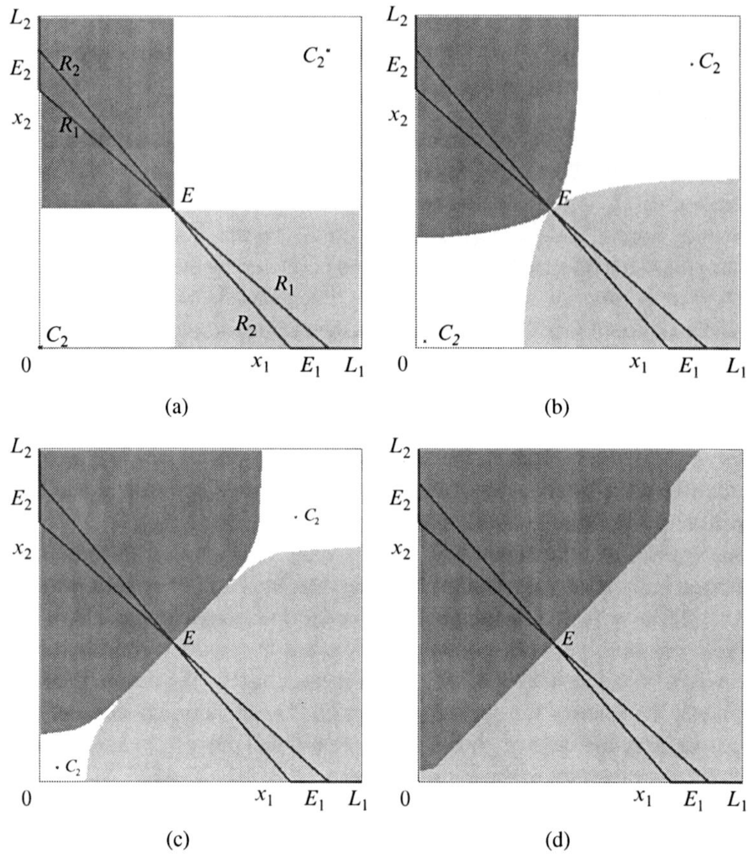


**Fig. 1.11** Basins of attraction for the Cournot duopoly when firms use partial adjustment towards the best response with linear demand/quadratic cost. *Light grey* basin of $E_1$; *dark grey* basin of $E_2$; *white* basin of the 2-cycle $C_2$. (**a**) Full adjustment, $a_1 = a_2 = 1$. The basins are rectangular. (**b**) Partial adjustment, $a_1 = 0.97$, $a_2 = 0.98$. The basins lose their rectangular shape. (**c**) Partial adjustment, $a_1 = 0.93$, $a_2 = 0.95$. The basin of $C_2$ shrinks. (**d**) Partial adjustment, $a_1 = 0.9$, $a_2 = 0.92$. The 2-cycle $C_2$ has become unstable, and its basin has disappeared

As a final remark we note that although the cyclic outcome $C_2$ is an attractor from a mathematical point of view, it has several shortcomings as a potential description of real-world economic behavior. First, whereas convergence to a steady state implies that the players' naive expectations are fulfilled at least in the long run, a sustained low-periodic oscillation implies that the players' expectations are permanently wrong. It seems plausible that in such a situation the players would learn how to improve their forecasts. Second, although profits are always positive in all Nash equilibria, this is not necessarily true in general for the cycle $C_2$. As an example consider again the best reply dynamics, where $C_2 = \{(0,0)\,;\left(x_1^M, x_2^M\right)\}$. The corresponding profits along the 2-cycle are $\varphi_k\,(0,0) = 0$ for firm $k$, with

$\varphi_1\left(x_1^M, x_2^M\right) = (A - c_1)\left[e_2(A - c_1) + B(c_2 - c_1)\right]/(4(B + e_1)(B + e_2))$, and $\varphi_2\left(x_1^M, x_2^M\right) = (A - c_2)\left[e_1(A - c_2) + B(c_1 - c_2)\right]/\ (4(B + e_1)(B + e_2))$. This shows that for at least one of the firms, profits are negative along the cycle. Moreover, if $-B < e_k < 0$ and $c_1 = c_2$, then we have negative profits for both firms, a situation which is not sustainable for any firm. As a consequence of these considerations, what our analysis of the global dynamics reveals is that for some initial production choices an economically infeasible situation will emerge for the firms. Notice that this important result can only be obtained through a global study of the structure of the basins of attraction.

We also would like to draw the reader's attention to a global bifurcation which is responsible for the drastic change in the dynamics obtained in this simple duopoly model. In a situation where marginal costs are decreasing strongly and $x_k^M < x_k^L$, we obtain three coexisting attractors: two boundary equilibria and a 2-cycle. Notice that the limiting quantities $x_k^L$ are located on a line where the map is not differentiable. Consider now what happens if marginal costs increase. At a certain point, a boundary equilibrium $x_k^M$ will collide with $x_k^L$, and if marginal costs are increased even further, then the interior equilibrium becomes globally stable. This is actually a first example of a border collision bifurcation, a global bifurcation occurring whenever a qualitative change in the phase diagram (that is, creation/destruction of invariant sets and/or stability change of existing ones) is due to a contact (and crossing) of an invariant set with a border where the map is not differentiable separating regions where it is differentiable. In this case the boundary that separates regions $\mathbb{D}^{(5)}$ and $\mathbb{D}^{(1)}$ is the one involved in the contact, and such a border is due to the presence of non-negativity constraint. This kind of global (or contact) bifurcations, specific to piece-wise differentiable dynamical systems, will be examined in more detail in Chap. 2, in particular in Examples 2.3 and 2.4.

### *1.3.2 A Cournot Oligopoly Game*

In his seminal paper, Theocharis (1960) studied the asymptotic stability of the Cournot–Nash equilibrium under discrete-time best reply dynamics with naive expectations. For this quantity-setting model with linear demand and linear costs, he found that the (unique) equilibrium is asymptotically stable only in the case of two competitors. It is marginally stable (see definition (A.1) in Appendix A) for three firms and unstable for more than three firms. Among others, McManus and Quandt (1961) and Fisher (1961) demonstrated that this result depends on the type of adjustment process the firms use to determine their production quantities. They showed that for certain adjustment processes in continuous-time the equilibrium is stable no matter what the number of firms is. These facts will be later discussed in Chap. 2. Despite this result Fisher (1961, p.125) notes that "... the tendency to instability does rise with the number of sellers for most of the processes considered". These early papers gave rise to a lively discussion that has endured until the present day. One of the main topics in this body of literature is the relation between the following issues: the quasi-competitiveness of the economy, that is the question as to whether output increases and the market price decreases with an increasing number of firms in the industry; the asymptotic stability of the equilibrium if entry occurs; the question as to whether perfect competition is obtained in the limit as the number of competitors is increased. The interested reader should consult for example Frank (1965), Ruffin (1971), Howrey and Quandt (1968), Okuguchi (1976), or more recently, Seade (1980) and Amir and Lambson (2000) to get an impression of the variety of interesting results obtained concerning this issue. In this section we focus on asymptotic stability issues and we try to answer the question: is local asymptotic stability obtained when the number of firms increases? Furthermore, we also address the topic of global dynamics, that is we look at the changes in the basins of attraction of the stable equilibria. Clearly, a discussion of these issues becomes more complicated when the model is nonlinear, since increasing the number of players means increasing the dimension of the dynamical system. This is so since such increases lead to greater complexity in the dynamics of nonlinear systems, whereas in the case of linear systems no new dynamic phenomena arise.

In order to keep the mathematical analysis tractable, but at the same time to also shed some light on the relation between asymptotic stability and the number of firms, in what follows we will consider both the symmetric and semi-symmetric models. Recall that in the *symmetric case* it is assumed that all firms are identical, so that they have identical cost functions and all firms start from the same initial production quantities. Since the cost and demand parameters are identical for all firms, the reaction functions $R_k$ will be identical, say $R_k = R$ for each $k$. Consequently, the quantities will be identical for all periods, and the dynamics are governed by a 1-dimensional system. If we let $x(t)$ denote the common output of the representative firm, then the one-dimensional model in the symmetric case is obtained by setting $Q_k = (N-1)x$ for each $k$. It is worth noting that the symmetric case may be structurally unstable, that is the outcome obtained for the representative firm in the symmetric case may be completely different from the outcome of the model with almost identical, but nevertheless heterogeneous firms (the firms might differ in their production costs or might select slightly different initial quantities). Therefore, the insights obtained from the symmetric model need to be accepted with some caution. In order to derive some results which can be compared with the existing literature, we reconsider the partial adjustment towards the best response process given by (1.23).

The symmetric case is obtained if we assume $N$ players with identical quadratic cost functions (as in Example 1.2), that is $c_1 = c_2 = \cdots = c_N = c$ and $e_1 = e_2 = \cdots = e_N = e$, identical adjustment speeds, that is $a_1 = a_2 = \ldots, a_N = a$, and identical capacity limits $L_1 = L_2 = \cdots = L_N = L$. It is also assumed that $B+e > 0$, so the payoff functions of the firms are strictly concave in their strategies. Then from (1.23) the 1-dimensional model which summarizes the common behavior of all identical firms starting from identical initial condition $x_1(0) = x_2(0) = \cdots = x_N(0) = x(0)$ is

$$x(t+1) = T(x(t)) \equiv (1-a)\,x(t) + aR\left((N-1)\,x(t)\right),$$

where (see the reaction function in case (i) of Example 1.2)

$$R((N-1)x) = \begin{cases} 0 & \text{if} \quad z^* < 0, \\ L & \text{if} \quad z^* > L, \\ z^* & \text{otherwise}, \end{cases}$$

with $z^* = \big(A - c - B(N-1)x\big)/\big(2(B+e)\big)$.

Observe that the number of firms $N$ enters as a parameter, so we can study the stability conditions as $N$ is increased. The positive equilibrium is given by

$$\bar{x} = \frac{A-c}{B(N+1)+2e}$$

and the map $T$ is a contraction provided that $|T'(x)| < 1$, that is

$$0 < a\frac{BN+B+2e}{2(B+e)} < 2.$$

This implies that the positive equilibrium is always asymptotically stable for sufficiently small values of the adjustment speed $a$. Moreover, given $0 < a \leq 1$, asymptotic stability is obtained for

$$N < \frac{(4-a)\,B + 2\,(2-a)\,e}{a\,B}.$$

In the case of best reply dynamics, $a = 1$, the stability condition reads $N < (3B + 2e)/B$. In the case of linear costs, $e = 0$, we obtain the result by Theocharis stating that asymptotic stability is obtained for $N < 3$.

In the *semi-symmetric* case $(N-1)$ firms are assumed to be identical, whereas one firm differs with regard to its production costs and/or initial production quantity. Let firms $2, \ldots, N$ be identical, then their production choices will coincide in each period, that is $x_k = x_2$ for all $k \geq 2$. Let us denote the production quantity of firm 1 by $x_1$, then

$$Q_1 = (N-1)x_2 \text{ and } Q_2 = x_1 + (N-2)x_2. \tag{1.44}$$

By using the reaction functions $R_1$ and $R_2 = \cdots = R_N$, we obtain a two-dimensional system with state variables $x_1$ and $x_2$. In (1.23) we set $c_2 = \cdots = c_N$, $e_2 = \cdots = e_N$, $a_2 = \cdots = a_N$, and $L_2 = \cdots = L_N$. Then the 2-dimensional model that governs the behavior of firm 1 and the common behavior of the identical firms $2, \ldots, N$ becomes

$$T_N\colon \begin{cases} x_1(t+1) = (1-a_1)\,x_1(t) + a_1 R_1\left((N-1)\,x_2(t)\right), \\ x_2(t+1) = (1-a_2)\,x_2(t) + a_2 R_2\left(x_1(t) + (N-2)\,x_2(t)\right), \end{cases}$$

where (again refer to the reaction function in case (i) of Example 1.2)

$$R_1((N-1)x_2) = \begin{cases} 0 & \text{if} \quad z_1^* < 0, \\ L_1 & \text{if} \quad z_1^* > L_1, \\ z_1^* & \text{otherwise,} \end{cases}$$

with $z_1^* = \big(A - c_1 - B(N-1)x_2\big)/\big(2(B+e_1)\big)$ and

$$R_2(x_1 + (N-2)x_2) = \begin{cases} 0 & \text{if} \quad z_2^* < 0, \\ L_2 & \text{if} \quad z_2^* > L_2, \\ z_2^* & \text{otherwise,} \end{cases}$$

with $z_2^* = \big(A - c_2 - B\,(x_1 + (N-2)x_2)\big)/\big(2(B+e_2)\big)$.

The interior equilibrium is independent of $a_k$, $k = 1, 2$, but depends on the number of firms $N$. It is given by $E = (\bar{x}_1(N), \bar{x}_2(N))$ with

$$\bar{x}_1(N) = \frac{A(B+2e_2) - 2c_1e_2 + B\,(c_2(N-1) - c_1N)}{2B(N-2)(B+e_1) + 4(B+e_1)(B+e_2) - B^2(N-1)},$$

$$\bar{x}_2(N) = \frac{2(B+e_1)(A-c_2) - B\,(A-c_1)}{2B(N-2)(B+e_1) + 4(B+e_1)(B+e_2) - B^2(N-1)}.$$

The Jacobian matrix computed at the interior equilibrium is

$$\begin{pmatrix} 1-a_1 & -a_1\frac{B(N-1)}{2(B+e_1)} \\ -a_2\frac{B}{2(B+e_2)} & 1-a_2-a_2\frac{B(N-2)}{2(B+e_2)} \end{pmatrix},$$

from which the stability conditions can be obtained by applying conditions (1.40). Interesting stability results are obtained for the boundary equilibria, in the case when $B^2 > 4(B+e_1)(B+e_2)$ (illustrated in Fig. 1.3 for one possible situation). The Jacobian evaluated in the neighborhood of $E_1$ is either

$$\begin{pmatrix} 1-a_1 & -a_1\frac{B(N-1)}{2(B+e_1)} \\ 0 & 1-a_2 \end{pmatrix} \quad \text{or} \quad \begin{pmatrix} 1-a_1 & 0 \\ 0 & 1-a_2 \end{pmatrix}$$

or both, if the equilibrium is on the boundary between the two regions, since $R_2 \equiv 0$ here. The Jacobian evaluated in the neighborhood of $E_2$ is either

$$\begin{pmatrix} 1-a_1 & 0 \\ -a_2\frac{B}{2(B+e_2)} & 1-a_2-a_2\frac{B(N-2)}{2(B+e_2)} \end{pmatrix} \quad \text{or} \quad \begin{pmatrix} 1-a_1 & 0 \\ 0 & 1-a_2-a_2\frac{B(N-1)}{2(B+e_2)} \end{pmatrix}$$

or both, because $R_2 \equiv 0$ here.

As before, let $\boldsymbol{P} = \begin{pmatrix} x & 0 \\ 0 & 1 \end{pmatrix}$, then the row norms of the Jacobians around $E_1$ generated by the matrix $\boldsymbol{P}$ are bounded by the row norm of the matrix

$$\begin{pmatrix} x & 0 \\ 0 & 1 \end{pmatrix} \begin{pmatrix} 1-a_1 & a_1 \frac{B(N-1)}{2(B+e_1)} \\ 0 & 1-a_2 \end{pmatrix} \begin{pmatrix} \frac{1}{x} & 0 \\ 0 & 1 \end{pmatrix} = \begin{pmatrix} 1-a_1 & a_1 \frac{Bx(N-1)}{2(B+e_1)} \\ 0 & 1-a_2 \end{pmatrix} \tag{1.45}$$

which is below one if

$$1 - a_1 + a_1 \frac{Bx(N-1)}{2(B+e_1)} < 1,$$

that is, when

$$x < \frac{2(B+e_1)}{B(N-1)}.$$

Hence the equilibrium $E_1$ is locally asymptotically stable for all values of $N$. Similarly, $E_2$ is locally asymptotically stable if there is a positive $x$ such that

$$\frac{a_2 B}{2x(B+e_2)} + \left| 1 - a_2 - a_2 \frac{B(N-2)}{2(B+e_2)} \right| < 1$$

which occurs if

$$-1 < 1 - a_2 \left( 1 + \frac{B(N-2)}{2(B+e_2)} \right) < 1.$$

Therefore, $E_2$ is stable provided that

$$0 < a_2 \frac{B(N-2) + 2(B+e_2)}{2(B+e_2)} < 2.$$

From this stability condition we can now derive several interesting results. First, as already shown before, in the case of duopoly ($N = 2$) the boundary equilibrium $E_2$ is also always stable, like $E_1$. Moreover, the boundary equilibrium $E_2$ is stable provided that $a_2$ is sufficiently small, which means that firms $2, \ldots, N$ have a high inertia in adjusting their quantities toward the best responses. Finally, increasing the number of firms has a destabilizing role. In fact the stability condition can be written as

$$N < 2 + \frac{2\,(2-a_2)\,(B+e_2)}{Ba_2},$$

so that for given cost parameters and adjustment speeds asymptotic stability is lost when the number of firms reaches a certain size.

To conclude this section, we study the global dynamics of the semi-symmetric model. Consider again the parameter values $A = 450$, $B = 30$ and $c_1 = c_2 = \ldots = c_N = 275$, $e_1 = e_2 = \cdots = e_N = -17$. For the adjustment speeds of the two firms we select $a_1 = 0.6$ and $a_2 = \cdots = a_N = 0.45$. For these parameter values the stability condition derived in the previous paragraph tells us that the boundary equilibrium $E_2$ is asymptotically stable if $N < 4$. In Fig. 1.12a we depict the basins

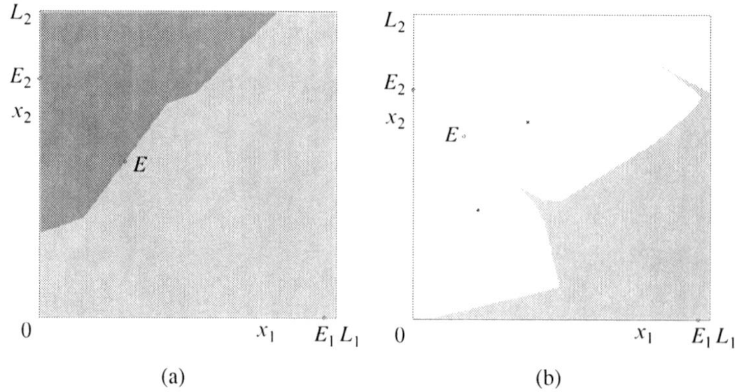


**Fig. 1.12** The Cournot oligopoly with linear demand/quadratic cost. Firms use partial adjustment towards the best response. Basins of attraction of the various equilibria for different values of the number of firms $N$. (**a**) The 3-firm case. Here both $E_1$ and $E_2$ are stable. *Dark grey* basin of $E_2$; *light grey* basin of $E_1$. (**b**) The 5-firm case. Now $E_1$ is stable, $E_2$ is unstable. *Light grey* basin of $E_1$; *white* basin of the two cycle

of the two boundary equilibria $E_1$ and $E_2$ for $N = 3$ firms. To guarantee non-negative prices, we have selected $L_1 = 7$ and $L_2 = L_3 = 4$. Both boundary equilibria are asymptotically stable, each with its own basin of attraction represented by the different shadings of grey. In Fig. 1.12b we show the situation for $N = 5$ firms where $L_1 = 7$ and $L_2 = \cdots = L_5 = 2$. Now only the boundary equilibrium $E_1$ is asymptotically stable, and its basin is represented by the light grey region. Points located in the white region converge to the 2-cycle represented by the two dots.

### 1.3.3 Cournot Duopoly Revisited: A Gradient Type Adjustment Process

The local stability of an equilibrium and the global dynamics depend on the adjustment mechanism the firms use to update their production choices. We now reconsider the duopoly case analyzed in Sect. 1.3.1, but instead of assuming partial adjustment towards the best response, we now consider a discrete time adjustment process based on marginal profits, similar to the gradient adjustment process discussed in Sect. 1.2 (1.32). However we assume now that the *relative* variation in production quantities is proportional to the marginal profits, that is firm $i$ adjusts its output according to

$$\frac{x_i(t+1) - x_i(t)}{x_i(t)} = a_i \left(\frac{\partial \varphi_i}{\partial x_i}\right)$$

with $a_i > 0$. With these assumptions, the dynamics are now governed by the discrete time system

$$T_g: \begin{cases} x_1(t+1) = x_1(t) + a_1 x_1(t)\left[A - c_1 - 2(B+e_1)x_1(t) - Bx_2(t)\right], \\ \\ x_2(t+1) = x_2(t) + a_2 x_2(t)\left[A - c_2 - 2(B+e_2)x_2(t) - Bx_1(t)\right]. \end{cases} \tag{1.46}$$

It is easy to see that the interior steady state of the adjustment process based on marginal profits coincides with the unique interior Nash equilibrium $E = (\bar{x}_1, \bar{x}_2)$ given in (1.12). To study the local asymptotic stability of $E$, we consider the Jacobian matrix of (1.46). Since the Nash equilibrium is located at the intersection of the two reaction functions given in (1.34) and (1.35), we have $B\bar{x}_i = A - c_j - 2\left(B + e_j\right)\bar{x}_j \quad (i, j = 1, 2, \ i \neq j)$. Therefore, the Jacobian matrix evaluated at the interior equilibrium $E$ can be written as

$$\begin{pmatrix} 1 - 2a_1(B+e_1)\bar{x}_1 & -a_1 B\bar{x}_1 \\ -a_2 B\bar{x}_2 & 1 - 2a_2(B+e_2)\bar{x}_2 \end{pmatrix}. \tag{1.47}$$

We can check the stability conditions by use of the relations (1.40) with

$$q = (1 - 2a_1(B+e_1)\bar{x}_1)(1 - 2a_2(B+e_2)\bar{x}_2) - a_1 a_2 B^2 \bar{x}_1 \bar{x}_2,$$

and

$$p = -2 + 2a_1(B+e_1)\bar{x}_1 + 2a_2(B+e_2)\bar{x}_2.$$

By assuming that $B + e_k > 0$ for $k = 1, 2$, clearly $q < 1$. Notice that

$$p + q + 1 = 4a_1 a_2 (B+e_1)(B+e_2)\bar{x}_1\bar{x}_2 - a_1 a_2 B^2 \bar{x}_1 \bar{x}_2,$$

which is positive if $B^2 < 4(B+e_1)(B+e_2)$. Similarly,

$$-p+q+1 = 4 - 4a_1(B+e_1)\bar{x}_1 - 4a_2(B+e_2)\bar{x}_2 + 4a_1 a_2 \bar{x}_1 \bar{x}_2 (B+e_1)(B+e_2),$$

so this is positive, if

$$(4(B+e_1)(B+e_2) - B^2)\bar{x}_1\bar{x}_2 a_1 a_2 - 4(B+e_1)\bar{x}_1 a_1 - 4(B+e_2)\bar{x}_2 a_2 + 4 < 0. \tag{1.48}$$

If $B^2 < 4(B+e_1)(B+e_2)$ and the equilibrium $E$ is positive, then this additional condition can be used to determine a region of stability in the $(a_1, a_2)$-plane. In contrast to the adjustment process where firms partially adjust their quantities towards the best reply, here the speeds of adjustment are crucial for local asymptotic stability of the Nash equilibrium. As remarked earlier, the stabilizing role of sufficiently small values of the adjustment speeds has been observed before by many authors (see for example Fisher (1961), McManus and Quandt (1961), and Flam (1993)). In Fig. 1.13 we depict the stability region (shaded) in the $(a_1, a_2)$ plane obtained for the parameter values $A = 450$, $B = 30$, $c_1 = c_2 = 275$, $e_1 = e_2 = -11$. For values

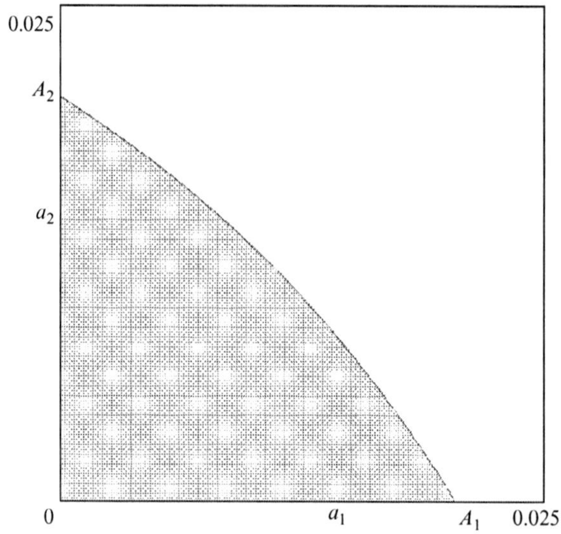


**Fig. 1.13** The Cournot duopoly with a gradient type adjustment process and linear demand/quadratic cost. The *hashed area* indicates the stability region of the interior Nash equilibrium $E$ in the $(a_1, a_2)$ plane of adjustment speeds

of $(a_1, a_2)$ inside the stability region, the Nash equilibrium $E$ is an asymptotically stable node. The boundary of this region represents a bifurcation curve at which $E$ loses asymptotic stability through a flip (or period doubling) bifurcation (see for example Guckenheimer and Holmes (1983), or Lorenz (1995)). This bifurcation curve intersects the axes in the points

$$A_1 = \left(\frac{1}{(B+e_1)\bar{x}_1}, 0\right) \quad \text{and} \quad A_2 = \left(0, \frac{1}{(B+e_2)\bar{x}_2}\right),$$

from which further information on the effects of the model's parameters on the local asymptotic stability of $E$ could be derived by further analysis.

So far we have only considered questions related to local asymptotic stability of the interior equilibrium. But what can we say about the global dynamics? That is, given that the interior Nash equilibrium is locally asymptotically stable, what can be said about its basin of attraction, defined as the set of feasible initial conditions which generate bounded and positive trajectories converging to $E$? In Fig. 1.14, obtained with parameters $A = 450$, $B = 30$, $c_1 = c_2 = 275$, $e_1 = e_2 = -11$ and speeds of adjustment $a_1 = 0.01, a_2 = 0.012$, the Nash equilibrium $E = (2.57, 2.57)$ is locally asymptotically stable and its basin of attraction (or feasible set) is represented by the white area. The region in grey represents the basin of infinity, denoted $B(\infty)$, that is the set of initial conditions that generates unbounded (and negative), therefore "infeasible", trajectories. The interior Nash equilibrium is not globally asymptotically stable since not all initial conditions in the strategy space

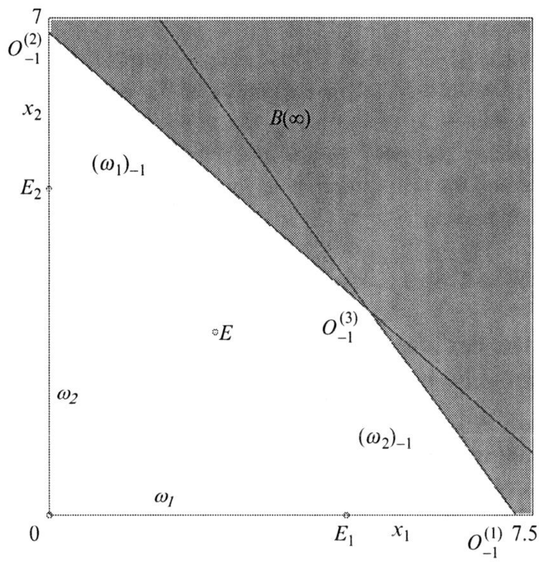


**Fig. 1.14** The Cournot duopoly with a gradient type adjustment process and linear demand/quadratic cost. The *white* region is the basin of attraction of the Nash equilibrium $E$, the *dark grey* region is $B(\infty)$. The basin of $E$ is bounded by the two segments $\omega_1$, $\omega_2$ and their rank-1 preimages $(\omega_1)_{-1}$, $(\omega_2)_{-1}$

are economically feasible. For all quantity choices in the basin of $E$, we obtain $x_1 + x_2 < A/B$. Therefore, non-negativity of prices is guaranteed. Note that for the set of parameters we have selected here, the interior equilibrium would be globally stable with respect to partial adjustment towards the best response.

For the set of parameters used to obtain Fig. 1.14, the set of initial conditions which lead to convergence to the Nash equilibrium $E$ is the interior of the quadrilateral $OO_{-1}^{(1)}O_{-1}^{(3)}O_{-1}^{(2)}$, where $O = (0,0)$ denotes the origin and the other three vertexes are the rank-1 preimages of $O$, meaning that for these points $T_g(O_{-1}^{(i)}) = O$ holds for $i = 1,2,3$ (Note that the mapping $T_g$ was defined in (1.46)). These points are given by

$$O_{-1}^{(1)} = \left(\frac{1 + a_1(A - c_1)}{2a_1(B + e_1)}, 0\right), \quad O_{-1}^{(2)} = \left(0, \frac{1 + a_2(A - c_2)}{2a_2(B + e_2)}\right) \tag{1.49}$$

and

$$O_{-1}^{(3)} = \left(\frac{2a_2(B + e_2)\,(1 + a_1(A - c_1)) - a_1 B\,(1 + a_2(A - c_2))}{3B^2 a_1 a_2 + 4a_1 a_2(e_1 + e_2) + 4a_1 a_2 e_1 e_2},\right.$$
$$\left.\frac{2a_1(B + e_1)\,(1 + a_2(A - c_2)) - a_2 B\,(1 + a_1(A - c_1))}{3B^2 a_1 a_2 + 4a_1 a_2(e_1 + e_2) + 4a_1 a_2 e_1 e_2}\right), \tag{1.50}$$

which can be obtained by solving the fourth degree algebraic system (1.46) for $x_i(t)$, upon setting $x_i(t+1) = 0$, $(i = 1, 2)$. A simple strategy for obtaining the preimages of $O$ is to start from the dynamics of $T_g$ restricted to the axes. Since $x_i(t) = 0$ implies $x_i(t+1) = 0$, starting from an initial condition on a coordinate axis, the dynamics are "trapped" on this axis for all $t$. In other words, a monopoly prevails over time and the one-dimensional "monopoly dynamics" is obtained from (1.46) with $x_i = 0$, namely

$$x_j(t+1) = (1 + a_j(A - c_j))x_j(t) - 2(B + e_j)a_j x_j^2(t). \tag{1.51}$$

We also note that this map is conjugate to the standard logistic map $x(t+1) = \mu x(t)(1 - x(t))$ through the linear transformation $x_j = \frac{1+a_j(A-c_j)}{2a_j(B+e_j)}x$, from which the relation $\mu = 1 + a_j(A - c_j)$ can be obtained. The following results for our map can be directly derived from the properties of the logistic map, which is well-studied in the literature; see for example, Devaney (1989). The rank-1 preimages $O_{-1}^{(j)}$ given in (1.49) can now be easily derived from (1.51). Along the $x_j$-axis $(j = 1, 2)$, the one-dimensional restriction (1.51) gives bounded dynamics for $a_j(A - c_j) \le 3$ provided that the initial conditions are taken inside the segment $\omega_j = OO_{-1}^{(j)}$. Observe that divergent trajectories along the invariant $x_j$ axis are obtained if the initial condition is out of the segment $\omega_j$ $(j = 1, 2)$. Let us now turn to the quadrilateral region bounded by the two segments $\omega_1$ and $\omega_2$ and their rank-1 preimages, say $(\omega_1)_{-1}$ and $(\omega_2)_{-1}$ respectively (see Fig. 1.14). The preimages $(\omega_1)_{-1}$ and $(\omega_2)_{-1}$ can be analytically computed as follows. Let $X = (x, 0)$ be a point of $\omega_1$. Its preimages are the real solutions $(x_1, x_2)$ of the algebraic system

$$\begin{cases} x_1\left[1 + a_1(A - c_1) - 2a_1(B + e_1)x_1 - a_1 B x_2\right] = x, \\ \\ x_2\left[1 + a_2(A - c_2) - a_2 B x_1 - 2a_2(B + e_2)x_2\right] = 0. \end{cases} \tag{1.52}$$

From the second equation it is easy to see that the preimages of the points of $\omega_1$ are either located on the same invariant axis $x_2 = 0$ or on the line represented by the equation

$$a_2 B x_1 + 2a_2(B + e_2)x_2 = 1 + a_2(A - c_2). \tag{1.53}$$

Analogously, the preimages of a point of $\omega_2$ belong to the same invariant axis $x_1 = 0$ or to the curve represented by equation

$$2a_1(B + e_1)x_1 + a_1 B x_2 = 1 + a_1(A - c_1). \tag{1.54}$$

It is now straightforward to see that the line (1.53) intersects the $x_2$ axis in the point $O_{-1}^{(2)}$ and the line (1.54) intersects the $x_1$ axis in the point $O_{-1}^{(1)}$. Moreover, the two lines intersect at the point $O_{-1}^{(3)}$. A summary of these observations leads to the following description of the basin of the asymptotically stable Nash equilibrium $E$ as shown in Fig. 1.14. The rank-1 preimages of the origin are the vertexes of the quadrilateral $OO_{-1}^{(1)}O_{-1}^{(3)}O_{-1}^{(2)}$. The sides of this region are given by $\omega_1$, $\omega_2$ and their respective rank-1 preimages $(\omega_1)_{-1}$ and $(\omega_2)_{-1}$ respectively. All points inside this quadrilateral region lead to convergence, all points outside cannot generate feasible trajectories. Points located to the right of $(\omega_2)_{-1}$ are mapped into points with negative value of $x_1$ after one iteration, as can be easily deduced from the first component of (1.46). Points located above $(\omega_1)_{-1}$ are mapped into points with negative value of $x_2$ after one iteration, as can be deduced from the second component of (1.46). The expressions in (1.53) and (1.54) can be used to determine the impact of parameter changes on the basin. Finally, observe that for these values of the parameters the basin of the unique interior Nash equilibrium is a rather simple and connected set.

### *1.3.4 Simple Basins and Critical Curves*

In this subsection we introduce the concept of critical curves (see also Appendix C). This subsection uses many concepts about dynamical systems that may not be familiar to some readers (such as noninvertible maps, critical sets, preimages of various ranks and so on). These concepts are reviewed in Appendix C, which the reader may need to study before working through this subsection.

Recall that in the previous subsection we have demonstrated how to obtain the boundaries of the feasible region by taking the preimages $(\omega_i)_{-1}$ $(i = 1, 2)$ of the coordinate axes. Since the map $T_g$ in (1.46) is a noninvertible map, as can be readily deduced from the fact that the origin has four preimages, there might be further preimages of $(\omega_i)_{-1}$ $(i = 1, 2)$, which have to be also considered in order to obtain the whole boundary of the feasible region. In order to determine if $(\omega_i)_{-1}$ $(i = 1, 2)$ have further preimages, we can use the critical curves of the map which can be used to identify regions in the feasible set (or strategy space) with a different number of preimages.

To begin with, let us consider a given point $(x_1', x_2')$ in the strategy space. Then its preimages can be calculated by setting $x_1(t+1) = x_1', x_2(t+1) = x_2'$ in (1.46) and solving with respect to $x_1$ and $x_2$ the fourth degree algebraic system,

$$\begin{cases} x_1 \left[1 + a_1 \left(A - c_1 - 2(B + e_1)x_1 - Bx_2\right)\right] = x_1', \\ \\ x_2 \left[1 + a_2 \left(A - c_2 - 2(B + e_2)x_2 - Bx_1\right)\right] = x_2'. \end{cases} \tag{1.55}$$

Clearly, this algebraic system may have up to four real solutions, which are the rank-1 preimages of $(x_1', x_2')$. We can now use this information to subdivide the strategy space into regions characterized by a different number of preimages. This is shown in Fig. 1.15a, which is obtained with the same parameters as Fig. 1.14. The regions $Z_k$ denote the sets of points which have $k$ real and distinct rank-1 preimages. For example, as shown above, the origin $O = (0, 0) \in Z_4$, because it has four

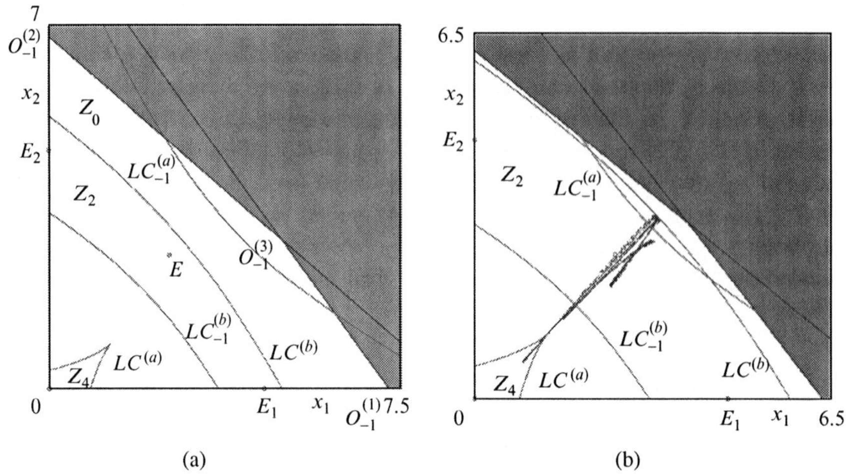


**Fig. 1.15** The Cournot duopoly with a gradient type adjustment process and linear demand/quadratic cost. Illustrating the regions of preimages of different ranks, the sets of points where the Jacobian vanishes ($LC_{-1}^{(a)}$ and $LC_{-1}^{(b)}$) and the critical curves $LC^a$ and $LC^b$. (**a**) The parameters are the same as in Fig. 1.13. (**b**) The speeds of adjustment are slightly higher, $E$ becomes unstable and a strange attractor emerges, but the basic structure of the basin remains the same as in (**a**). Note however that the critical curve $LC^{(b)}$ in now quite close to the boundary of the *white* and *grey* regions

rank-1 preimages, given by $O$ itself (since $T_g(0,0) = (0,0)$) and $O_{-1}^{(i)}$, $i = 1,2,3$ (since $T_g(O_{-1}^{(i)}) = (0,0)$ as well). The regions $Z_k$ are separated by segments of critical curves denoted as $LC^{(a)}$ and $LC^{(b)}$ in Fig. 1.15a.

An intuitive understanding of the importance of critical curves can be obtained by referring to the folding or unfolding mechanism of a map. The map (1.46) is noninvertible, which means that distinct points in the action set can be mapped into the same point by $T_g$. This can be geometrically envisioned by imagining a process which folds the action space onto itself (so that points which are in different locations are folded onto each other). A result from algebraic geometry tells us that the folding process can be characterized by a change of sign of the determinant of the Jacobian of the map: if the sign is positive, then the map is orientation preserving, whereas it is orientation reversing otherwise.[4] The folding curves where the sign change occurs is the locus of points where the determinant of the Jacobian of the map vanishes. Its *image* gives the so-called critical curve, which separates zones or regions with different numbers of *preimages* (this indicates the importance of the unfolding action of the map). To sum up, the following numerical procedure

[4] Consider a one-dimensional, continuously differentiable map $g(y)$. If $g'(y) > 0$, then for $x < y$, it follows that $g(x) < g(y)$. If, on the other hand, $g'(y) < 0$, the orientation is reversed. Obviously, the change of signs occurs exactly at the point where the derivative vanishes.

(see also Appendix C) can be used to obtain the critical curves (for a given set of parameters):

1. The map (1.46) is continuously differentiable, so the (folding) set $LC_{-1}$ can be obtained numerically as the locus of points $(x_1, x_2)$ for which the Jacobian determinant of $T_g$ vanishes.
2. The critical curves $LC$, which separate the regions $Z_k$, are obtained by computing the images of the points belonging to $LC_{-1}$, that is $LC = T_g(LC_{-1})$.

In Fig. 1.15a the set of points at which the Jacobian vanishes gives the curves denoted by $LC_{-1}^{(a)}$ and $LC_{-1}^{(b)}$. It is formed by the union of the two branches of a hyperbola. Also the critical curve $LC = T_g(LC_{-1})$ is formed by two branches, denoted by $LC^{(a)} = T_g(LC_{-1}^{(a)})$ and $LC^{(b)} = T_g(LC_{-1}^{(b)})$. The curve $LC^{(b)}$ separates the region $Z_0$, whose points have no preimages, from the region $Z_2$, whose points have two distinct rank-1 preimages. The curve $LC^{(a)}$ separates the region $Z_2$ from $Z_4$, whose points have four distinct preimages.

Our analysis based on the critical curves of the map now reveals why the set of initial conditions that lead to convergence to the Nash equilibrium, bounded by $\omega_1, \omega_2$ and its preimages $(\omega_1)_{-1}$ and $(\omega_2)_{-1}$, is a rather simple set. It is due to the fact that only preimages of rank-1 of $\omega_1$ and $\omega_2$ exist. Note that $(\omega_1)_{-1}$ and $(\omega_2)_{-1}$ are entirely included in $Z_0$, that is a region of the feasible set whose points have no preimages. Therefore, the preimages $(\omega_i)_{-1}$ $(i = 1, 2)$ of the invariant axes, have no preimages of higher rank. Consequently, the whole boundary that separates the basin $B(E)$ and the infeasible set $B(\infty)$ is

$$\mathcal{F} = \left(\cup_{n=0}^{\infty} T_g^{-n}(\omega_1)\right) \bigcup \left(\cup_{n=0}^{\infty} T_g^{-n}(\omega_2)\right), \tag{1.56}$$

that is, the union of all the preimages of the segments $\omega_1$ and $\omega_2$ (see Appendix C), which is a rather simple set.

To conclude this subsection, we would like to stress the fact that the properties of the basin boundaries are related to the global dynamics of our duopoly model. Such a simple structure of the basin may be also maintained when the Nash equilibrium loses stability due to local (period-doubling) bifurcations. In Fig. 1.15b, obtained with the same parameters as before except that $a_1 = 0.015$ and $a_2 = 0.0165$, we depict a situation where (after the usual period-doubling sequence) a chaotic attractor describes the long run evolution of the production decisions of the duopolists. Despite the fact that the dynamic behavior can be considered as complex, the basin boundaries are still given by the same quadrilateral.

The reader should notice, however, that basins are not always as simple as in the examples presented so far in this book. Indeed, a closer look at Fig. 1.15b reveals that the critical curve $LC^{(b)}$ is rather close to a basin boundary. This indicates that a small shift of this curve due to a parameter variation may cause a contact, after which a portion of the set of infeasible points $B(\infty)$ crosses the critical curve and, consequently, enters the region $Z_2$. In the next subsection we will show that such contact bifurcations may have a considerable impact on the topological structure of the feasible set.

### 1.3.5 Disconnected Basins

In all the examples encountered up to now, the basins of the corresponding attractor were rather simple and were connected sets. As we shall now demonstrate, basins can have a quite complicated structure. For example, they can be pierced by many holes or may consist of areas without any connection. In such situations predicting the long run outcome of the duopoly game where players use certain adjustment processes to determine their production quantities over time is quite difficult. This becomes particularly relevant when stochastic influences play a role.

In Fig. 1.16a we depict the situation after an increase in the adjustment speeds from $a_1 = 0.015$, $a_2 = 0.0165$ (the values in Fig. 1.15b) to $a_1 = 0.015$, $a_2 = 0.017$. After the contact of the curve $LC^{(b)}$ with the boundary of $B(\infty)$, a set indicated as $H_0$ which belongs to the infeasible set $B(\infty)$ enters $Z_2$ (see the region indicated by the arrow in Figs. 1.16a, b).

This means that points belonging to $H_0$ have two distinct preimages, say $H_{-1}^{(1)}$ and $H_{-1}^{(2)}$, which are located on opposite sides of the curve $LC_{-1}^{(b)}$ (the preimages of points exactly on the curve $LC^{(b)}$ inside $B(\infty)$ are located on $LC_{-1}^{(b)}$). Obviously, since $H_0$ belongs to the set $B(\infty)$, initial conditions belonging to $H_{-1}^{(1)}$ and $H_{-1}^{(2)}$ also lead to infeasible trajectories, since they are mapped into the infeasible set after one iteration. The rank-1 preimages of $H_0$ constitute a so-called *hole* of $B(\infty)$ which is located entirely inside the feasible set (this hole is also called a "lake" in Mira et al. (1996)). Since this hole, also referred to as the *main hole*, again lies inside the region $Z_2$, it also has two preimages. These smaller holes, denoted as $H_{-2}^{(1)}$ and

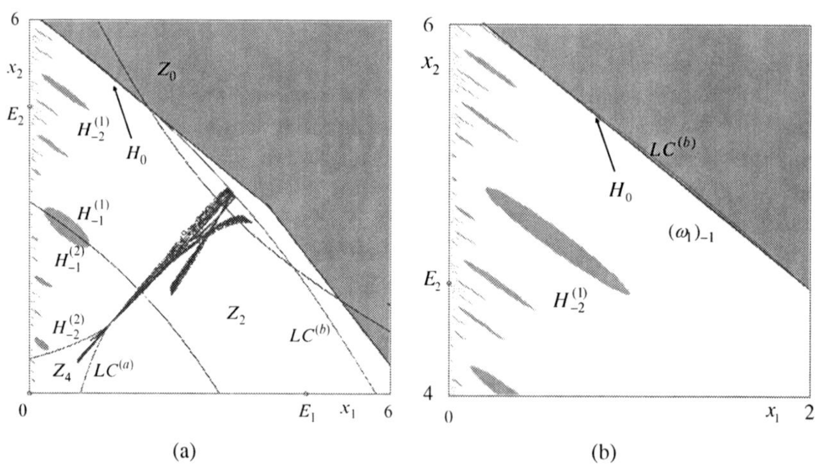


**Fig. 1.16** The Cournot duopoly with a gradient type adjustment process and linear demand/quadratic cost. Slightly higher speeds of adjustment than in the case of Fig. 1.15. The critical curve $LC^{(b)}$ has crossed the basin boundary and a disconnected basin of attraction now results. (**a**) The entire region. (**b**) A close up of the set $H_0$ and its preimages

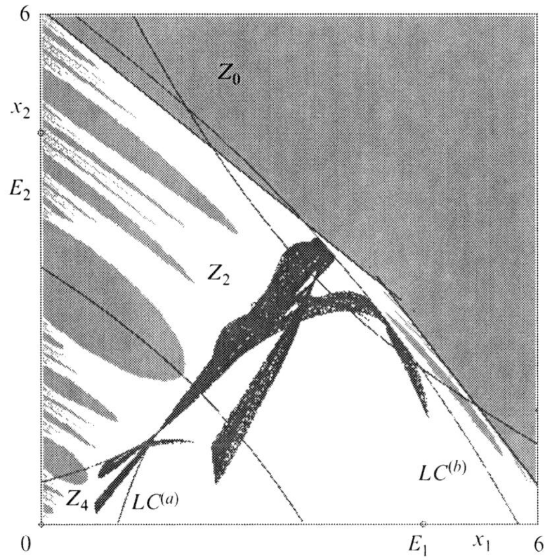


**Fig. 1.17** The Cournot duopoly with a gradient type adjustment process and linear demand/quadratic cost. The same situation as in Fig. 1.16, but with slightly higher speeds of adjustment. Note how the holes have become larger and connected along the vertical axis

$H_{-2}^{(2)}$, contain initial conditions which are mapped into the main hole and then into the infeasible set. The sets $H_{-2}^{(1)}$ and $H_{-2}^{(2)}$ are bounded by preimages of rank 3 of $\omega_1$. Since these smaller holes are again both inside $Z_2$, each of them has again two further preimages inside $Z_2$, and so on. Summarizing, we can conclude that the global bifurcation which we have just described transforms a *simply connected* basin into a *multiply connected* basin. The latter set has a countably infinite number of holes, called an *arborescent sequence of holes*, which belong to the infeasible set $B(\infty)$. As the speeds of adjustment are further increased, the holes become more pronounced and they become connected along the vertical axis as shown in Fig. 1.17.

Our numerical results show that the structure of the basins may become considerably more complex as the adjustment speeds are increased. The transition between qualitatively different structures of the boundary occur through so called *contact bifurcations* (see for example Mira et al. (1996)) and these bifurcations can be described in terms of contacts between the basin boundaries and arcs of the *critical curves*. To conclude this chapter, we would like to stress that in general there is no relation between the bifurcations which change the qualitative properties of the basins (global bifurcations) and the bifurcations which change the qualitative properties of the attractor (sequences of local bifurcations). The former is related to the global dynamics, whereas the latter focuses on the local (stability) properties. In later chapters we will encounter situations where the attractor is a rather simple set (that is, an equilibrium), but the structure of its basin is quite complex. As demonstrated above, in other situations exactly the opposite might be the case.

# Chapter 2
# Concave Oligopolies

In the previous chapter we have seen that except in very special cases oligopoly models have nonlinear features and therefore can generally exhibit a vast array of dynamical behavior ranging from simple to complicated. Under special conditions however the uniqueness of the equilibrium can be guaranteed, simple conditions can be derived for the local asymptotic stability of the equilibrium with both discrete and continuous time scales, and the global dynamics are less complicated and can be handled with some of the standard tools of nonlinear dynamical systems. In this chapter we will consider concave oligopolies, which are the straightforward generalizations of linear oligopolies and are the most frequently discussed cases in the literature (see for example, Okuguchi and Szidarovszky (1999) and the references therein).

In the first section we consider oligopolies both without and with cost externalities, derive the best response functions in both cases and their various properties that will be invoked in the ensuing analysis of the dynamics. In Sect. 2.2 we examine the local stability of discrete time oligopolies using the adjustment processes introduced in Sect. 1.2. In Sect. 2.3 we then consider the global stability of the discrete time oligopoly, bringing to bear the tools developed in Sect. 1.3. Section 2.4 gives a brief description of the local stability of the dynamics in both discrete and continuous time where firms use gradient adjustment processes. The local stability of continuous time oligopolies using certain types of best response dynamics is studied in Sect. 2.5. Finally in Sect. 2.6 we study the impact of various kinds of information delays on the local stability of continuous time oligopolies using best response dynamics.

## 2.1 Introduction

We will first consider oligopolies without cost externalities. As in the previous chapter, let $N$ be the number of firms, $x_k$ the output of firm $k$ $(k = 1, 2, \dots, N)$, and $Q = \sum_{k=1}^{N} x_k$ the total output of the industry. If $p = f(Q)$ denotes the inverse demand function and $C_k(x_k)$ is the cost of firm $k$, then the profit of this firm can be

G.I. Bischi et al., *Nonlinear Oligopolies*, DOI 10.1007/978-3-642-02106-0_2, 

written as (1.1) that we repeat here for the sake of convenience,

$$\varphi_k(x_1, \ldots, x_N) = x_k f(Q) - C_k(x_k).$$

Assume that the price function $f$ and all cost functions are twice continuously differentiable and satisfy the conditions

(A) $f'(Q) < 0$,

(B) $x_k f''(Q) + f'(Q) \leq 0$,

(C) $f'(Q) - C_k''(x_k) < 0$,

for all $k$ and feasible values of $x_k$ and $Q$.

Condition (A) means that $f(Q)$ is strictly decreasing in $Q$, that is, a larger total output can only be sold for a lower price. Condition (B) is called the decreasing marginal revenue condition, it states that marginal revenue for firm $k$ decreases for higher levels of output of the rest of the industry (see for example, Vives (1999)). Condition (C) relates the lower bound on the convexity/concavity of the cost function to the degree of negativity of the slope of the price function. It is assumed by many authors that $f(Q)$ is concave and $C_k(x_k)$ is convex for all $k$. In this case $f' < 0$, $f'' \leq 0$, $C_k'' \geq 0$, and naturally $C_k' > 0$, since a larger output level requires higher cost. Conditions (B) and (C) are then clearly satisfied. In fact these conditions are slightly more general, since they can be also satisfied if $f$ is slightly convex and/or $C_k$ is slightly concave provided that $-f'$ is large enough. We have to mention as well, that conditions (A)–(C) are more restrictive than the simple condition that the profit functions be concave.

Notice that

$$\frac{\partial}{\partial x_k}\varphi_k(x_1, \ldots, x_N) = f(x_k + Q_k) + x_k f'(x_k + Q_k) - C_k'(x_k), \tag{2.1}$$

and under these conditions

$$\frac{\partial^2}{\partial x_k^2}\varphi_k(x_1, \ldots, x_N) = 2f'(x_k + Q_k) + x_k f''(x_k + Q_k) - C_k''(x_k) < 0, \tag{2.2}$$

where $Q_k = \sum_{l \neq k} x_l$ as in the previous chapter. Hence $\varphi_k$ is strictly concave in $x_k$.

In order to prove the existence of a unique equilibrium under conditions (A)–(C) and develop dynamic models we have to determine first the best response functions of the firms.

The concavity of the profit functions implies that the best response functions can be obtained in the form

$$R_k(Q_k) = \begin{cases} 0 & \text{if} \quad f(Q_k) - C_k'(0) \le 0, \\ L_k & \text{if} \quad f(L_k + Q_k) + L_k f'(L_k + Q_k) - C_k'(L_k) \ge 0, \\ z_k^* & \text{otherwise,} \end{cases} \tag{2.3}$$

where $z_k^*$ is the unique solution of the strictly monotonic equation

$$f(z_k + Q_k) + z_k f'(z_k + Q_k) - C_k'(z_k) = 0 \tag{2.4}$$

in the interval $(0, L_k)$. Observe that due to our assumptions the left hand side of (2.4) strictly decreases and is continuous in $z_k$, positive at $z_k = 0$ and negative at $z_k = L_k$, therefore there is a unique solution.

In order to analyze the asymptotic behavior of any one of the discrete time and continuous time dynamical systems (1.19), (1.28)–(1.31) emerging from partial adjustment and best reply behavior, we will need to examine the Jacobian of the systems, and to do so, we have to determine the derivatives of the best response functions. These derivatives can be obtained by implicitly differentiating equation (2.4). Assuming that $z_k^*$ is interior (that is, $0 < z_k^* < L_k$), then we have

$$f'(1 + R_k') + R_k' f' + R_k f''(1 + R_k') - C_k'' R_k' = 0,$$

implying that

$$R_k' = -\frac{f' + R_k f''}{2f' + R_k f'' - C_k''}. \tag{2.5}$$

Note that the same result could be obtained directly by using

$$R_k' = -\frac{\partial^2 \varphi_k / \partial x_k \partial Q_k}{\partial^2 \varphi_k / \partial x_k^2}. \tag{2.6}$$

Conditions (B) and (C) imply that

$$-1 < R_k'(Q_k) \le 0 \tag{2.7}$$

for all $k$ and $Q_k$. In the first two cases of (2.3), the derivative of $R_k$ is zero, except at two possible break-points, so (2.7) is always satisfied. This property will play a crucial role later in the stability analysis.

Notice that relation (2.4) shows that $R_k(Q_k)$ decreases in $Q_k$, that is, larger total output of the rest of the industry requires smaller output responses from the firms. Since $R_k'(Q_k)$ is larger than $-1$, best responses cannot decrease very rapidly.

We notice that if $f(0) > C_k'(0)$, that is the reservation price is higher than the marginal costs at $x_k = 0$, then the monopoly quantity (ignoring capacity limits) of firm $k$ is the solution of equation (2.4) with $Q_k = 0$. If it is below $L_k$, then it is the best response of firm $k$ at $Q_k = 0$.

We can also rewrite the best responses of the firms in terms of the total output of the industry. This idea will be very helpful in proving the existence and uniqueness of the equilibrium and it can be also used to derive a simple computational method to find the equilibrium. From (2.3) we have

$$\widetilde{R}_k(Q) = \begin{cases} 0 & \text{if} \quad f(Q) - C_k'(0) \leq 0, \\ L_k & \text{if} \quad f(Q) + L_k f'(Q) - C_k'(L_k) \geq 0, \\ z_k & \text{otherwise,} \end{cases} \tag{2.8}$$

where $z_k$ is the unique solution of the equation

$$f(Q) + z_k f'(Q) - C_k'(z_k) = 0 \tag{2.9}$$

inside the interval $(0, L_k)$. We point out that in the case when we consider the best response as a function of $Q$ (rather than $Q_k$) we denote it by $\widetilde{R}_k$. Notice that in the third case of (2.8), the left hand side is positive at $z_k = 0$, negative at $z_k = L_k$, and strictly decreasing, since it has a negative derivative given by

$$\frac{\partial}{\partial z_k}\{f(Q) + z_k f'(Q) - C_k'(z_k)\} = f'(Q) - C_k''(z_k) < 0.$$

The derivative of $\widetilde{R}_k(Q)$ can be obtained by implicit differentiation, so that

$$f' + \widetilde{R}_k' f' + \widetilde{R}_k f'' - C_k'' \widetilde{R}_k' = 0,$$

from which

$$\widetilde{R}_k' = -\frac{f' + \widetilde{R}_k f''}{f' - C_k''} \leq 0.$$

Since $\widetilde{R}_k(Q)$ is continuous in the interval $[0, \sum_{l=1}^N L_l]$, it is non-increasing in $Q$ for all $Q \in [0, \sum_{l=1}^N L_l]$. Finally, consider the single-variable equation

$$\sum_{k=1}^N \widetilde{R}_k(Q) - Q = 0, \tag{2.10}$$

which must hold at the equilibrium. The left hand side of (2.10) is strictly decreasing in $Q$, it is non-negative at $Q = 0$ and non-positive at $Q = \sum_{k=1}^N L_k$. Therefore there is a unique solution $\bar{Q}$, and the corresponding equilibrium outputs are $\bar{x}_k = \widetilde{R}_k(\bar{Q})$.

*Example 2.1.* In our earlier Example 1.1 we introduced oligopolies with linear price and cost functions,

$$f(Q) = A - BQ \quad \text{and} \quad C_k(x_k) = d_k + c_k x_k, \quad (1 \leq k \leq N). \tag{2.11}$$

We also proved that the best response function of firm $k$ is a piece-wise linear function

$$R_k(Q_k) = \begin{cases} 0 & \text{if} \quad A - BQ_k - c_k \le 0, \\ L_k & \text{if} \quad A - 2BL_k - BQ_k - c_k \ge 0, \\ -\frac{1}{2}Q_k + (A - c_k)/(2B) & \text{otherwise} \end{cases} \tag{2.12}$$

by assuming that $\sum_{k=1}^{N} L_k \le \frac{A}{B}$.
In the first two cases $R'_k(Q_k) = 0$ and in the third case $R'_k(Q_k) = -\frac{1}{2}$ showing that (2.7) is always satisfied. ▼

In Example 1.2 we examined oligopolies with linear price and quadratic cost functions. In case (ii) of that example we observed the possibility of multiple equilibria, however it is easy to see that under the stated assumptions condition (C) is violated.

Let us turn our attention next to the general case when the cost of firm $k$ is $C_k(x_k, Q_k)$, perhaps because of the presence of externalities as discussed in Sect. 1.1. In this more general case the profit of firm $k$ is given as

$$\varphi_k(x_1, \ldots, x_N) = x_k f(x_k + Q_k) - C_k(x_k, Q_k)$$

with derivatives

$$\frac{\partial \varphi_k}{\partial x_k} = f(x_k + Q_k) + x_k f'(x_k + Q_k) - C'_{kx}(x_k, Q_k)$$

and

$$\frac{\partial^2 \varphi_k}{\partial x_k^2} = 2f'(x_k + Q_k) + x_k f''(x_k + Q_k) - C''_{kxx}(x_k, Q_k),$$

where $C'_{kx}$ and $C''_{kxx}$ denote the first and second order partial derivatives of $C_k$ with respect to $x_k$. Assume that conditions $(A)$, $(B)$ are satisfied, furthermore assume

(C') $f'(x_k + Q_k) - C''_{kxx}(x_k, Q_k) < 0$,

for all $k$ and feasible values of $x_k$ and $Q_k$.

Under conditions (A), (B) and (C'), the profit $\varphi_k$ of firm $k$ is strictly concave in $x_k$, therefore there is a unique best response function of firm $k$ given by

$$R_k(Q_k) = \begin{cases} 0 & \text{if} \quad f(Q_k) - C'_{kx}(0, Q_k) \le 0, \\ L_k & \text{if} \quad f(L_k + Q_k) + L_k f'(L_k + Q_k) - C'_{kx}(L_k, Q_k) \ge 0, \\ z_k^* & \text{otherwise,} \end{cases}$$

where $z_k^*$ is the unique solution of the equation

$$f(z_k + Q_k) + z_k f'(z_k + Q_k) - C'_{kx}(z_k, Q_k) = 0,$$

where the left hand side is strictly decreasing in $z_k$. The derivative of the best response function can be determined by implicitly differentiating this equation to obtain

$$f'(1+R_k') + R_k' f' + R_k f''(1+R_k') - C''_{kxx} R_k' - C''_{kxQ} = 0,$$

implying that

$$R_k' = -\frac{f' + R_k f'' - C''_{kxQ}}{2f' + R_k f'' - C''_{kxx}},$$

where $C''_{kxQ}$ is the mixed second order partial derivative of $C_k$. If in addition to conditions (A), (B) and (C') we further assume for all $k$ and feasible values of $x_k$ and $Q_k$ that

$$\text{(D)} \quad f'(x_k + Q_k) + x_k f''(x_k + Q_k) \le C''_{kxQ}(x_k, Q_k) < C''_{kxx}(x_k, Q_k) - f'(x_k + Q_k),$$

then relation (2.7) remains valid even in this more general case. It can be proved, similarly to the special case without cost externalities, that under conditions (A), (B), (C') and (D) there is always a unique Nash equilibrium.

The existence and uniqueness of the Nash equilibrium has been examined by many authors. Some earlier results used the Brouwer or Kakutani fixed point theorem, which unfortunately is an approach that does not offer computational methods to find the equilibria, and this would be required in the situation of general price and cost functions. A comprehensive summary of the most important earlier results is given in Okuguchi (1976). Okuguchi and Szidarovszky (1999) provide some extensions of the earlier results that do lead to computational methods to find the equilibria. The existence and uniqueness proof presented in this section is taken from Szidarovszky and Yakowitz (1977). Uniqueness and existence results for Nash equilibria can also be found in Vives (1999), using arguments based on the Tarski fixed point theorem.

## 2.2 Discrete Time Models and Local Stability

Consider first the best reply dynamics with adaptive expectations which are governed by (1.28) and (1.29). A vector $(\bar{x}_1, \ldots, \bar{x}_N, \bar{Q}_1^E, \ldots, \bar{Q}_N^E)$ is a steady state of this system if and only if

$$\bar{Q}_k^E = \sum_{l \neq k} \bar{x}_l, \tag{2.13}$$

and

$$\bar{x}_k = R_k(\bar{Q}_k^E). \tag{2.14}$$

In other words, $(\bar{x}_1, \ldots, \bar{x}_N)$ is a Nash equilibrium. The same is true obviously for partial adjustment towards the best response, which are modeled by (1.30). If the sign-preserving function $\alpha_k$ is a homogeneous linear function of the form $\alpha_k(z) = a_k z$, then the system (1.28)–(1.29) reduces to (1.20)–(1.21), so we will not discuss system (1.20)–(1.21) directly, but only as a special case. Similarly, system (1.30) also reduces to (1.23) in this case.

The local and global stability properties of an equilibrium depend on the particular adjustment process which is used by the firms to update their quantity choices. So any stability result to be introduced and proved in this book is always applicable to the particular dynamical system for which it is proved. In this section best reply dynamics with adaptive expectations and partial adjustment towards the best response will be examined. We will return to gradient adjustments later in this chapter.

The asymptotic stability of the equilibrium will be examined by the technique of linearization around the equilibrium, which is summarized briefly in Appendix B. Here we assume that the equilibrium is interior, otherwise the best response functions are not differentiable. In such cases we have to assume that the conditions of Theorem B.3 are satisfied in a neighborhood of the equilibrium. First we show that as far as local asymptotic stability is concerned, the conditions for the best reply dynamics with adaptive expectations and partial adjustments are equivalent since it turns out that the Jacobians of the two processes have identical nonzero eigenvalues. The Jacobian of the the best reply dynamics[1] (1.28)–(1.29) has a special structure, namely

$$\begin{pmatrix} \bar{\boldsymbol{J}}_{11} & \bar{\boldsymbol{J}}_{12} \\ \bar{\boldsymbol{J}}_{21} & \bar{\boldsymbol{J}}_{22} \end{pmatrix} \tag{2.15}$$

with

$$\bar{\boldsymbol{J}}_{11} = \begin{pmatrix} 0 & r_1 a_1 & \ldots & r_1 a_1 \\ r_2 a_2 & 0 & \ldots & r_2 a_2 \\ \vdots & \vdots & & \vdots \\ r_N a_N & r_N a_N & \ldots & 0 \end{pmatrix}, \bar{\boldsymbol{J}}_{12} = \begin{pmatrix} r_1(1-a_1) & & & 0 \\ & r_2(1-a_2) & & \\ & & \ddots & \\ 0 & & & r_N(1-a_N) \end{pmatrix},$$

$$\bar{\boldsymbol{J}}_{21} = \begin{pmatrix} 0 & a_1 & \ldots & a_1 \\ a_2 & 0 & \ldots & a_2 \\ \vdots & \vdots & & \vdots \\ a_N & a_N & \ldots & 0 \end{pmatrix}, \quad \text{and } \bar{\boldsymbol{J}}_{22} = \begin{pmatrix} 1-a_1 & & & 0 \\ & 1-a_2 & & \\ & & \ddots & \\ 0 & & & 1-a_N \end{pmatrix},$$

where for all $k$,

[1] See Appendix B for a definition of the Jacobian of a dynamical system. We stress that unless indicated otherwise the elements of this matrix are evaluated at the steady state of the system, which is indicated by the overbar.

$$r_k = R'_k(\bar{Q}_k^E) \quad \text{and} \quad a_k = \alpha'_k(0). \tag{2.16}$$

If we use $(u_1, u_2, \ldots, u_N, v_1, v_2, \ldots, v_N)$ to denote the typical eigenvector and $\lambda$ the associated eigenvalue of the matrix (2.15) then it is relatively straightforward to see that the $u_k, v_k$ (for $k = 1, 2, \ldots, N$) are given by the two sets of equations

$$r_k a_k \sum_{l \neq k} u_l + r_k(1 - a_k)v_k = \lambda u_k, \quad (k = 1, 2, \ldots, N), \tag{2.17}$$

$$a_k \sum_{l \neq k} u_l + (1 - a_k)v_k = \lambda v_k, \quad (k = 1, 2, \ldots, N). \tag{2.18}$$

By subtracting the $r_k$-multiple of (2.18) from (2.17) we find that

$$\lambda(u_k - r_k v_k) = 0. \tag{2.19}$$

Since a zero eigenvalue does not destroy the asymptotic stability of the system, we will consider only nonzero eigenvalues of the Jacobian. If $\lambda \neq 0$, then (2.19) implies that $u_k = r_k v_k$ and if we substitute this condition into (2.17) we see that the $u_k$ values are determined by

$$r_k a_k \sum_{l \neq k} u_l + (1 - a_k)u_k = \lambda u_k, \quad (1 \leq k \leq N),$$

which is readily shown to be the eigenvalue equation of the $N \times N$ matrix

$$\bar{H} = \begin{pmatrix} 1 - a_1 & r_1 a_1 & \ldots & r_1 a_1 \\ r_2 a_2 & 1 - a_2 & \ldots & r_2 a_2 \\ \vdots & \vdots & \ddots & \vdots \\ r_N a_N & r_N a_N & \ldots & 1 - a_N \end{pmatrix}. \tag{2.20}$$

Observe that this matrix coincides with the Jacobian of the partial adjustment dynamics (1.30). Therefore, if local asymptotic stability is our concern, then the conditions for the process (1.28)–(1.29) of best reply dynamics with adaptive expectations is equivalent to the process (1.30) of partial adjustment towards the best response with naive expectations. This means that best reply dynamics with adaptive expectations and best reply dynamics with partial adjustments share the same local asymptotic stability properties, and the eigenvalue structure of matrix (2.20) determines whether an equilibrium is locally asymptotically stable or not. In the case of $N = 2$ (duopoly) or very special response functions with arbitrary value of $N$, the two processes are even equivalent as was shown earlier in Sect. 1.2. The following theorem presents conditions for the local asymptotic stability of the equilibrium. It allows us to assert that if the initial outputs of the firms are sufficiently close to the equilibrium, then as $t \to \infty$, the outputs converge to the equilibrium.

**Theorem 2.1.** *Assume that* $a_k = \alpha_k'(0) > 0$ *for all* $k = 1, 2, \ldots, N$.

*(i) The equilibrium is locally asymptotically stable if for all* $k$,

$$a_k(1 + r_k) < 2 \tag{2.21}$$

*and*

$$\sum_{k=1}^{N} \frac{r_k a_k}{2 - a_k(1 + r_k)} > -1. \tag{2.22}$$

*(ii) The equilibrium is unstable if for at least one* $k$,

$$a_k(1 + r_k) \geq 2$$

*or*

$$\sum_{k=1}^{N} \frac{r_k a_k}{2 - a_k(1 + r_k)} < -1.$$

*Proof.* Notice that the structure of matrix $\bar{\boldsymbol{H}}$ is the same as matrix $\boldsymbol{A}$ given in equation (E.4) of Appendix E. Therefore we can use relation (E.5) to determine that its characteristic equation has the form

$$\prod_{k=1}^{N} (1 - a_k(1 + r_k) - \lambda) \cdot \left[ 1 + \sum_{k=1}^{N} \frac{r_k a_k}{1 - a_k(1 + r_k) - \lambda} \right] = 0. \tag{2.23}$$

In order to make the mathematical analysis easier assume that $a_k > 0$ for all $k$ and the firms are numbered in such a way that the different $a_k(1 + r_k)$ values satisfy $a_1(1 + r_1) > a_2(1 + r_2) > \cdots > a_s(1 + r_s)$ and their values are repeated $m_1, m_2, \ldots, m_s$ times. By adding the terms with identical denominators in the bracketed expression and denoting by $\theta_j$ the sum of the corresponding numerators $r_k a_k$, we can rewrite (2.23) as

$$\prod_{j=1}^{s} (1 - a_j(1 + r_j) - \lambda)^{m_j} \cdot \left[ 1 + \sum_{j=1}^{s} \frac{\theta_j}{1 - a_j(1 + r_j) - \lambda} \right] = 0, \tag{2.24}$$

where $\theta_j \leq 0$. So we conclude that if $\theta_j = 0$ or $m_j \geq 2$, then $1 - a_j(1 + r_j)$ is an eigenvalue of $\bar{\boldsymbol{H}}$. This eigenvalue is always less than 1, so it is inside the unit circle if and only if $a_j(1 + r_j) < 2$. All other eigenvalues are the roots of the equation

$$g(\lambda) \equiv 1 + \sum_{j=1}^{s} \frac{\theta_j}{1 - a_j(1 + r_j) - \lambda} = 0,$$

where we can assume that all $\theta_j$ values are nonzero. This last equation is equivalent to a polynomial equation of degree $s$, so there are $s$ real or complex roots. Clearly,

$$\lim_{\lambda \to \pm\infty} g(\lambda) = 1, \qquad \lim_{\lambda \to 1 - a_j(1+r_j) \pm 0} g(\lambda) = \pm\infty$$

and

$$g'(\lambda) = \sum_{j=1}^{s} \frac{\theta_j}{(1 - a_j(1 + r_j) - \lambda)^2} < 0.$$

Using these properties we can graph $g(\lambda)$ as shown in Fig. 2.1. This figure indicates that the structure of the roots is such that there is one root before $1 - a_1(1 + r_1)$, and one root between each pair of poles $1 - a_j(1 + r_j)$ and $1 - a_{j+1}(1 + r_{j+1})$ for $j = 1, 2, \dots, s - 1$. So all roots have been found and they are real. Furthermore all are inside the unit circle if and only if

$$1 - a_1(1 + r_1) > -1, \quad \text{and} \quad g(-1) > 0.$$

At least one eigenvalue is outside the unit circle if either $1 - a_1(1 + r_1) \le -1$ or $g(-1) < 0$. ■

In the case of constant speeds of adjustment we usually assume that $0 < a_k \le 1$ for all $k$, and from relation (2.7) we know that $-1 < r_k \le 0$. So condition (2.21) is usually satisfied in this case.

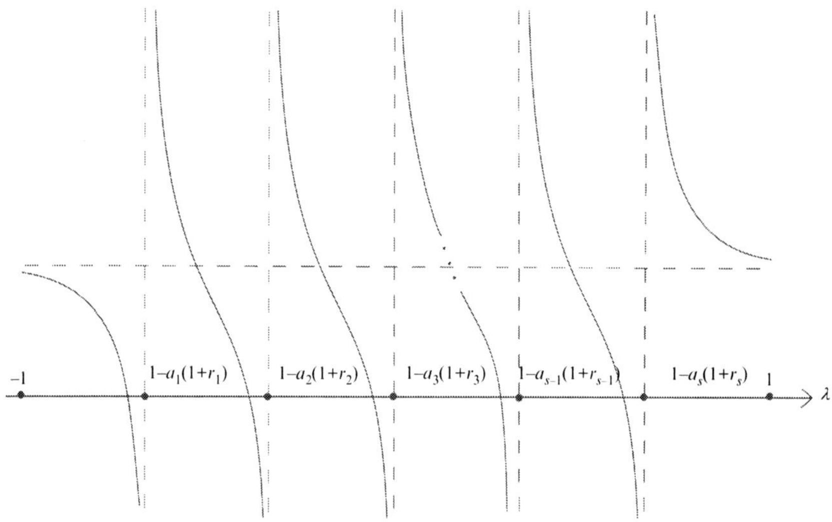


**Fig. 2.1** Graph of $g(\lambda)$, the roots of which are eigenvalues of the Jacobian of the system describing the dynamics of the discrete time oligopoly under best reply dynamics

Notice that conditions (A)–(C) (or (A),(B),(C') and (D)) were assumed for all feasible values of $x_k$ and $Q_k$, and they imply the existence of a Nash equilibrium. However they need to be satisfied only in a neighborhood of an interior equilibrium in order to guarantee the local asymptotic stability of that equilibrium. Observe first that with given price and cost functions the best responses are fixed, so the $r_k$ values are uniquely determined. The firms' choices are only the adjustment mechanisms, which are characterized by the functions $\alpha_k$. Clearly, conditions (2.21) and (2.22) are satisfied, if all $a_k = \alpha_k'(0)$ values are sufficiently small.

*Example* 2.2. Consider again the case of a linear inverse demand and linear cost functions as in Example 2.1, where $r_k = -\frac{1}{2}$ for all $k$. Then (2.21) holds if $a_k < 4$ for all $k$, and (2.22) holds if

$$\sum_{k=1}^{N} \frac{a_k}{4 - a_k} < 1. \tag{2.25}$$

In the further special case when the firms select identical adjustment schemes (that is, when $\alpha_k'(0) = a_k \equiv a$), then (2.25) can be rewritten in the form

$$a < \frac{4}{N+1}. \tag{2.26}$$

If $a \in (0, 1]$, then this condition always holds for duopolies ($N = 2$). If $N \geq 3$, then this condition is violated with naive expectations ($a = 1$). This is the result derived by Theocharis (1960). The equilibrium can still be stabilized however by selecting sufficiently small values of $a$. ▼

Consider next the nonlinear case with identical firms. In this case $a_k \equiv a$ and $r_k \equiv r$. Condition (2.22) can now be rewritten as

$$a < \frac{2}{1 - r(N-1)}. \tag{2.27}$$

In this case we do not assume that the initial outputs of the firms are the same, so the system cannot be reduced to a one-dimensional one. Notice that in the special linear case with $r = -\frac{1}{2}$, (2.27) reduces to (2.26).

It is also interesting to analyze condition (2.22) from the point of view of a single firm $k$. If for any other firm $l$, $a_l(1 + r_l) \geq 2$, or

$$\sum_{l \neq k} \frac{r_l a_l}{2 - a_l(1 + r_l)} \leq -1,$$

then the equilibrium becomes unstable regardless of the adjustment scheme of firm $k$. Firm $k$ is able to stabilize the equilibrium alone merely by selecting an adjustment function $\alpha_k$ such that its derivative at zero is sufficiently small. That is, the equilibrium becomes locally asymptotically stable when $a_k$ satisfies the two relations

$$a_k < \frac{2}{1 + r_k}$$

and

$$\frac{r_k a_k}{2 - a_k(1 + r_k)} > -1 - \sum_{l \neq k} \frac{r_l a_l}{2 - a_l(1 + r_l)}.$$

Simple algebra shows that this last inequality holds if and only if

$$a_k < \frac{2(1 + S_k)}{1 + S_k(1 + r_k)}, \tag{2.28}$$

where

$$S_k = \sum_{l \neq k} \frac{r_l a_l}{2 - a_l(1 + r_l)} \in (-1, 0].$$

It is easy to see that the right hand side of relation (2.28) is always positive, since $r_k \in (-1, 0]$.

Consider next the case of a duopoly when $N = 2$. In this special case conditions (2.21) and (2.22) can be rewritten as

$$a_1(1 + r_1) < 2, \quad a_2(1 + r_2) < 2$$

and

$$\frac{a_1 r_1}{2 - a_1(1 + r_1)} + \frac{a_2 r_2}{2 - a_2(1 + r_2)} > -1.$$

With fixed values of $a_1 \in (0, 2/(1 + r_1))$, firm 2 has to select a sign-preserving adjustment function with $\alpha_2'(0) = a_2$ satisfying condition (2.28) in order to stabilize the system. In the case of a duopoly this condition has the special form

$$a_2 < \frac{2\left(1 + \dfrac{a_1 r_1}{2 - a_1(1 + r_1)}\right)}{1 + \dfrac{a_1 r_1(1 + r_2)}{2 - a_1(1 + r_1)}} = \frac{2(2 - a_1)}{2 - a_1(1 - r_1 r_2)}. \tag{2.29}$$

Notice that for $a_1, a_2 \in (0, 1]$ this relation is always satisfied, so the equilibrium is always locally asymptotically stable. The stability region of this condition in the $(a_1, a_2)$ plane is illustrated in Fig. 2.2.

Several generalizations of the above analysis, including multiproduct models, are discussed in Okuguchi and Szidarovszky (1999). In addition, the existence and uniqueness of the equilibrium is proved without imposing the conditions of differentiability of the price and cost functions, and in the linear cases several alternative sufficient and necessary stability conditions are derived. The very first stability result in discrete time dynamic oligopolies dates back to Theocharis (1960) and following in his footsteps many researchers have worked intensively on this topic, a task which continues even to the present day. For an extensive literature review, see Kopel (2009).

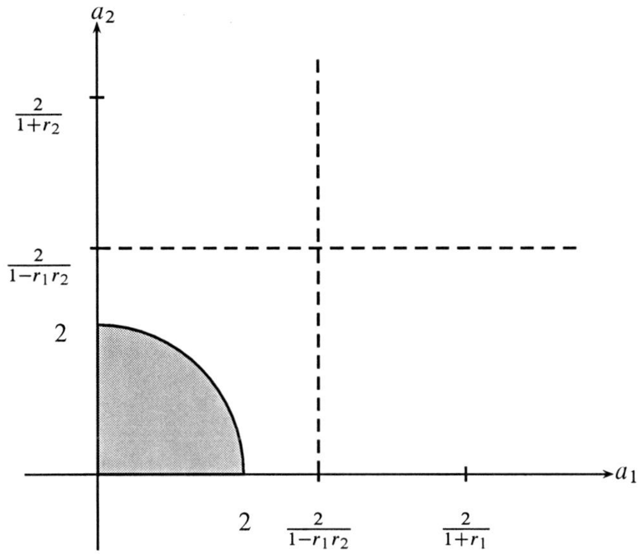


**Fig. 2.2** Stability region of a discrete time duopoly under best reply dynamics in the $(a_1, a_2)$ plane

## 2.3 Discrete Time Oligopolies and Global Stability

We start this section with a simple discussion of global asymptotic stability. Our analysis will be based on the sufficient condition presented in Appendix B which states that if there exists a matrix norm such that the norms of the Jacobians of a discrete time dynamical system in all regions are less than some $q < 1$ everywhere in the phase space, then the system is globally asymptotically stable. Here we present the case of partial adjustment towards the best response, the case of best reply with adaptive expectations can be discussed in a similar way. The feasible output set is divided into subregions depending on the different cases in the best response function (2.3). In each subregion the Jacobian of the partial adjustment dynamics (1.30) (that we shall denote by **H**) has the special structure (2.20), where

$$r_k = R'_k\left(\sum_{l\neq k} x_l\right) \quad \text{and} \quad a_k = \alpha'_k\left(R_k\left(\sum_{l\neq k} x_l\right) - x_k\right),$$

where $r_k$ is either given by (2.5) or equals zero. Assume that $0 < a_k \leq 1$ for all $k$ and for all feasible output levels $x_1, x_2, \ldots, x_N$. Under conditions (A)–(C), inequality (2.7) holds for all $k$ and all feasible output levels. Therefore with the choice of the row norm, we have

$$\begin{aligned}\|\mathbf{H}\|_\infty &= \max_k \{|1 - a_k| + (N-1)|r_k a_k|\} \\ &= \max_k \{1 - a_k(1 + r_k(N-1))\}.\end{aligned} \tag{2.30}$$

Therefore the equilibrium is globally asymptotically stable if for all $k$ and all feasible output levels,

$$1 - a_k \left(1 + r_k(N-1)\right) \le q < 1 \tag{2.31}$$

for some positive $q$. All subregions of the feasible output set are compact and on each of them the functions $R_k$ and $\alpha_k$ are continuously differentiable, so the function $1 - a_k \left(1 + r_k(N-1)\right)$ attains its maximum value in each subregion. Since there are only finitely many subregions, condition (2.31) can be weakened by assuming that for all $k$, and all feasible output levels,

$$1 - a_k \left(1 + r_k(N-1)\right) < 1,$$

that is,

$$r_k > -\frac{1}{N-1}. \tag{2.32}$$

Notice that this is a *sufficient* condition for global asymptotic stability.

By assuming no cost externalities and using (2.5), this sufficient condition holds for all feasible output levels if and only if

$$(N-2)(f' + R_k f'') + (C_k'' - f') > 0. \tag{2.33}$$

Conditions (B) and (C) imply that the first term is always non-positive and the second term always positive. In the case of duopoly $N = 2$, then (2.33) holds, so the equilibrium is globally asymptotically stable. If we have a triopoly $N = 3$, then (2.33) can be written as

$$R_k f'' + C_k'' > 0,$$

which is not guaranteed to be satisfied. If $N$ becomes larger, then the first term on the left hand side of (2.33) converges to negative infinity if $f' + R_k f''$ is not identically zero, so with a larger number of firms condition (2.33) is violated.

In such cases we might try to apply different matrix norms, for example row or column norms generated by special diagonal matrices, similar to the examples discussed in the previous chapter. The choice of an appropriate norm depends on the problem and its existence is not guaranteed. This fact raises the need to develop and apply more sophisticated methods for the global analysis of the nonlinear oligopoly models that we will encounter in this book. The need for more advanced methods combining numerical, analytical, and geometrical arguments is also underlined by the fact that neither local analysis, nor the above described sufficient global asymptotic stability condition can be used in the case of non-differentiable best responses.

In order to illustrate our general approach to piecewise differentiable dynamic models, we reconsider the simplest oligopoly model introduced in Example 1.1 and further studied in Examples 2.1 and 2.2. We assume a linear inverse demand function, $p = f(Q) = A - BQ$, and linear cost functions, $C_k(x_k) = c_k x_k$, $k = 1, \ldots, N$, where for simplicity we let the fixed costs be zero. The presence of non-negativity and capacity constraints makes the resulting dynamical system non-differentiable. In what follows we investigate the asymptotic dynamics for an increasing number of firms in the industry and we explore the role of the capacity constraints, the presence of which leads to non-differentiability of the dynamical system considered. Moreover, we will explain some peculiar dynamic properties of piecewise linear maps, consider a particular type of bifurcation which causes the loss of stability of the unique equilibrium, and illustrate which kind of non-equilibrium dynamics might occur.

*Example 2.3.* As a first step let us consider the symmetric case of $N$ identical firms with $c_1 = c_2 = \ldots = c_N = c$, linear adjustment functions with $a_1 = a_2 = \ldots, a_N = a$, and $L_1 = L_2 = \ldots = L_N = L$. We assume that $A > c$ and that firms use partial adjustment towards their best responses. Then, given that firms start from the identical initial condition $x_1(0) = x_2(0) = \ldots = x_N(0) = x(0)$, the dynamics are captured by the one-dimensional model

$$x(t+1) = T(x(t)) = (1-a)\,x(t) + aR\left((N-1)\,x(t)\right),$$

where

$$R((N-1)x) = \begin{cases} L & \text{if} \quad x \leqslant \frac{A-c}{(N-1)B} - \frac{2L}{N-1}, \\ \frac{A-c}{2B} - \frac{N-1}{2}x & \text{if} \quad \frac{A-c}{(N-1)B} - \frac{2L}{N-1} \leqslant x \leqslant \frac{A-c}{(N-1)B} \\ 0 & \text{if} \quad x \geqslant \frac{A-c}{(N-1)B}. \end{cases}$$

Obviously, the function $T$ is a piecewise linear map, characterized by three regions where it is differentiable. These regions are separated by two kinks (points of non-differentiability), so that the map can be written in detail as

$$T(x) = \begin{cases} (1-a)\,x + aL & \text{if} \quad x \leqslant \frac{A-c}{(N-1)B} - \frac{2L}{N-1}, \\ \left(1 - \frac{a(N+1)}{2}\right)x + a\frac{A-c}{2B} & \text{if} \quad \frac{A-c}{(N-1)B} - \frac{2L}{N-1} \leqslant x \leqslant \frac{A-c}{(N-1)B}, \\ (1-a)\,x & \text{if} \quad x \geqslant \frac{A-c}{(N-1)B}. \end{cases}$$

Figure 2.3 depicts a typical graph of the map $T(x)$ together with the graph of the reaction function (dashed). It should be clear that the exact shape of the graph and the locations of the kinks depend on the market and cost parameters $A, B, c$, on the capacity level $L$ of the firms and the number of firms $N$ in the industry, and in particular on the adjustment speed $a$. For larger values of $a$ the graph of $T(x)$ is closer to that of $R(x)$, for smaller values of $a$ the graph of $T(x)$ is closer to the diagonal. Furthermore, as we now show, these parameters determine if the

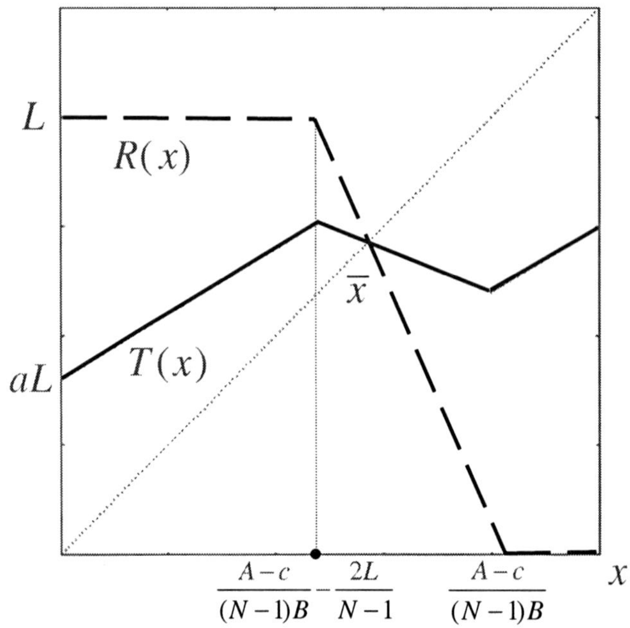


**Fig. 2.3** Example 2.3; linear inverse demand and cost functions and identical capacity constrained firms. The piece-wise linear map $T(x)$ and reaction function $R(x)$

equilibrium is either at the boundary of the feasible set $[0, L]$ or in the interior, and if the equilibrium is stable. The definition of $R((N-1)x)$ implies that 0 cannot be equilibrium, so the equilibrium $\bar{x}$ is either interior or equals $L$.

As usual, the steady states of the adaptive adjustment process are the equilibria of the underlying game, since $T(x) = x$ if and only if $R((N-1)x) = x$. However, the equilibrium might be located on the boundary. We account for this possibility by writing the unique equilibrium point as $\bar{x} = \min\left\{\frac{A-c}{B(N+1)}, L\right\}$. Its stability, under the dynamic adjustment process governed by the iteration of the map $T$, depends on the derivative of $T$, which has three segments with two different derivatives: $1-a$ and $1 - \frac{a(N+1)}{2}$. From the results of Appendix B we know that the equilibrium is globally asymptotically stable of both $|1-a|$ and $\left|1 - \frac{a(N+1)}{2}\right|$ are less than one, which is the case if $0 < a < \frac{4}{N+1}$.

We now turn to the asymptotic dynamics of the production sequences generated by $T$ if, (1) the number of firms in the industry changes, and (2) each firm is capacity-constrained. Figure 2.4 depicts a bifurcation diagram of output $x$ with respect to the number of firms $N$ obtained with the parameters $A = 16$, $B = 1$, $a = 0.5$, $c = 6$ and $L = 1$ (in all the numerical simulations in this subsection we select the parameters such that $NL \leq A/B$ in order to ensure non-negative prices). Observe that as long as $(A-c)/(B(N+1)) > L$, that is $N < 9$, each firm produces

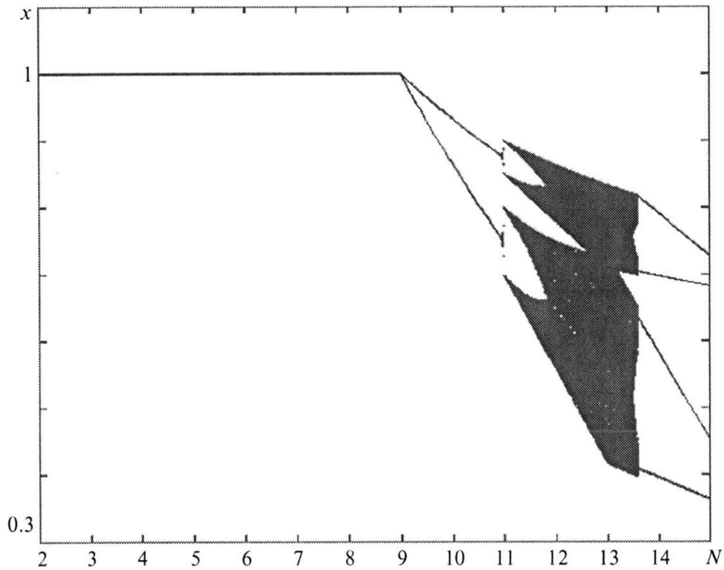


**Fig. 2.4** Example 2.3; linear inverse demand and cost functions and identical capacity constrained firms. Bifurcation diagram of output as the number of firms in the industry increases. Parameter values are $A = 16$, $B = 1$, $a = 0.5$, $c = 6$ and $L = 1$

at capacity. In these cases $\bar{x} = L$ is stable. Note that this holds even if the stability condition $a(N + 1) < 4$ is violated due to $N > 4/a - 1 = 7$ (which yields a slope of the decreasing branch of $T(x)$ less than $-1$).

The bifurcation diagram reveals that for an increasing number of firms cyclic and even chaotic behavior of the production sequences can be observed. The first qualitative change occurs when the number of firms goes from $N = 9$ up to $N = 10$ and it involves a particular kind of global bifurcation known as a *border collision bifurcation*. This kind of bifurcation is specific to piecewise differentiable dynamical systems (see Nusse and Yorke (1995) and Zhanybai and Mosekilde (2003)), hence we describe it in more detail (see the sequence of pictures in Fig. 2.5). In some of the figures to follow we depict portions of the phase space outside the strategy space $[0, L]$. We do this to illustrate and emphasize that in order to understand global bifurcations sometimes it is not sufficient to focus on the local properties around the equilibrium. Let us start with a situation where $N = 8$ firms are in the market. In this case we have $(A - c)/(B(N + 1)) > L$ and the stable equilibrium $\bar{x} = L$ is located on the left upward-sloping branch of the graph of $T(x)$ (Fig. 2.5a). Now, if an additional firm enters the market, $N = 9$, then $\bar{x} = (A - c)/(B(N + 1)) = L$. Additionally, since $((A - c)/B(N - 1)) - 2L(N - 1) = L$, the kink of the graph of the map $T$ also is located exactly on the boundary (see Fig. 2.5b). As the number of firms is even further increased to $N = 10$, the steady state of the map enters the decreasing branch. The derivative $T'(\bar{x})$ of the map for this increasing sequence of $N$ crosses a jump discontinuity, since it assumes the value $(1 - a) = 0.5$ on the left

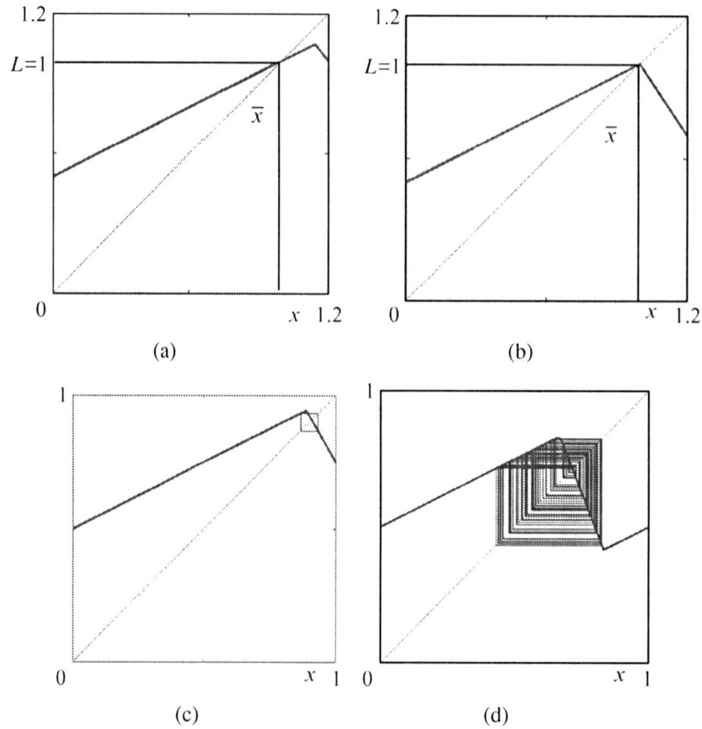


**Fig. 2.5** Example 2.3; linear inverse demand and cost functions and identical capacity constrained firms. Illustrating the border collision bifurcation that occurs as the number of firms varies from 8 to 13. (**a**) $N = 8$; (**b**) $N = 9$; (**c**) $N = 10$; (**d**) $N = 13$. Note how the derivative of the map $T$ at $\bar{x}$ crosses a jump discontinuity as $N$ passes through the value 9

branch and suddenly attains the value $1 - \frac{a(N+1)}{2} < -1$ on the right branch without passing through the bifurcation value $T'(\bar{x}) = -1$.

In general it is not easy to predict which kind of attractor will emerge from such a type of bifurcation. In our case, for $N > 9$ a stable cycle of period 2 is created around the unstable equilibrium $\bar{x}$ (see Fig. 2.5c). The points of the stable 2-cycle are located on different branches of the piecewise linear map. Hence, the multiplier of this 2-cycle is given by the product of its derivatives, that is

$$\lambda_2(C_2) = (1 - a)\left(1 - \frac{a(N+1)}{2}\right).$$

For increasing values of $N$ the multiplier $\lambda(C_2)$ crosses the critical value $-1$ for $N = 11$ after which a chaotic attractor suddenly appears (Fig. 2.5d).

Border collision bifurcations may occur in the presence of piecewise smooth reaction functions for example, due to non-negativity and capacity constraints or even discontinuous reaction functions. As demonstrated above, they are related to (1) the crossing of the equilibria (or points of a cycle) through sets where the dynamical system is not differentiable, and (2) these sets of non-differentiability separate regions where the maps that represent the dynamical system are differentiable, but different. From the viewpoint of economics and oligopoly theory, these bifurcations seem to be important since they may cause sudden stability switches and/or the appearance or disappearance of equilibria, cycles, or chaotic attractors. Although the study of border collision bifurcations is quite new even in the mathematical literature, some studies report this phenomenon for economic models with constraints (see for example Hommes (1991, 1995), Hommes and Nusse (1991), Hommes et al. (1995), Puu and Sushko (2002), Puu et al. (2005), Puu and Sushko (2006), Sushko et al. (2005, 2006). Some of the main results on this subject can be found in Nusse and Yorke (1992, 1995), Maistrenko et al. (1993, 1995, 1998), Di Bernardo et al. (1999), Banerjee et al. (2000a, b), Halse et al. (2003), Zhanybai and Mosekilde (2003), Zhusubaliyev et al. (2002, 2007). However, we should stress that the study of the global dynamical properties of piecewise differentiable dynamical systems is still at a pioneering stage. Many open problems still await a systematic approach, even in the case of one-dimensional maps, see for example the recent papers by Avrutin and Schanz (2006), Avrutin et al. (2006).

To show that after an equilibrium has lost its stability via a border collision bifurcation any kind of attractor may be created, in Fig. 2.6 we present a bifurcation diagram of production $x$ obtained for $A = 16$, $B = 1, a = 0.5, c = 6$ and $N = 15$ and increasing values of the capacity limit $L$ in the range $[0.2, 1]$. In this case at the bifurcation value $L = \frac{A-c}{(N+1)B} = 0.625$ the stable equilibrium $\bar{x} = L$ becomes unstable by crossing the kink of the map $T$, and a chaotic attractor is suddenly created.

Dynamical systems generated by one-dimensional differentiable maps are among the most frequently studied in the literature. It is well-known, for example, that the critical point of the map, where the derivative vanishes, and its images play an important role in deriving the bounds of the attractors in a bifurcation diagram as well as the regions of higher density of points (see for instance Gumowski and Mira (1980), Collet and Eckmann (1980), Mira et al. (1996)). In higher-dimensional dynamical systems based on noninvertible maps the critical curves introduced earlier assume this important role. For a piecewise differentiable map, like the one encountered in the present example, the two kink points, the relative maximum and minimum point, can be used to obtain the upper and lower boundaries of the (chaotic) attractors. These points assume the role of critical (that is, folding) points in our noninvertible map. Indeed, the piecewise linear map $T$ is a noninvertible map of $Z_1 - Z_3 - Z_1$ kind (see Appendix C). However, in contrast to the situations studied before, these critical points are not found by looking for points of vanishing derivative, as for differentiable maps, but they are the points where the map is non-differentiable. In Fig. 2.7a, obtained with $L = 0.8$ and the other parameters as in Fig. 2.6, the upper boundary of the attractor is the maximum value denoted by $m$, and the lower boundary is its image $m_1 = T(m)$. In Fig. 2.7b, obtained with $L = 0.9$, the chaotic interval is $[m_1, m]$. The property that the dynamics are trapped between the critical points and their iterates is useful to bound the chaotic attractors in the

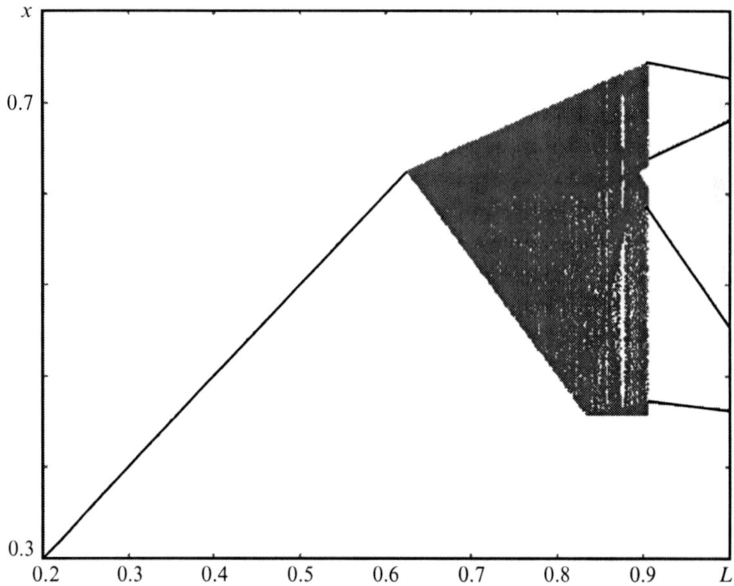


**Fig. 2.6** Example 2.3; linear inverse demand and cost functions and identical capacity constrained firms. Bifurcation diagram of output with respect to capacity $L$ with $N = 15$ firms. A border collision occurs as the bifurcation value $L = 0.625$ is crossed

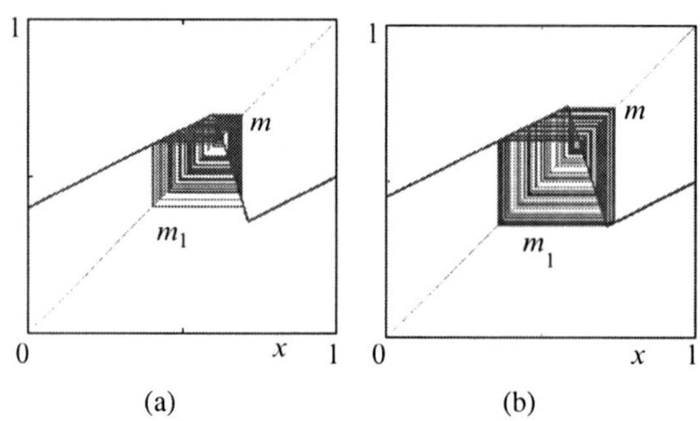


**Fig. 2.7** Example 2.3; linear inverse demand and cost functions and identical capacity constrained firms. The determination of the bounds for the chaotic attractor as capacity $L$ varies. The critical point $m$ is determined at a point where the map $T$ is not differentiable. (**a**) $L = 0.8$; (**b**) $L = 0.9$

case of two-dimensional discrete-time dynamical systems represented by piecewise differentiable maps, like the one obtained in the semi-symmetric case, to which we turn in the next example.

To conclude our analysis of the mathematical properties of the piecewise linear symmetric model, we investigate what kind of bifurcation occurs when the interior equilibrium $\bar{x}$ loses stability for increasing values of $N$ because the derivative $T'(\bar{x})$ become less than $-1$. As we demonstrate below, a quite particular kind of bifurcation, sometimes called *degenerate* or *critical flip bifurcation*, occurs in our piecewise linear map. Although the cycle emerging after the equilibrium has lost its stability is probably not a plausible description of agents' behavior from an economic point of view (see below), we present this case mainly to discuss some mathematical properties of dynamical systems involving piecewise differentiable maps. Since piecewise linear functions are often used in models of economic systems, a study of their peculiar dynamic features may be useful as a reference in other circumstances.

Let us consider the set of parameters $A = 16$, $B = 1$, $a = 0.4$, $c = 8$ and $L = 1$, and take $N$ as a bifurcation parameter. If $N < 7$ then $A - c > (N + 1)BL$, so the equilibrium is $\bar{x} = L$, and it is stable as was shown earlier. For $N > 7$ the equilibrium $\bar{x} = \frac{A-c}{B(N+1)}$ is stable for $N < 4/a - 1 = 9$, and for $N > 9$ it is unstable. For a linear map instability of the equilibrium means divergence of all the trajectories starting arbitrarily close to it, however this is not the case for our piecewise linear model, as its trajectories are bounded. Indeed, as shown in the numerically computed bifurcation diagram of Fig. 2.8, after the bifurcation occurring at $N = 9$, a stable cycle $C_2$ of period 2 suddenly appears. We note that the amplitude of the oscillations along the newly born stable cycle is of finite amplitude *from the moment of its creation.*

This suggests that such a "hard" bifurcation is different from what is usually called a flip (or period doubling) bifurcation. The appearance of the stable cycle is not due to local properties around the equilibrium but is related to the global shape of the map generating the dynamical system. It should be mentioned, however, that the non-equilibrium dynamics emerging after the equilibrium has lost its stability do

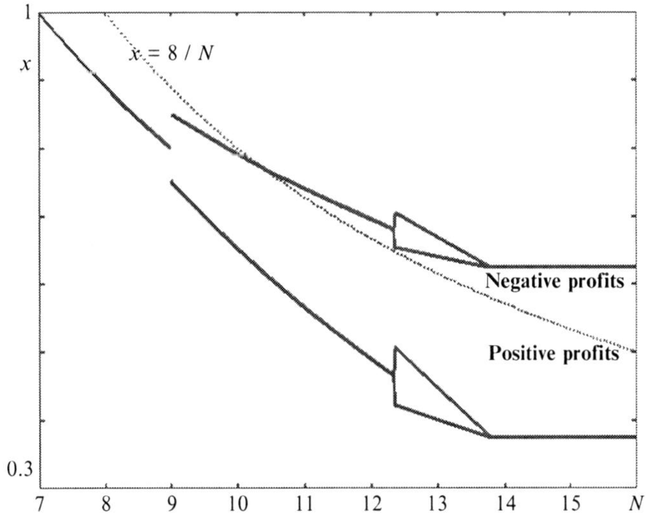


**Fig. 2.8** Example 2.3; linear inverse demand and cost functions and identical capacity constrained firms. Bifurcation of output with respect to number of firms $N$ when stability of equilibrium is lost because $T'(\lambda)$ becomes less than $-1$, and a 2-cycle emerges. Note that profits become negative above the *dotted line*

not seem to be a plausible description of real-world economic behavior for at least two reasons. First, it seems reasonable to assume that such low-periodic dynamics would be detected by the firms. As a result, they most probably would change their expectations. Second, for profits to be positive, $A - c - BNx > 0$ has to hold. However, for the parameter values considered here, profits are negative for $x > 8/N$, that is above the dotted line depicted in Fig. 2.8. Therefore, for $N > 10$ the production at the upper periodic points involve negative profits, that is all firms would cyclically make a loss every other time period. Again firms would most likely change their expectations.

In what follows we give a brief mathematical description of the mechanism which leads to such a bifurcation. Let us consider Figs. 2.9a, b, where the graph of the map $T(x)$ is shown for $N = 8$ and $N = 10$ respectively (the other parameters have the same values as in Fig. 2.5). In Fig. 2.9a the equilibrium is stable (a typical trajectory is shown), whereas in Fig. 2.9b it is unstable, with a stable 2-cycle around it and having periodic points located on different branches of the piecewise linear map. The multiplier of this 2-cycle is given by the product of the derivatives

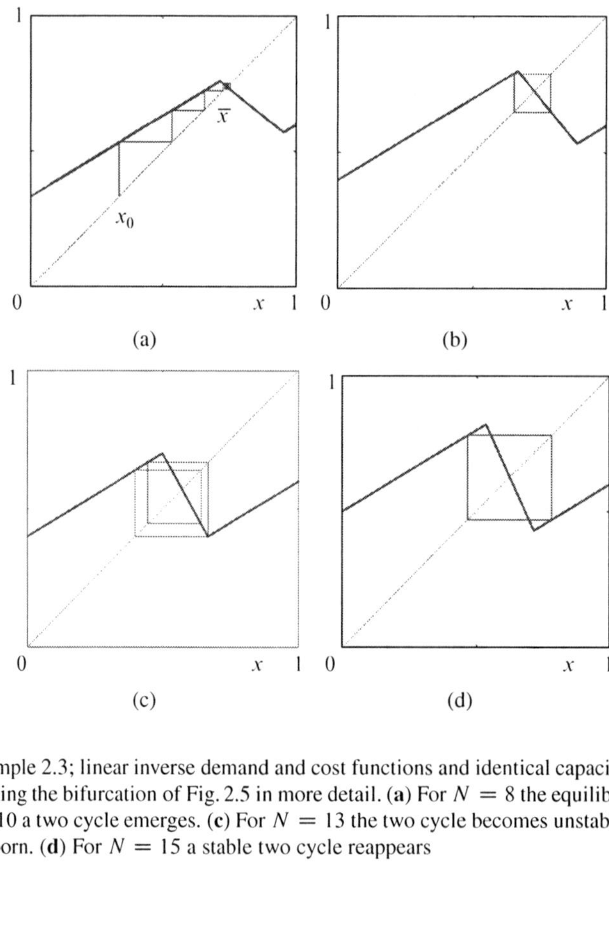


**Fig. 2.9** Example 2.3; linear inverse demand and cost functions and identical capacity constrained firms. Examining the bifurcation of Fig. 2.5 in more detail. (**a**) For $N = 8$ the equilibrium is stable. (**b**) For $N = 10$ a two cycle emerges. (**c**) For $N = 13$ the two cycle becomes unstable and a stable four cycle is born. (**d**) For $N = 15$ a stable two cycle reappears

computed at the two periodic points : $\lambda(C_2) = (1-a)\left(1 - \frac{a(N+1)}{2}\right)$. For the set of parameters used in Fig. 2.9(b) we obtain $\lambda(C_2) = -0.72$, hence it is stable, however $\lambda(C_2)$ decreases for increasing values of $N$ and it reaches the bifurcation value $\lambda(C_2) = -1$ at $N = \frac{2}{a}\left(\frac{2-a}{1-a}\right) - 1 = 12.33$ for $a = 0.4$ (see Fig. 2.5). When the cycle $C_2$ becomes unstable, a stable cycle of period 4, say $C_4$, appears (see Fig. 2.9(c), obtained for $N = 13$) and a further increase of $N$ leads to a period halving bifurcation, typical of bimodal maps (one-dimensional maps with a local maximum and a local minimum), at which $C_4$ is replaced by another stable cycle of period 2, with periodic points located on the first and third branches (as in Fig. 2.9(d) for $N = 15$). This means that the multiplier associated with this cycle is $(1-a)^2$, independent of $N$, from which we deduce that this 2-cycle will remain stable for each value of $N$.

To sum up, this example has illustrated that in a piecewise linear map, even if the loss of stability of an equilibrium point or of a periodic cycle is related to the local values of their multipliers, the effects of such bifurcations, as well as the location of the emerging attractors in the phase space, is related to the global shape of the iterated map. In particular, the periodic cycles emerging from such bifurcations have periodic points belonging to branches of the map that are far from the bifurcating equilibrium. This property also holds for piecewise linear dynamical systems of dimension greater than one, as we shall see in the next example.

*Example 2.4.* In this example we consider the *semi-symmetric* oligopoly, which is obtained by assuming $c_2 = \ldots = c_N$, $a_2 = \ldots = a_N$, and $L_2 = \ldots = L_N$. Further let $x_1(0)$ and $x_2(0) = \ldots = x_N(0)$ denote the initial production quantities of the firms. If the firms partially adjust their production quantities towards the best replies with linear adjustment functions, then the decisions made by firm 1 and the identical firms $2, \ldots, N$ are captured by the two-dimensional dynamical system

$$T : \begin{cases} x_1(t+1) = (1-a_1)\,x_1(t) + a_1 R_1\left((N-1)\,x_2(t)\right), \\ x_2(t+1) = (1-a_2)\,x_2(t) + a_2 R_2\left(x_1(t) + (N-2)\,x_2(t)\right), \end{cases} \tag{2.34}$$

where

$$R_1((N-1)x_2) = \begin{cases} 0 \text{ if } x_2 \geqslant \frac{A-c_1}{B(N-1)}, \\ L_1 \text{ if } x_2 \leqslant \frac{A-c_1}{B(N-1)} - \frac{2L_1}{N-1}, \\ \frac{A-c_1}{2B} - \frac{1}{2}(N-1)x_2 \quad \text{otherwise}, \end{cases}$$

and

$$R_2(x_1 + (N-2)x_2) = \begin{cases} 0 \text{ if } x_1 + (N-2)x_2 \geqslant \frac{A-c_2}{B}, \\ L_2 \text{ if } x_1 + (N-2)x_2 \leqslant \frac{A-c_2}{B} - 2L_2, \\ \frac{A-c_2}{2B} - \frac{1}{2}\left[x_1 + (N-2)x_2\right] \quad \text{otherwise}. \end{cases}$$

Therefore

$$\frac{\partial R_1}{\partial x_1} = 0 \quad \text{and} \quad \frac{\partial R_1}{\partial x_2} \quad \text{is either} \quad 0 \quad \text{or} \quad -\frac{N-1}{2}.$$

Similarly,

$$\frac{\partial R_2}{\partial x_1} \quad \text{is either} \quad 0 \quad \text{or} \quad -\frac{1}{2},$$

and

$$\frac{\partial R_2}{\partial x_2} \quad \text{is either} \quad 0 \quad \text{or} \quad -\frac{N-2}{2}.$$

Therefore the Jacobians in the different regions have the forms

$$\begin{pmatrix} 1-a_1 & \frac{-a_1(N-1)}{2} \\ -\frac{a_2}{2} & 1-a_2-\frac{a_2(N-2)}{2} \end{pmatrix} = \begin{pmatrix} 1-a_1 & \frac{-a_1(N-1)}{2} \\ -\frac{a_2}{2} & 1-\frac{a_2 N}{2} \end{pmatrix},$$

$$\begin{pmatrix} 1-a_1 & \frac{-a_1(N-1)}{2} \\ -\frac{a_2}{2} & 1-a_2 \end{pmatrix}$$

and a form in which one or both of the off-diagonal elements are equal to zero. Using the row norm generated by the diagonal matrix $\boldsymbol{P} = \begin{pmatrix} x & 0 \\ 0 & 1 \end{pmatrix}$ we see that this norm of all possible Jacobians is below one if

$$1 - a_1 + \frac{a_1(N-1)x}{2} < 1,$$

$$\frac{a_2}{2x} + \left|1 - \frac{a_2 N}{2}\right| < 1,$$

and

$$\frac{a_2}{2x} + 1 - a_2 < 1.$$

The first inequality can be rewritten as

$$x < \frac{2}{N-1},$$

and the third condition can be simplified to

$$x > \frac{1}{2}.$$

The second inequality is equivalent to

$$x > \frac{1}{N} \quad \text{and} \quad a_2 < x(4 - Na_2).$$

Therefore the equilibrium is globally asymptotically stable if

$$a_2 < \frac{4}{N} \quad \text{and} \quad \max\left\{\frac{1}{2}; \frac{a_2}{4 - Na_2}\right\} < x < \frac{2}{N-1}.$$

Notice that a feasible $x$ exists if

$$a_2 < \frac{4}{N} \quad \text{and} \quad \max\left\{\frac{1}{2}; \frac{a_2}{4 - Na_2}\right\} < \frac{2}{N-1}.$$

The second relation implies that $N \leq 4$. If $N = 2$, then $a_2 < 2$ and $a_2/(4 - 2a_2) < 2$ are the stability conditions, which can be rewritten as $a_2 < 1.6$. This condition always holds since we assume that $a_2 \leq 1$. If $N = 3$, then the conditions reduce to $a_2 < 4/3$. Finally, if $N = 4$, then $a_2 < 1$ and $a_2/(4 - 2a_2) < 2/3$ are the conditions which can be summarized as $a_2 < 1$. As in the earlier examples, the Jacobian matrix assumes different forms in the different regions (see Fig. 2.10), where the different regions refer to the different regions of the best response functions based on the non-negativity and capacity constraints, similarly to Sect. 1.3.1. Therefore, whenever a variation of parameters causes a displacement of the equilibrium point (or of a periodic point of a cycle) into a different region by crossing the

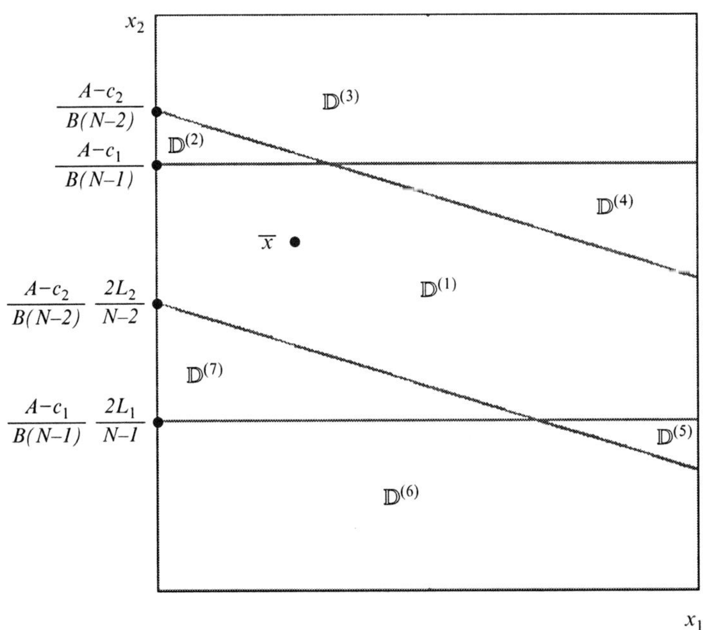


**Fig. 2.10** Example 2.4; linear inverse demand and cost functions, the case of semi-symmetric capacity constrained firms. The regions for the different expressions for the quantity dynamics map

border where the map is not differentiable, the eigenvalues of the equilibrium (or of the periodic cycle) may suddenly change. Such transitions are again accompanied by border collision bifurcations, which we shall demonstrate below. The equilibrium $\bar{\boldsymbol{x}} = (\bar{x}_1, \bar{x}_2)$ in the semi-symmetric case can be obtained from the result of Sect. 1.3.2 with $e_1 = e_2 = 0$, and has the form

$$\bar{x}_1 = \frac{A - Nc_1 + (N-1)c_2}{B(N+1)}, \quad \bar{x}_2 = \frac{A - 2c_2 + c_1}{B(N+1)}. \tag{2.35}$$

If this equilibrium is interior, then its local asymptotic stability is determined by the eigenvalues of the Jacobian matrix

$$\boldsymbol{J}^{(1)} = \begin{pmatrix} 1 - a_1 & -\frac{a_1(N-1)}{2} \\ -\frac{a_2}{2} & 1 - \frac{Na_2}{2} \end{pmatrix}.$$

The characteristic polynomial of this matrix is the quadratic equation

$$\lambda^2 + p\lambda + q = 0$$

with

$$p = -2 + a_1 + \frac{Na_2}{2},$$

and

$$q = (1 - a_1)\left(1 - \frac{Na_2}{2}\right) - \frac{a_1 a_2 (N-1)}{4}.$$

Simple calculation shows that the stability conditions $q < 1$, $p + q + 1 > 0$ and $-p + q + 1 > 0$ (see Appendix F) are satisfied if and only if

$$N < N_b(a_1, a_2) = \frac{16 - a_1(8 - a_2)}{a_2(4 - a_1)}.$$

The right hand side is decreasing in $a_1$, so the global asymptotic stability condition is obtained by selecting the smallest right hand side value for $a_1 = 1$. As expected, also in this case, an increasing number of firms in the oligopoly leads to instability, and the bifurcation value $N_b$ depends on the speeds of adjustment. As for the one-dimensional symmetric case in Example 2.3, it is not easy to predict what kind of asymptotic dynamics are obtained when the equilibrium point is unstable. The attracting sets created after the bifurcation of a piecewise linear map depend on the global properties of the map and are influenced by the borders between the different regions $\mathbb{D}^{(i)}$, where the map is not differentiable. In order to illustrate this point, we consider a case of border collision bifurcation where the border crossing has a remarkable qualitative effect. The bifurcation diagram of outputs in Fig. 2.11 is obtained for $N = 21$, $A = 16$, $B = 1$, $a_1 = 0.2$, $a_2 = 0.3$, $c_1 = 6$, $c_2 = 6$, $L_1 = 2$ and the capacity limit $L_2$ is the bifurcation parameter in the range $[0.4, 0.6]$. At a capacity level of $L_2 \simeq 0.45$ the equilibrium $\bar{x}$ crosses the boundary from region

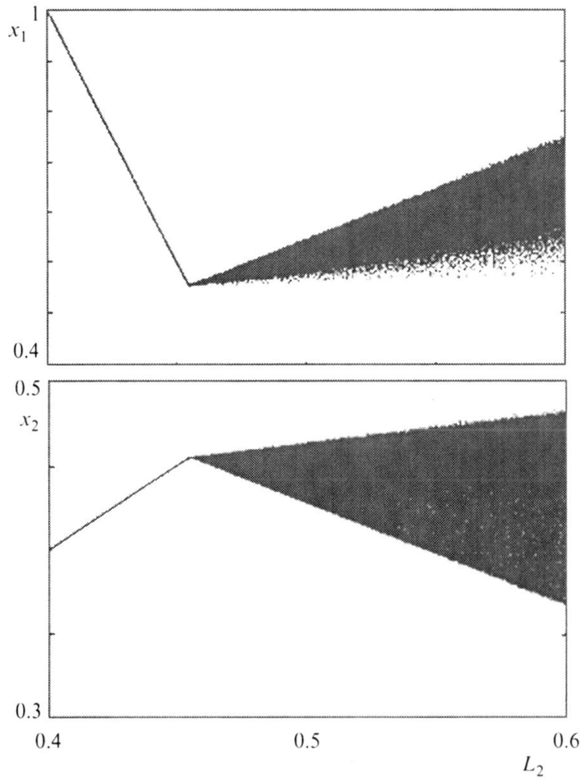


**Fig. 2.11** Example 2.4; linear inverse demand and cost functions, the case of semi-symmetric capacity constrained firms. Border collision bifurcations of $x_1$ (the output of firm 1) and $x_2$ (the output of the other firms) as a function of $L_2$ (the capacity constraint of the other firms). Parameter values are $N = 21$, $A = 16$, $B = 1$, $a_1 = 0.2$, $a_2 = 0.3$, $c_1 = 6$, $c_2 = 6$ and $L_1 = 2$

$\mathbb{D}^{(7)}$, where it is always stable, to $\mathbb{D}^{(1)}$, where the equilibrium is unstable (since $N = 21 > N_b(0.2, 0.3) = 12.6$). The effect of this border collision is the *sudden creation of a chaotic attractor* that becomes larger and larger as the capacity limit $L_2$ increases. Therefore, if firms in this industry invest in capacity, this can cause quite dramatic effects in the asymptotic dynamics of the output sequences. Whereas smaller capacity levels stabilized the industry, a small increase may lead to complex dynamics. What about non-negativity of prices and profits in this situation? The profits of firm $k$ are positive as long as $x_1 + (N-1)x_2 < (A - c_k)/B$, and for the set of parameters used in Fig. 2.11 this means that all the profits are positive as long as $x_1 + 20x_2 < 10$. Non-negativity of prices is ensured because $Q_{\max} = L_1 + (N-1)L_{2\max} = 2 + 20 \cdot 0.6 = 14 < 16 = A/B$. Of course, with these values of the parameters we could even consider capacities $L_2$ up to 0.7, however this would lead to chaotic oscillations of greater amplitude. Consequently, a larger proportion of the chaotic attractor in the regions of the strategy space would be characterized by negative profits. For the one-dimensional model of Example 2.3

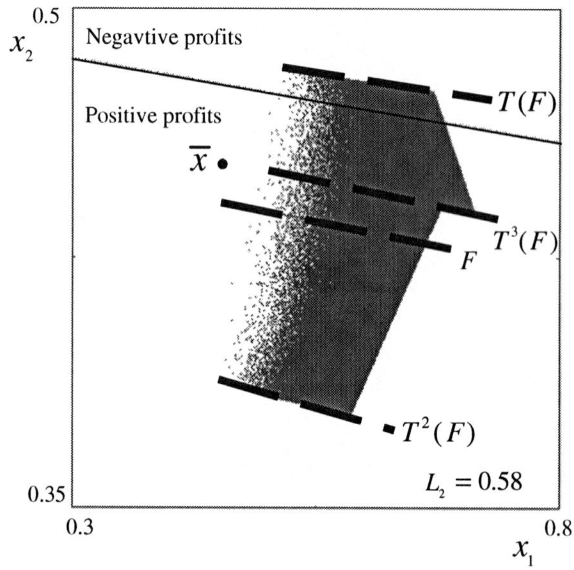


**Fig. 2.12** Example 2.4; linear inverse demand and cost functions, the case of semi-symmetric capacity constrained firms. Computing the bounds on the chaotic attractor in the space of $x_1$ (output of firm 1) and $x_2$ (output of all other firms). Also shown is the line above which profits are negative

obtained in the symmetric case we noted that the kinks (the local maxima and minima, where the map is non-differentiable) can be used to determine bounds for the asymptotic dynamics of the dynamical system. In the two-dimensional model describing the semi-symmetric case the lines of non-differentiability which separate the different regions may play the role of folding curves. That is, they may act as critical curves of noninvertible maps. As explained in Appendix C, the images of the curves where the Jacobian determinant changes sign can be used to bound trapping regions, within which the asymptotic dynamics are confined. Indeed, if we represent the chaotic attractor obtained for a capacity level of $L_2 = 0.58$ (see Fig. 2.12), we notice that it is crossed by the line $F$ of non-differentiability, the equation of which is $x_1 + (N-2)x_2 = \frac{A-c_2}{B} - 2L_2$, separating the regions $\mathbb{D}^{(7)}$ and $\mathbb{D}^{(1)}$. This line acts as a folding line and its images of increasing rank, say $F_1 = T(F)$, $F_2 = T(F_1) = T^2(F)$, give the upper and lower boundaries of the output sequences along the chaotic attractor. In Fig. 2.12 the line $x_1 + 20x_2 = 10$ is also displayed (thin line). Above this line profits of all firms are negative.

Obviously, as the trajectory of production quantities evolves along the chaotic attractor, some time periods exist in which the profits are negative. For the set of parameters used to obtain the bifurcation diagram of Fig. 2.11, this problem only occurs for $L_2 > 0.5$, so that when $L_2 < 0.5$ the chaotic attractor is entirely included in the region of the strategy space where profits are positive. Also in this case, it is not easy to prove what kind of asymptotic dynamics are obtained when

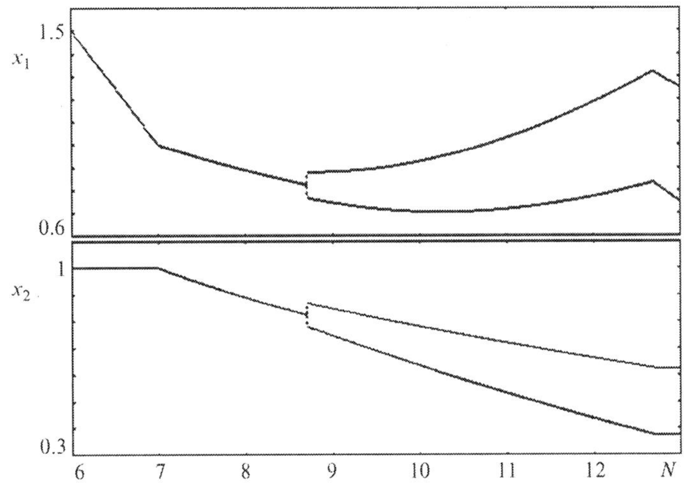


**Fig. 2.13** Example 2.4; linear inverse demand and cost functions, the case of semi-symmetric capacity constrained firms. Bifurcations of $x_1$ (output of firm 1) and $x_2$ (output of all other firms) with respect to $N$ (the number of firms)

the equilibrium point is unstable. In fact, as already noted in the study of the symmetric case, the attracting sets created after the bifurcation of a piecewise linear map depend on the global properties of the map and are influenced by the borders between the different regions $\mathbb{D}^{(i)}$, where the map is not differentiable. In order to illustrate this point, we consider in Fig. 2.13 the numerically computed bifurcation diagram of outputs obtained with parameters $A = 14$, $B = 1, a_1 = 0.5, a_2 = 0.4$, $c_1 = 6$, $c_2 = 6$, $L_1 = 2$, $L_2 = 1$ and values of the bifurcation parameter $N$ in the range $[6, 13]$.

Analogously to the symmetric one-dimensional model in Example 2.3 we give a mathematical description of the bifurcations involved in the present situation in order to understand the peculiar properties of piecewise linear dynamical systems. However, also here we should point out that asymptotic behavior characterized by a low-periodic stable cycle is not realistic from an economic point of view, because presumably firms would detect such a simple periodicity and change their naive expectations. Moreover, also in this case the profits are negative in the upper periodic point. For $N < 7$ the point $\bar{x}$ given in (2.35) is outside the region $\mathbb{D}^{(1)}$, with $\bar{x}_2 > L_2$, hence it is not an equilibrium of the dynamical system (2.34). For $N = 7$, $\bar{x} = (1, 1)$ (see Fig. 2.14a) and it then enters the region $\mathbb{D}^{(1)}$ as $N$ is further increased (Fig. 2.14b, obtained with $N = 8$), and it is locally asymptotically stable for $N < N_b\ (0.5, 0.4) = 8.714$. At this bifurcation value the equilibrium $\bar{x}$ becomes unstable and a stable cycle of period 2 appears, clearly visible in the bifurcation diagram of Fig. 2.13, with periodic points located in different regions (see Fig. 2.14c, obtained with $N = 10$, where the two stable periodic points are labelled by $c_2^{(i)}$).

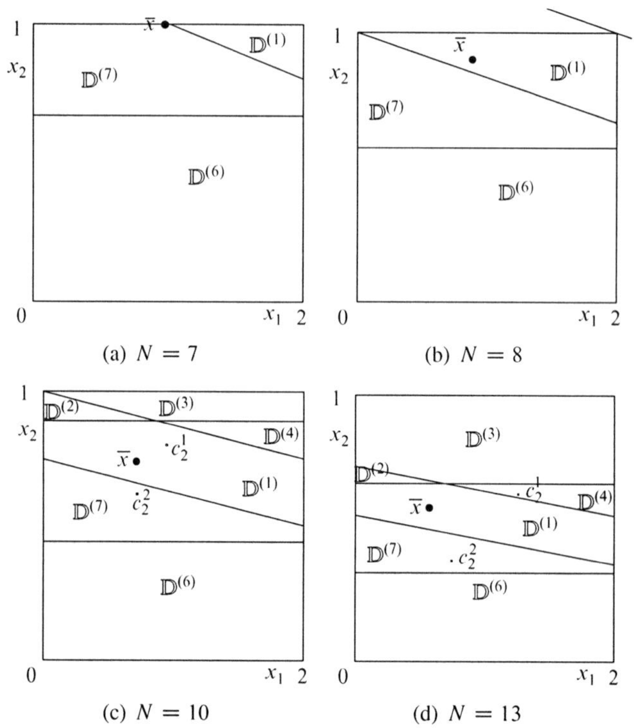


**Fig. 2.14** Example 2.4; linear inverse demand and cost functions, the case of semi-symmetric capacity constrained firms. A more detailed study of the bifurcations with respect to $N$. Note the changing structure of the different regions of the map as $N$ increases, and the border collision that occurs as $N$ increases from 8 to 10

As $N$ is further increased the periodic points may cross the boundaries that separate different regions, giving rise to border collision bifurcations that may change the stability of the cycle involved and create new attractors. For example, in Fig. 2.14d, obtained with $N = 13$, we can see that the periodic point $c_2^1$ has crossed the border, moving from region $\mathbb{D}^{(1)}$ to $\mathbb{D}^{(4)}$. However this border collision did not cause a change of stability of the 2-cycle, because after the border crossing the two periodic points are in regions $\mathbb{D}^{(4)}$ and $\mathbb{D}^{(7)}$, so the 2-cycle remains stable with its multipliers given by $\lambda_1(C_2) = (1 - a_1)^2$, $\lambda_2(C_2) = (1 - a_2)^2$ (the two Jacobian matrices $J^{(4)} = J^{(7)}$ are triangular matrices). Nevertheless, in the bifurcation diagram of Fig. 2.13 the occurrence of this border crossing can be easily detected around the value $N \simeq 12.7$.

*Example 2.5.* In this example we return to the case of a quadratic price function

$$f(Q) = \begin{cases} A - Q^2 & \text{if } 0 \leqslant Q \leqslant \sqrt{A}, \\ 0 & \text{if } Q > \sqrt{A}, \end{cases}$$

and linear costs $C_k(x_k) = c_k x_k$, which was introduced in Example 1.4. We assume that $A > c_k$ for all $k = 1, \ldots, N$. As shown in Example 1.4, the best response of firm $k$ is given by the continuous and piecewise differentiable function

$$R_k(Q_k) = \begin{cases} 0 \text{ if } x_k^* \leqslant 0 \text{ i.e., } & Q_k \geqslant \sqrt{A - c_k}, \\ L_k \text{ if } x_k^* \geqslant L_k \text{ i.e., } & Q_k \leqslant \sqrt{L_k^2 + A - c_k} - 2L_k, \\ z_k^* \quad \text{otherwise i.e., } & \sqrt{L_k^2 + A - c_k} - 2L_k < Q_k < \sqrt{A - c_k}, \end{cases}$$

where

$$z_k^* = \frac{1}{3}\left(\sqrt{Q_k^2 + 3(A - c_k)} - 2Q_k\right).$$

It is easy to see that $\sqrt{L_k^2 + A - c_k} - 2L_k < \sqrt{A - c_k}$ holds given our assumption that $A > c_k$ and $L_k > 0$. In the case of $N$ firms, the unique equilibrium, see (1.14), is

$$\bar{x}_k = \frac{2A + \sum_{l=1}^N c_l - (N+2)c_k}{2\sqrt{(N+2)(NA - \sum_{l=1}^N c_l)}},$$

under the assumption that it is interior.

Let us first consider the case of duopoly, that is $N = 2$, with partial adjustment towards the best response. The sequence of production quantities in this case is obtained by the repeated application of the piecewise differentiable map

$$T: \begin{cases} x_1(t+1) = (1 - a_1)\,x_1(t) + a_1 R_1(x_2(t)), \\ x_2(t+1) = (1 - a_2)\,x_2(t) + a_2 R_2(x_1(t)), \end{cases}$$

where the reaction functions are given by

$$R_1(x_2) = \begin{cases} 0 & \text{if } \; x_2 \geqslant \sqrt{A - c_1}, \\ L_1 & \text{if } \; x_2 \leqslant \sqrt{L_1^2 + A - c_1} - 2L_1, \\ \frac{1}{3}\left(\sqrt{x_2^2 + 3(A - c_1)} - 2x_2\right) & \text{otherwise}, \end{cases}$$

and

$$R_2(x_1) = \begin{cases} 0 & \text{if } \; x_1 \geqslant \sqrt{A - c_2}, \\ L_2 & \text{if } \; x_1 \leqslant \sqrt{L_2^2 + A - c_2} - 2L_2, \\ \frac{1}{3}\left(\sqrt{x_1^2 + 3(A - c_2)} - 2x_1\right) & \text{otherwise}. \end{cases}$$

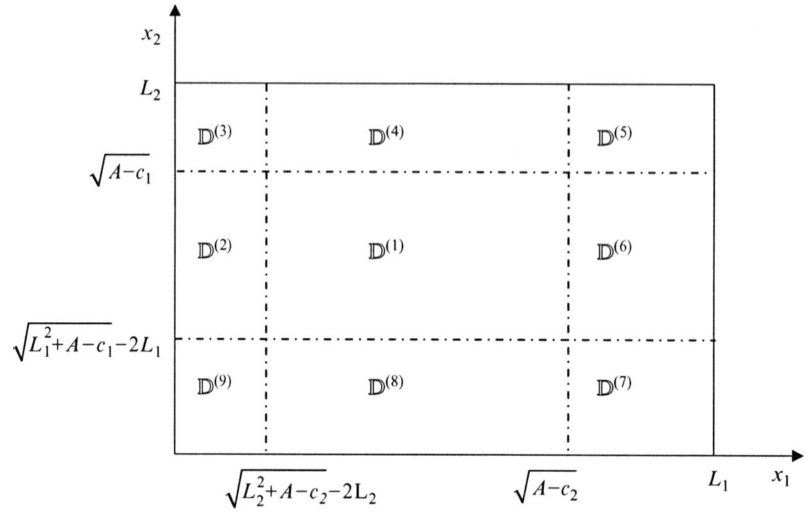


**Fig. 2.15** Example 2.5; quadratic price and linear cost functions. The regions of the piece-wise map in the duopoly case

The strategy space, given by the trapping region $\mathbb{D} = [0, L_1] \times [0, L_2]$, can be subdivided into nine different regions $\mathbb{D}^{(i)}$, $i = 1, \ldots, 9$, (see Fig. 2.15), similar to the previous examples.

Some of these regions may be empty when one or both the capacity limits are too small. For example $\mathbb{D}^{(3)}$, $\mathbb{D}^{(4)}$ and $\mathbb{D}^{(5)}$ do not exist if $L_2 \leqslant \sqrt{A - c_1}$. Moreover, the profit of firm $k$ is positive as long as $x_1 + x_2 < \sqrt{A - c_k}$, hence some of the regions in Fig. 2.15 may involve negative profits. For example, if $c_1 = c_2$ then regions $\mathbb{D}^{(3)}$, $\mathbb{D}^{(4)}$, $\mathbb{D}^{(5)}$, $\mathbb{D}^{(6)}$ and $\mathbb{D}^{(7)}$ all involve negative profits and if an attractor is completely included inside these regions, it should be considered as economically infeasible. Instead, trajectories that pass through such regions and then exit it to enter other regions characterized by positive profits can be considered as economically feasible. The Jacobians of the regions are

$$\boldsymbol{J}^{(1)} = \begin{pmatrix} 1 - a_1 & a_1 \left( \frac{x_2}{3\sqrt{x_2^2 + 3(A - c_1)}} - \frac{2}{3} \right) \\ a_2 \left( \frac{x_1}{3\sqrt{x_1^2 + 3(A - c_2)}} - \frac{2}{3} \right) & 1 - a_2 \end{pmatrix},$$

and matrices which can be obtained from this Jacobian by changing one or both off-diagonal elements to zero. That is,

$$\boldsymbol{J}^{(2)} = \boldsymbol{J}^{(6)} = \begin{pmatrix} 1 - a_1 & a_1 \left( \frac{x_2}{3\sqrt{x_2^2 + 3(A - c_1)}} - \frac{2}{3} \right) \\ 0 & 1 - a_2 \end{pmatrix},$$

$$J^{(3)} = J^{(5)} = J^{(7)} = J^{(9)} = \begin{pmatrix} 1-a_1 & 0 \\ 0 & 1-a_2 \end{pmatrix},$$

$$J^{(4)} = J^{(8)} = \begin{pmatrix} 1-a_1 & 0 \\ a_2\left(\frac{x_1}{3\sqrt{x_1^2+3(A-c_2)}} - \frac{2}{3}\right) & 1-a_2 \end{pmatrix}.$$

The interior equilibrium is

$$\bar{\boldsymbol{x}} = (\bar{x}_1, \bar{x}_2) = \left(\frac{2A-3c_1+c_2}{4\sqrt{2A-c_1-c_2}}, \frac{2A-3c_2+c_1}{4\sqrt{2A-c_1-c_2}}\right)$$

and it is an equilibrium of the dynamical system provided it belongs to the region $\mathbb{D}^{(1)}$. Unfortunately, even in the duopoly case, explicit conditions for the local stability of the interior equilibrium are not easy to obtain. However, numerical simulations indicate that whenever $\bar{\boldsymbol{x}}$ exists then it appears to be globally asymptotically stable. Now the questions arises under which conditions is stability lost in the oligopoly case, that is for $N > 2$. Clearly, the general case is hard to analyze. However, to get some insight into the effect of an increasing number of firms or an increase in the speeds of adjustment on the stability of the equilibrium we can consider the *semi-symmetric* case. Hence, we assume $c_2 = \ldots = c_N$, $a_2 = \ldots = a_N$, and $L_2 = \ldots = L_N$. Under the further assumption of identical initial conditions for firms $2, \ldots, N$, that is $x_2(0) = \ldots = x_N(0)$, the production decisions of firm 1 and the identical firms $2, \ldots, N$ are governed by the two-dimensional dynamical system

$$T\colon \begin{cases} x_1(t+1) = (1-a_1)\,x_1(t) + a_1 R_1\left((N-1)\,x_2(t)\right), \\ x_2(t+1) = (1-a_2)\,x_2(t) + a_2 R_2\left(x_1(t) + (N-2)\,x_2(t)\right), \end{cases} \tag{2.36}$$

where

$$R_1((N-1)x_2) = \begin{cases} 0 \quad \text{if} \quad x_2 \geqslant \frac{\sqrt{A-c_1}}{N-1}, \\ L_1 \text{ if} \quad x_2 \leqslant \frac{\sqrt{L_1^2+A-c_1-2L_1}}{N-1}, \\ \frac{1}{3}\left(\sqrt{(N-1)^2x_2^2+3\,(A-c_1)} - 2(N-1)x_2\right) \quad \text{otherwise,} \end{cases}$$

and

$$R_2(x_1+(N-2)x_2)$$
$$= \begin{cases} 0 \quad \text{if} \quad x_1+(N-2)x_2 \geqslant \sqrt{A-c_2}, \\ L_2 \text{ if} \quad x_1+(N-2)x_2 \leqslant \sqrt{L_2^2+A-c_2-2L_2}, \\ \frac{1}{3}\left(\sqrt{(x_1+(N-2)x_2)^2+3\,(A-c_2)} - 2\,(x_1+(N-2)x_2)\right) \quad \text{otherwise.} \end{cases}$$

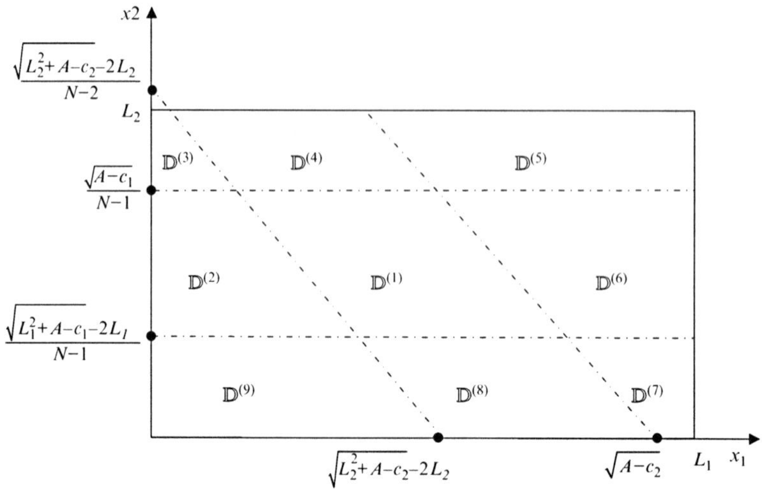


**Fig. 2.16** Example 2.5; quadratic price and linear cost functions. The regions of the piece-wise map in the semi-symmetric case

Also in this case the strategy space $\mathbb{D} = [0, L_1] \times [0, L_2]$ can be subdivided into up to nine different regions $\mathbb{D}^{(i)}$, as shown in Fig. 2.16. In each of these regions the map $T$ is then differentiable. For example, in regions $\mathbb{D}^{(1)}$ and $\mathbb{D}^{(2)}$ we have

$$T|_{\mathbb{D}^{(1)}} : \begin{cases} x_1(t+1) = (1-a_1)x_1(t) \\ \qquad\qquad + a_1\frac{1}{3}\left[\sqrt{(N-1)^2x_2^2 + 3(A-c_1)} - 2(N-1)x_2\right], \\ x_2(t+1) = (1-a_2)x_2(t) \\ \qquad\qquad + a_2\frac{1}{3}\Big[\sqrt{(x_1+(N-2)x_2)^2 + 3(A-c_2)} \\ \qquad\qquad -2(x_1+(N-2)x_2)\Big], \end{cases}$$

$$T|_{\mathbb{D}^{(2)}} : \begin{cases} x_1(t+1) = (1-a_1)x_1(t) \\ \qquad\qquad + a_1\frac{1}{3}\left[\sqrt{(N-1)^2x_2^2 + 3(A-c_1)} - 2(N-1)x_2\right], \\ x_2(t+1) = (1-a_2)x_2(t) + a_2L_2, \end{cases}$$

and the corresponding Jacobian matrices are given by

$$J^{(1)} = \begin{pmatrix} 1-a_1 & a_1(N-1)\left(\frac{(N-1)x_2}{3\sqrt{(N-1)^2x_2^2+3(A-c_1)}} - \frac{2}{3}\right) \\ a_2\left(\frac{x_1+(N-2)x_2}{3\sqrt{(x_1+(N-2)x_2)^2+3(A-c_2)}} - \frac{2}{3}\right) & j_{22}^{(1)} \end{pmatrix},$$

where $j_{22}^{(1)} = 1 - a_2 + a_2(N-2)\left(\frac{x_1+(N-2)x_2}{3\sqrt{(x_1+(N-2)x_2)^2+3(A-c_2)}} - \frac{2}{3}\right)$.

and

$$J^{(2)} = J^{(6)} = \begin{pmatrix} 1 - a_1 & j_{12}^{(2)} \\ 0 & 1 - a_2 \end{pmatrix}.$$

where $j_{12}^{(2)} = a_1(N-1)\left(\frac{(N-1)x_2}{3\sqrt{(N-1)^2x_2^2+3(A-c_1)}} - \frac{2}{3}\right)$.

The other Jacobians are obtained by changing the off-diagonal elements of $J^{(1)}$ to zero. The interior equilibrium $\bar{x} = (\bar{x}_1, \bar{x}_2)$ in the semi-symmetric case is

$$\begin{aligned} \bar{x}_1 &= \frac{2A - (N+1)c_1 + (N-1)c_2}{2\sqrt{(N+2)(NA - c_1 - (N-1)c_2)}}, \\ \bar{x}_2 &= \frac{2A - 3c_2 + c_1}{2\sqrt{(N+2)(NA - c_1 - (N-1)c_2)}}, \end{aligned} \tag{2.37}$$

provided it is in region $\mathbb{D}^{(1)}$. Due to the algebraic complexity of the expressions involved, the study its local stability analytically is quite difficult. Moreover, when the equilibrium $\bar{x}$ crosses the boundaries of the region $\mathbb{D}^{(1)}$ (or other periodic points move across different regions $\mathbb{D}^{(i)}$) border collision bifurcations may occur that cause the creation, destruction, or modification of the qualitative properties of the attractors. In what follows we employ a combination of analytical and numerical methods to gain some information about the global dynamical behavior of this nonlinear piece-wise differentiable model and about the global bifurcations occurring as some parameters are varied.

Let us assume that firm 1 has higher unit costs than the rest of the industry, so that $c_1 > c_2$, and we shall study the properties of the equilibrium as the number of firms varies. Note that $2A + c_1 - 3c_2 > 0$ is guaranteed by the assumptions $c_1 > c_2$ and $A > c_k$ $(k = 1, 2)$. As long as $2A + (N-1)c_2 - (N+1)c_1 > 0$, all firms are active in the market. Profits are then given by

$$\begin{aligned} \bar{\varphi}_1 &= \frac{(2A + (N-1)c_2 - (N+1)c_1)^2}{2(N+2)\sqrt{(N+2)(NA - c_1 - (N-1)c_2)}}, \\ \bar{\varphi}_2 &= \frac{(2A + c_1 - 3c_2)^2}{2(N+2)\sqrt{(N+2)(NA - c_1 - (N-1)c_2)}}. \end{aligned}$$

However, if $N > (2A - c_2 - c_1)/(c_1 - c_2)$ then firm 1 stops producing. In this case $\bar{x}_1 = 0$ and the other $N - 1$ identical firms select their symmetric equilibrium

$$\bar{x}_2 = \frac{A - c_2}{\sqrt{(N+1)(N-1)(A - c_2)}}, \tag{2.38}$$

which can be obtained from relation (1.14) with $N$ being replaced by $N - 1$ and all $c_k$ by $c_2$. The profit of the active firms reads

$$\bar{\varphi}_2 = \frac{2(A-c_2)^2}{(N+1)\sqrt{(N+1)(N-1)(A-c_2)}}.$$

As a numerical example, consider $A = 16, c_1 = 10 > 8 = c_2 = ... = c_N$. Then using the expressions given in (2.37) we obtain the unique equilibrium

$$\bar{x}_1 = \frac{7-N}{\sqrt{8N^2+14N-4}}, \quad \bar{x}_2 = ... = \bar{x}_N = \frac{9}{\sqrt{8N^2+14N-4}},$$

which shows that for $N > 7$ firm 1 stops producing and we have a boundary equilibrium

$$\bar{x}_1 = 0 \quad \text{and} \quad \bar{x}_2 = ... = \bar{x}_N = \sqrt{\frac{8}{N^2-1}}.$$

For $N = 7$, total equilibrium industry output becomes $\bar{Q} = \sqrt{6}$ and the corresponding equilibrium price is $f(\bar{Q}) = 10$, which obviously equals the marginal cost of firm 1. However this is not the end of the story. Figure 2.17 shows a bifurcation diagram of outputs obtained for the model (2.36) with speeds of adjustment $a_1 = 0.5$, $a_2 = 0.4$, capacity limits $L_1 = L_2 = 0.4$ and bifurcation parameter $N$ in the range $[3, 10]$ (notice that $L_1 + (N-1)L_2 \leqslant A$ in the whole range, so that non-negativity of prices is ensured). The bifurcation diagram of Fig. 2.17 confirms that $\bar{x}_1$ goes to zero for $N > 7$. However, for $N = 10$ a positive stable cycle of period 2 characterizes the long-run dynamics and it appears that firm 1 resumes production. Mathematically, this stable cycle is created through a border collision bifurcation between $N > 9$ and $N > 10$, and at its creation it coexists with the stable boundary equilibrium. So,

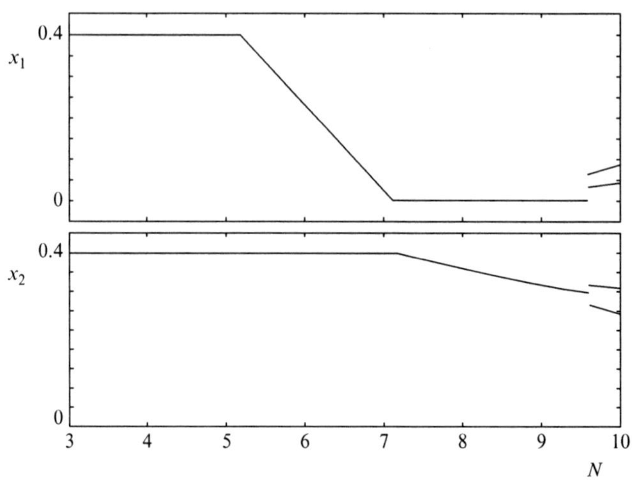


**Fig. 2.17** Example 2.5; quadratic price and linear cost function. The semi-symmetric case. Bifurcation diagrams of $x_1$, $x_2$ with respect to the number of firms $N$. Illustrating how $x_1$ can go to zero for some values of $N$

for $N > 9$ the boundary equilibrium point $\left(0, \sqrt{8/(N^2-1)}\right)$ is the unique (global) attractor. Then, as $N$ is increased, a 2-cycle appears with periodic points located in the regions $\mathbb{D}^{(2)}$ and $\mathbb{D}^{(4)}$ respectively, coexisting with the stable boundary equilibrium with the stable boundary equilibrium point. Each attractor has its own basin of attraction, and then the basin of the boundary equilibrium point shrinks, until the equilibrium becomes unstable. For $N = 10$ the stable 2-cycle remains the unique (global) attractor. However, this cycle is not meaningful as a description of a long run solution of the game that we are considering here. The reason is that a careful analysis should check if profits, given by $\varphi_k = x_1\left[A - c_1 - (x_1 + (N-1)x_2)^2\right]$, are positive. The two periodic points of the stable cycle for $N > 10$ have coordinates $c_2^1 \simeq (0.0889, 0.3058)$ and $c_2^1 \simeq (0, 0445, 0.2429)$. Hence, the profits of firm 1 along this cycle are $\varphi_1\left(c_2^1\right) = -0.18$ and $\varphi_1\left(c_2^2\right) = 0.046$, and the corresponding profits of firm 2 are $\varphi_1\left(c_2^1\right) = -0.006$ and $\varphi_1\left(c_2^2\right) = 0.13$. Consequently, as expected firm 1 has no incentive to resume production again since the average profit along the stable 2-cycle is negative. With a higher number of firms more complicated non-equilibrium dynamics can be observed. For example, let us consider the bifurcation diagram obtained with parameters $N = 12$, $A = 144$, $c_1 = 10$, $c_2 = 8$, $L_1 = L_2 = 1$, $a_2 = 0.7$ and increasing values of the speed of adjustment $a_1 \in (0.1]$ (Fig. 2.18). In this case chaotic oscillations of large amplitude dominate, a situation that may imply that firms have great difficulty in forecasting, so that naive expectations may in fact represent a reasonable assumption.

However, the shape of the chaotic attractor in the strategy space (see Fig. 2.19a, obtained with the same parameters as those used in Fig. 2.18 and $a_1 = 0.8$) reveals a certain degree of correlation among the production quantities of firm 1 and the

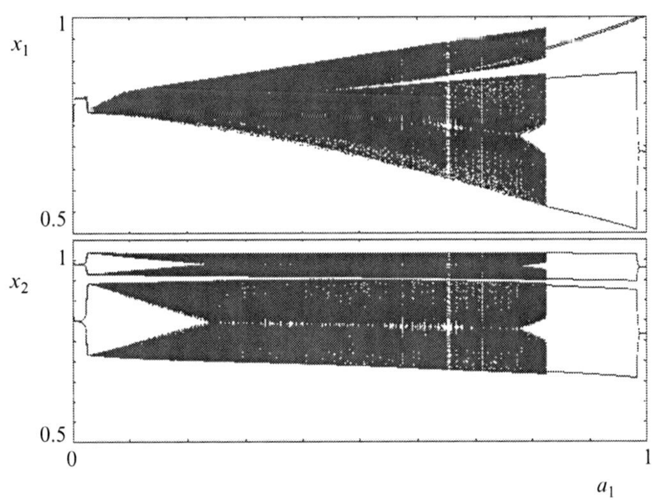


**Fig. 2.18** Example 2.5; quadratic price and linear cost function. The semi-symmetric case. Bifurcation diagrams of $x_1$, $x_2$ with respect to $a_1$, the speed of adjustment of firm 1. The number of firms is held equal to $N = 12$

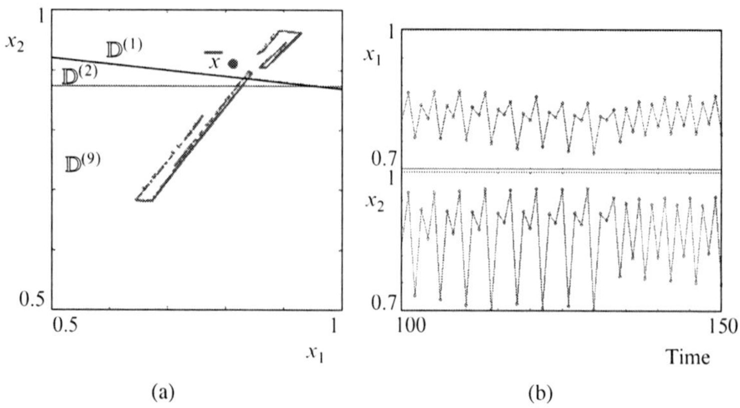


**Fig. 2.19** Example 2.5; quadratic price and linear cost functions. The semi-symmetric case. (**a**) The attractor associated with Fig. 2.18 for $a_1 = 0.8$. (**b**) The time series along the chaotic attractor. Note the correlation between $x_1$ and $x_2$ in both figures

quantities of the other (symmetric) firms in the following sense: periods where the output $x_1(t)$ of firm 1 is high are associated with periods where $x_2(t)$ is high. The same can be observed for periods with low output values (see also Fig. 2.19b, where the asymptotic values of typical time series for $x_1(t)$ and $x_2(t)$ moving along the chaotic attractor are shown). In this case, the firms' actual production choices and their expectations of the joint outputs of the other firms move jointly up and down and enable the observer to make a qualitative prediction of what to expect next, an increase in industry output or a decrease.

## 2.4 Gradient Adjustments

In this section we briefly examine the gradient adjustment processes introduced in Sect. 1.2. Further examples and applications of gradient dynamics will be presented in the following chapters. In the case of the classical Cournot model without cost externalities the discrete time gradient adjustment process (1.32) becomes

$$x_k(t+1) = x_k(t) + \alpha_k \left( x_k(t) f'\left(\sum_{i=1}^{N} x_i(t)\right) + f\left(\sum_{i=1}^{N} x_i(t)\right) - C_k'(x_k(t)) \right), \tag{2.39}$$

and the continuous time gradient adjustment process (1.33) simplifies to

$$\dot{x}_k(t) = \alpha_k \left( x_k(t) f'\left(\sum_{i=1}^{N} x_i(t)\right) + f\left(\sum_{i=1}^{N} x_i(t)\right) - C_k'(x_k(t)) \right). \tag{2.40}$$

Models including externalities can be discussed similarly, the final conclusions remain similar to those to be presented in this section. First the local asymptotic stability of the equilibrium is discussed. The Jacobian of the system (2.39) has the special form

$$\bar{H} = \begin{pmatrix} 1 + a_1(x_1 f'' + 2f' - C_1'') & a_1(x_1 f'' + f') & \ldots \; a_1(x_1 f'' + f') \\ a_2(x_2 f'' + f') & 1 + a_2(x_2 f'' + 2f' - C_2'') & \ldots \; a_2(x_2 f'' + f') \\ \vdots & \vdots & \ddots \\ a_N(x_N f'' + f') & a_N(x_N f'' + f') & \ldots \; 1 + a_N(x_N f'' \\ & & + 2f' - C_N'') \end{pmatrix}$$

where all derivatives are taken at the equilibrium and $a_k = \alpha_k'(0)$ for all $k$. Notice that this matrix has the special structure (E.4) introduced in Appendix E, therefore (E.5) can be used to write the characteristic polynomial as

$$\prod_{k=1}^{N}(1 + a_k(f' - C_k'') - \lambda) \cdot \left[1 + \sum_{k=1}^{N} \frac{a_k(x_k f'' + f')}{1 + a_k(f' - C_k'') - \lambda}\right].$$

Similarly to best response dynamics with adaptive expectations , the eigenvalues are $1 + a_k(f' - C_k'')$ and the roots of the equation

$$1 + \sum_{k=1}^{N} \frac{a_k(x_k f'' + f')}{1 + a_k(f' - C_k'') - \lambda} = 0. \tag{2.41}$$

Since assumptions (A)–(C) hold, the graph of the function on the left hand side of (2.41) is the same as shown in Fig. 2.1, so all eigenvalues are inside the unit circle if and only if for all $k$,

$$a_k(C_k'' - f') < 2, \tag{2.42}$$

and

$$\sum_{k=1}^{N} \frac{a_k(x_k f'' + f')}{2 + a_k(f' - C_k'')} > -1. \tag{2.43}$$

Notice that these conditions are very similar to conditions (2.21) and (2.22) given for the best reply dynamics with adaptive expectations. In order to compare the two cases substitute relation (2.5) (from which $r_k$ is calculated) into conditions (2.21) and (2.22) to obtain

$$a_k \frac{C_k'' - f'}{-(2f' + x_k f'' - C_k'')} < 2 \tag{2.44}$$

and

$$\sum_{k=1}^{N} \frac{a_k \dfrac{f' + x_k f''}{-(2f' + x_k f'' - C_k'')}}{2 - a_k \dfrac{C_k'' - f'}{-(2f' + x_k f'' - C_k'')}} > -1, \tag{2.45}$$

from which it is clear that if $a_k$ is replaced by $a_k/(-(2f' + x_k f'' - C_k''))$ for all firms, then (2.44) is the same as (2.42), and (2.45) is identical to (2.43).

If $2f' + x_k f'' - C_k'' = -1$, then the cases are exactly the same. If $2f' + x_k f'' - C_k'' > -1$ for all $k$, then (2.44) implies (2.42) and (2.45) implies (2.43), so the best reply dynamics are more stable. Notice that

$$2f' + x_k f'' - C_k'' = (f' + x_k f'') + (f' - C_k''),$$

where the first term is non-positive and the second term is negative. The condition that this quantity is larger than $-1$ requires that the absolute values of both terms be sufficiently small. For example, in the case of linear price and cost functions,

$$f'' = C_k'' = 0,$$

and the condition requires that $f' > -1/2$, so price cannot decrease very fast with increasing total output of the industry in order to guarantee stability. If $2f' + x_k f'' - C_k'' < -1$, then we reach the opposite conclusion.

The Jacobian of the continuous time system (2.40) also has the special form, $\bar{\boldsymbol{H}} - \boldsymbol{I}$, where $\bar{\boldsymbol{H}}$ is the Jacobian of the discrete system and $\boldsymbol{I}$ is the identity matrix. It is very easy to show that all eigenvalues of $\bar{\boldsymbol{H}} - \boldsymbol{I}$ are real and negative, so the gradient adjustment process is always locally asymptotically stable. Therefore, for continuous time scales there is no difference between best reply dynamics and gradient adjustments as long as our concern is only local asymptotic stability. The global asymptotic properties of gradient adjustment and adaptive adjustment processes are usually different.

## 2.5 Continuous Time Oligopolies and Local Stability

Consider now the continuous time model (1.31) of the dynamics of partial adjustment towards the best response with naive expectations. A vector $(\bar{x}_1, \ldots, \bar{x}_N)$ is a steady state of the system if and only if

$$\bar{x}_k = R_k\Big(\sum_{l \neq k} \bar{x}_l\Big),$$

in which case $(\bar{x}_1, \ldots, \bar{x}_N)$ is a Nash equilibrium.

The main result of this section is the following, in which we use the notation of Sect. 2.2.

**Theorem 2.2.** *Assume that* $a_k > 0$ *for all* $k = 1, 2, \ldots, N$. *Then the equilibrium with respect to the continuous adjustment process (1.31) is always locally asymptotically stable.*

*Proof.* Using linearization, the eigenvalues of the Jacobian have to be examined. Similarly to the discrete case, the Jacobian of system (1.31) has the special structure

$$\begin{pmatrix} -a_1 & a_1 r_1 & \cdots & a_1 r_1 \\ a_2 r_2 & -a_2 & \cdots & a_2 r_2 \\ \vdots & \vdots & & \vdots \\ a_N r_N & a_N r_N & \cdots & -a_N \end{pmatrix}, \tag{2.46}$$

which is a special case of the form (E.4) (studied in Appendix E). The characteristic equation of this matrix can also be given as a special case of equation (E.5), namely

$$\prod_{k=1}^{N} (-a_k(1+r_k) - \lambda) \cdot \left[ 1 + \sum_{k=1}^{N} \frac{a_k r_k}{-a_k(1+r_k) - \lambda} \right] = 0. \tag{2.47}$$

We will now proceed similarly to the discrete case examined earlier. Assume again that $a_k > 0$ for all $k$ and the firms are numbered in such a way that the different $a_k(1+r_k)$ values are

$$a_1(1+r_1) > a_2(1+r_2) > \cdots > a_s(1+r_s)$$

and these values are repeated $m_1, m_2, \ldots, m_s$ times, respectively, among the $N$ firms. By adding the terms with identical denominators in the bracketed expression and denoting by $\theta_j$ the sum of the corresponding numerators $a_k r_k$, we can rewrite (2.47) as

$$\prod_{j=1}^{s} (-a_j(1+r_j) - \lambda)^{m_j} \cdot \left[ 1 - \sum_{j=1}^{s} \frac{\theta_j}{a_j(1+r_j) + \lambda} \right] = 0, \tag{2.48}$$

with $\theta_j \leq 0 \ \ (1 \leq j \leq s)$. So we can reach the following conclusion. If $\theta_j = 0$ or $m_j \geq 2$, then $-a_j(1+r_j)$ is an eigenvalue, and this value is always negative. All other eigenvalues are the roots of the equation

$$1 - \sum_{j=1}^{s} \frac{\theta_j}{a_j(1+r_j) + \lambda} = 0, \tag{2.49}$$

where we assume that $\theta_j \neq 0$ for all $j$. This is equivalent to a polynomial equation of degree $s$, so there are $s$ real or complex roots. Let $g(\lambda)$ denote the left hand side, then clearly

$$\lim_{\lambda \to \pm\infty} g(\lambda) = 1, \qquad \lim_{\lambda \to -a_j(1+r_j)\pm 0} g(\lambda) = \pm\infty,$$

and

$$g'(\lambda) = \sum_{j=1}^{s} \frac{\theta_j}{(a_j(1+r_j)+\lambda)^2} < 0.$$

The graph of $g(\lambda)$ is the same as the one shown earlier in Fig. 2.1 with the only difference being that the poles are now the values $-a_j(1+r_j)$ $(j = 1, 2, \ldots, s)$. Since all poles are negative, all roots have to be real and negative. This observation implies the assertion. ■

The result of this theorem can also be obtained directly from the proof of Theorem 2.1. Notice that the Jacobian (2.46) can be written as $\bar{\boldsymbol{H}} - \boldsymbol{I}$, where $\bar{\boldsymbol{H}}$ is given in (2.20) and $\boldsymbol{I}$ is the identity matrix. Therefore the eigenvalues of the Jacobian of the continuous time case can be obtained by subtracting one from the eigenvalues of the discrete time case. Since the eigenvalues in the discrete time case are less than unity, all eigenvalues of the continuous time case have to be negative.

So far conditions (A)–(C) (or (A),(B),(C') and (D)) have been (see Sect. 2.1) assumed to hold in the entire feasible set of $x_k$ and $Q_k$, and they have implied the existence of a Nash equilibrium. If the oligopoly has an interior equilibrium, then these conditions need to be satisfied only in a neighborhood of this equilibrium in order to guarantee its local asymptotic stability. In comparing Theorems 2.1 and 2.2 we notice that in the discrete time case asymptotic stability can be lost if one or more firms change their adjustment schemes so that the conditions of Theorem 2.1 no longer hold. In the continuous case the equilibrium is always locally asymptotically stable, thus we see that the asymptotic behavior of the equilibrium is much richer in the discrete case. In the continuous case asymptotic stability cannot be lost by changing adjustment schemes, however – as we will demonstrate in the next session – it can be lost if the firms have only delayed information to which to react, or they respond to certain averaged past information.

Finally we mention that several linear extensions and modifications of the main result of this section can be found in Okuguchi and Szidarovszky (1999) especially for multiproduct oligopolies. Al-Nowaihi and Levine (1985), Dixit (1986) and Furth (1986) introduced adjustment processes based on the marginal profits of the firms by requiring that $\dot{x}_k$ for all times must have the same sign as the marginal profit. These gradient adjustment processes have been briefly discussed at the end of the previous chapter. Bellman (1969) offers a comprehensive background in the stability theory of ordinary differential equations. Global analysis of the asymptotic stability of continuous time systems is usually based on Lyapunov theory (see Appendix A) and on the construction of special Lyapunov functions. Hahn (1962) has shown with the special choice of

$$\mathbb{V}(\boldsymbol{x}) = \frac{1}{2}\sum_{k=1}^{N} \alpha_k x_k^2$$

that the equilibrium of the continuous time system (1.31) is globally asymptotically stable with symmetric firms and linear cost functions. This result has been generalized to non-symmetric firms by Okuguchi (1964). We also mention that Sect. 6.4 of Okuguchi and Szidarovszky (1999) discusses the multi-product case with special Lyapunov function selections and derives particular stability conditions.

## 2.6 Continuous Time Oligopolies with Continuously Distributed Time Lags

In examining the dynamic model (1.31) we assumed that at each time period $t$, each firm knew the simultaneous output levels $x_l(t)(l \neq k)$ of the competitors, so it was able to apply the adjustment scheme represented by the right hand side of the governing differential equation. This assumption is however unrealistic in real economic situations, since there is an inevitable time lag because of information collection and decision implementation. A similar situation occurs when the firms want to react to certain averaged past information rather than reacting to sudden market changes. In both cases the output of the rest of the industry as well as the firm's own output levels have to be replaced by averaged values of corresponding past information.

Therefore the differential equations (1.31) are modified to the form

$$\begin{aligned}\dot{x}_k(t) &= \alpha_k \left( R_k \left( \int_0^t w(t-s, T_k, m_k) \sum_{l \neq k} x_l(s) \mathrm{d}s \right) \right. \\ &\quad \left. - \int_0^t w(t-s, S_k, l_k) x_k(s) \mathrm{d}s \right). \end{aligned} \tag{2.50}$$

In the first term of (2.50) the firm reacts to a time weighted average (back to the beginning of the process) of the output of the rest of the industry. In the second term the firm computes its reaction to a time weighted average of its own output. In the ensuing analysis we select the weighting function given by

$$w(t-s, T, m) = \begin{cases} \dfrac{1}{T} \exp\{-(t-s)/T\} & \text{if} \quad m = 0, \\ \dfrac{1}{m!} \left(\dfrac{m}{T}\right)^{m+1} (t-s)^m \exp\{-m(t-s)/T\} & \text{if} \quad m \geq 1. \end{cases} \tag{2.51}$$

This weighting function has been frequently used in the analysis of time lagged dynamical system (see the Appendix D and Cushing (1977)) since as we shall see, it affords a great deal of mathematical tractability. The main properties of this weighting function are summarized in Appendix D, here we simply point out that $T$ may be interpreted as the average time delay (in fact for $m > 0$ the weighting function peaks at $t - s = T$), whilst the parameter $m$ plays the role of "squeezing" the weights around this average value. If we interpret $w$ as a distribution then $T$ is the mean and $m$ is related to the inverse of the standard deviation. We see from (2.50) that firm $k$ may apply a different average time lag ($T_k$) and "squeezing" factor ($m_k$) to information about output of the rivals than to information about its own output (denoted by $S_k$ and $l_k$ respectively). This reflects the fact that a firm should be better informed about its own production process than that of its rivals.

Equation (2.50) is a Volterra-type integro-differential equation, and as is also shown in Appendix D, it is equivalent to a system of ordinary differential equations. Therefore all known tools from the stability theory of ordinary differential equations can be used to analyze the asymptotic behavior of system (2.50), including linearization.

For $k = 1, 2, \ldots, N$, let $x_{k\delta}(t)$ denote the deviation of $x_k(t)$ from its equilibrium level, then the linearized system has the form

$$\dot{x}_{k\delta}(t)=a_k\left\{r_k\int_0^t w(t-s,T_k,m_k)\sum_{l\neq k}x_{l\delta}(s)\mathrm{d}s-\int_0^t w(t-s,S_k,l_k)x_{k\delta}(s)\mathrm{d}s\right\}, \tag{2.52}$$

where $1 \le k \le N$, $a_k = \alpha_k'(0)$ and $r_k = R_k'\left(\sum_{l\neq k}\bar{x}_l\right)$ as before. The characteristic equation of this linear system can be obtained with the same technique that is usually used in the case of linear differential equations (see Miller (1972)). We seek the solution in the form

$$x_{k\delta} = v_k e^{\lambda t} \quad (1 \le k \le N),$$

substitute it into (2.52) and let $t \to \infty$. The resulting equation becomes

$$\left[\lambda+a_k\int_0^\infty w(s,S_k,l_k)e^{-\lambda s}\mathrm{d}s\right]v_k-\left[a_k r_k\int_0^\infty w(s,T_k,m_k)e^{-\lambda s}\mathrm{d}s\right]\sum_{l\neq k}v_l=0.$$

By using the limiting values of the integral (D.3) derived in Appendix D we can further simplify this equation to

$$A_k(\lambda)v_k + B_k(\lambda)\sum_{l\neq k}v_l = 0, \tag{2.53}$$

with

$$A_k(\lambda) = \lambda + a_k \left( \frac{S_k \lambda}{q_k} + 1 \right)^{-(l_k+1)},$$

and

$$B_k(\lambda) = -a_k r_k \left( \frac{T_k \lambda}{p_k} + 1 \right)^{-(m_k+1)},$$

where

$$q_k = \begin{cases} 1 & \text{if} \quad l_k = 0, \\ l_k & \text{if} \quad l_k > 0, \end{cases}$$

and

$$p_k = \begin{cases} 1 & \text{if} \quad m_k = 0, \\ m_k & \text{if} \quad m_k > 0. \end{cases}$$

The set of equations (2.53) have non-trivial solution if and only if

$$det \begin{pmatrix} A_1(\lambda) & B_1(\lambda) & \cdots & B_1(\lambda) \\ B_2(\lambda) & A_2(\lambda) & \cdots & B_2(\lambda) \\ \vdots & \vdots & & \vdots \\ B_N(\lambda) & B_N(\lambda) & \cdots & A_N(\lambda) \end{pmatrix} = 0. \tag{2.54}$$

Notice that this determinant is the same as (E.2) discussed in Appendix E, where it is shown that this equation can be rewritten as

$$\prod_{k=1}^{N} (A_k(\lambda) - B_k(\lambda)) \cdot \left[ 1 + \sum_{k=1}^{N} \frac{B_k(\lambda)}{A_k(\lambda) - B_k(\lambda)} \right] = 0. \tag{2.55}$$

Since $A_k(\lambda)$ and $B_k(\lambda)$ are all rational functions, this equation is equivalent to a polynomial equation showing that there is a finite number of eigenvalues.

Equation (2.55) generally reduces to a very complicated high order polynomial equation, so no general analytic results can be derived. However in the special case of symmetric firms we will be able to derive simple stability conditions and examine the complex asymptotic behavior of the system. For this purpose assume that $a_k \equiv a, r_k \equiv r, T_k \equiv T, S_k \equiv S, m_k \equiv m, l_k \equiv l$, so $q_k \equiv q$ and $p_k \equiv p$ that is the firms are identical with respect to speeds of reaction and slopes of their reaction functions at the steady state and furthermore they use the same time weighting schemes. Assume in addition that the initial output levels of the firms are identical. Then system (2.52) reduces to a one-dimensional integro-differential equation, since the assumed symmetry implies that the output trajectories of the firms are also identical. Therefore in (2.53) we set, $v_k \equiv v$, $A_k(\lambda) \equiv A(\lambda)$ and $B_k(\lambda) \equiv B(\lambda)$, so that it simplifies to

$$A(\lambda) + (N-1)B(\lambda) = 0, \tag{2.56}$$

or

$$\lambda + a\left(\frac{S\lambda}{q} + 1\right)^{-(l+1)} - (N-1)ar\left(\frac{T\lambda}{p} + 1\right)^{-(m+1)} = 0,$$

which can be rewritten as the polynomial equation

$$\lambda\left(\frac{S\lambda}{q}+1\right)^{l+1}\left(\frac{T\lambda}{p}+1\right)^{m+1} + a\left(\frac{T\lambda}{p}+1\right)^{m+1} - (N-1)ar\left(\frac{S\lambda}{q}+1\right)^{l+1} = 0. \tag{2.57}$$

Assume first that the firm's own information lag $S$ is much smaller than $T$, the information lag about rival firms. Making the simplest assumption that $S = 0$, (2.57) becomes

$$(\lambda + a)\left(\frac{T\lambda}{p} + 1\right)^{m+1} - (N-1)ar = 0. \tag{2.58}$$

Consider first the special case of $T = 0$, when there is no information lag about rival firms. Then (2.58) reduces to the linear equation

$$(\lambda + a) - (N-1)ar = 0,$$

with solution

$$\lambda = (N-1)ar - a < 0,$$

so the equilibrium is locally asymptotically stable. This case was discussed under much more general conditions in Theorem 2.2 where the same conclusion was reached.

Consider next the case when $T > 0$ and $m = 0$. Then (2.58) becomes the quadratic,

$$\lambda^2 T + \lambda(1 + aT) + a(1 - (N-1)r) = 0.$$

Since all coefficients are positive, both roots are negative or have negative real parts (see Appendix F), so again the equilibrium is locally asymptotically stable.

In the case of $m = 1$, (2.58) reduces to the cubic equation

$$\lambda^3 T^2 + \lambda^2(aT^2 + 2T) + \lambda(1 + 2aT) + a(1 - (N-1)r) = 0. \tag{2.59}$$

All coefficients are positive and the Routh–Hurwitz criterion (see Szidarovszky and Bahill (1998)) implies that all roots have negative real parts if and only if

$$(aT^2 + 2T)(1 + 2aT) > T^2 a(1 - (N-1)r). \tag{2.60}$$

This inequality can be rewritten as a quadratic inequality in the variable $aT$ in the form

$$2(aT)^2 + aT(4 + r(N-1)) + 2 > 0. \tag{2.61}$$

The discriminant of the left hand side is

$$(4 + r(N - 1))^2 - 16 = r(N - 1)[r(N - 1) + 8].$$

The first factor, $r(N - 1)$, is negative, so we have the following cases.

Case 1. If $r(N - 1) + 8 > 0$, then the discriminant is negative, so (2.61) always holds.

Case 2. If $r(N-1)+8 = 0$, then (2.61) holds for all values of $aT$ except the single root of the quadratic polynomial. So the equilibrium is locally asymptotically stable unless

$$aT = -\frac{4 + r(N - 1)}{4} = -\frac{8 + r(N - 1)}{4} + 1 = 1.$$

Case 3. If $r(N-1)+8 < 0$, then the quadratic polynomial (2.61) has two real roots,

$$(aT)^*_{1,2} = \frac{-4 - r(N - 1) \pm \sqrt{r(N - 1)[r(N - 1) + 8]}}{4}. \tag{2.62}$$

Since $-4-r(N-1) = -(8+r(N-1))+4 > 0$, both roots are positive. Hence the equilibrium is locally asymptotically stable if

$$aT < (aT)^*_1 \quad \text{or} \quad aT > (aT)^*_2,$$

where $(aT)^*_1 < (aT)^*_2$. The equilibrium is unstable if

$$(aT)^*_1 < aT < (aT)^*_2.$$

Summarizing these results the stability region is shown as the shaded area in Fig. 2.20.

From the above analysis we can draw the following interesting conclusions. If $N \leq 9$, then $r(N-1)+8 > 0$, so Case 1 always occurs and the equilibrium is always locally asymptotically stable. Assume next that $N > 9$. Then Case 1 occurs if $r > -\frac{8}{N-1}$ resulting in the local asymptotic stability of the equilibrium. Case 2 occurs if $r = -\frac{8}{N-1}$, so the equilibrium is locally asymptotically stable unless $aT = 1$. Case 3 is obtained when $r < -\frac{8}{N-1}$, in which case local asymptotic stability occurs if $aT$ is either sufficiently small (less than $(aT)^*_1$) or sufficiently large (greater than $(aT)^*_2$). The asymptotic stability does not depend on the individual values of $a$ and $T$, it depends on only the product of $aT$. This property shows a certain kind of compensation between the speed of adjustment and the average information delay. If the average delay $T$ is given, then in Case 2 the firms must not select $a = \frac{1}{T}$, and in Case 3 they should select either a small $\left(a < \frac{(aT)^*_1}{T}\right)$ or a large $\left(a > \frac{(aT)^*_2}{T}\right)$ value of $a$ in order to stabilize the equilibrium.

Assume next that Case 3 occurs, that is, $-1 < r < -\frac{8}{N-1}$. If $aT < (aT)^*_1$ or $aT > (aT)^*_2$ then the equilibrium is locally asymptotically stable, and if $aT$ is

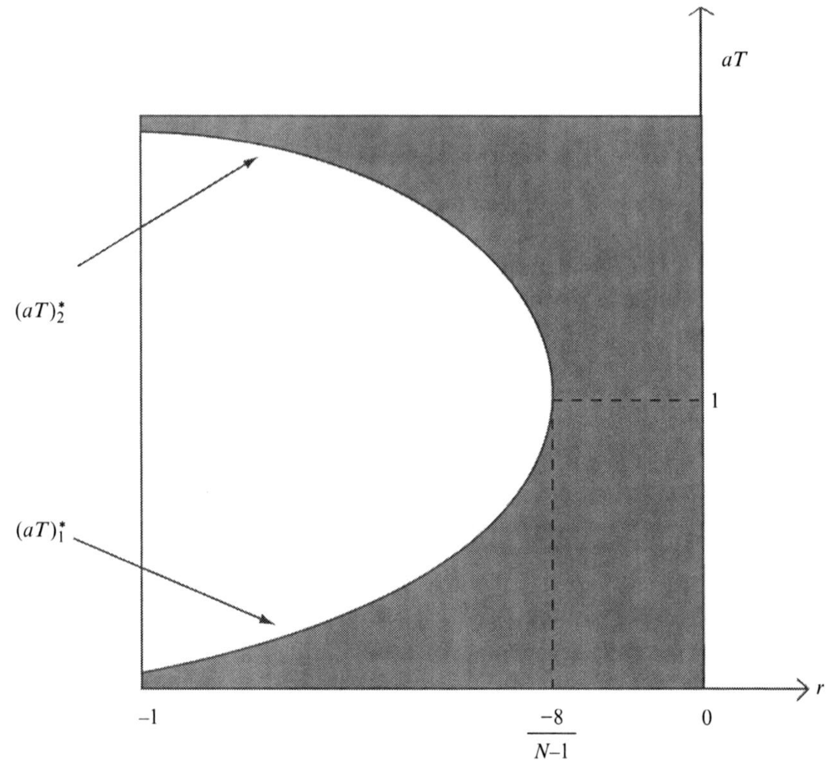


**Fig. 2.20** Stability region in the $(r, aT)$-space for continuous time symmetric oligopolies with time delay in information about rival firms

between $(aT)_1^*$ and $(aT)_2^*$, then it is unstable. So with fixed value of $r$, if $aT$ is gradually increased from a very small value and crosses $(aT)_1^*$, then asymptotic stability is lost. This instability holds until the value of $aT$ reaches $(aT)_2^*$, and on crossing this value, asymptotic stability is regained. It is very interesting to see what happens at these critical values $(aT)_1^*$ and $(aT)_2^*$. We will show that a Hopf bifurcation occurs (see for example, Guckenheimer and Holmes (1983)) giving the possibility of the birth of limit cycles around the equilibrium as $aT$ crosses these critical values.

In fact we may state the following theorem concerning a Hopf bifurcation in the $m = 1$ case:

**Theorem 2.3.** *In the case of $m = 1$ the dynamics of the symmetric oligopoly loses local asymptotic stability and a Hopf bifurcation occurs as $aT$ crosses the critical value $(aT)_1^*$ from below and the critical value $(aT)_2^*$ from above.*

*Proof.* We select $T$ as the bifurcation parameter, and consider the roots $\lambda$ of the eigenvalue equation (2.59) as functions of $T$, that is $\lambda = \lambda(T)$. In order to show that a Hopf bifurcation occurs we have to prove two facts. First, that at the critical value of $T$, we have a pair of pure complex roots while all other eigenvalues have negative real parts. Second, that the derivative of $\lambda(T)$ at the critical value has nonzero real part.

At the critical values the inequalities (2.60) as well as (2.61) become equalities, so the eigenvalue equation (2.59) reduces to

$$\begin{aligned} 0 &= \lambda^3 T^2 + \lambda^2(aT^2 + 2T) + \lambda \frac{T^2 a(1-(N-1)r)}{aT^2 + 2T} + a(1-(N-1)r) \\ &= \left[\lambda T^2 + (aT^2 + 2T)\right]\left[\lambda^2 + \frac{a(1-(N-1)r)}{aT^2 + 2T}\right]. \end{aligned}$$

Therefore the eigenvalues are

$$\lambda_{1,2} = \pm i\sqrt{\frac{a(1-(N-1)r)}{aT^2 + 2T}}, \tag{2.63}$$

and

$$\lambda_3 = -\frac{aT^2 + 2T}{T^2} < 0.$$

So the first condition is satisfied at the critical values. In order to show that the second condition is also satisfied we have to differentiate implicitly the eigenvalue equation (2.59) with respect to $T$. A simple calculation shows that (with the notation $\dot{\lambda} = \frac{d\lambda}{dT}$)

$$3\lambda^2\dot{\lambda}T^2 + 2\lambda^3 T + 2\lambda\dot{\lambda}(aT^2 + 2T) + \lambda^2(2aT + 2) + \dot{\lambda}(1 + 2aT) + 2\lambda a = 0,$$

implying that

$$\dot{\lambda} = \frac{-2\lambda^3 T - \lambda^2(2aT + 2) - 2\lambda a}{3\lambda^2 T^2 + 2\lambda(aT^2 + 2T) + (1 + 2aT)}. \tag{2.64}$$

For the sake of simplicity introduce the notation

$$\alpha^2 = \frac{a(1-(N-1)r)}{aT^2 + 2T} \left(= \frac{1 + 2aT}{T^2}\right).$$

Then $\lambda_{1,2} = \pm\alpha i$, and at these values

$$\begin{aligned} \dot{\lambda} &= \frac{\pm 2\alpha^3 iT + \alpha^2(2aT + 2) \mp 2a\alpha i}{-3\alpha^2 T^2 \pm 2\alpha i(aT^2 + 2T) + (1 + 2aT)} \\ &= \frac{\alpha^2(2aT + 2) + (\pm 2\alpha^3 T \mp 2a\alpha)i}{-2\alpha^2 T^2 \pm 2\alpha i(aT^2 + 2T)}. \end{aligned}$$

Multiplying both the numerator and denominator by the complex conjugate of the denominator, after some simple calculations we find that

$$Re\dot{\lambda} = \frac{-2\alpha^4 T^2(2aT+2) + 4\alpha^2 T(\alpha^2 T - a)(aT+2)}{(-2\alpha^2 T^2)^2 + 4\alpha^2(aT^2+2T)^2}.$$

The numerator can be simplified to $4\alpha^2 T(\alpha^2 T - Ta^2 - 2a)$. Here the first factor is positive and the second factor can be rewritten as

$$\frac{(1+2aT)T}{T^2} - Ta^2 - 2a = \frac{1-T^2a^2}{T} \neq 0$$

since it is easy to see that

$$(aT)_1^* < 1 < (aT)_2^*.$$

Hence the conditions for a Hopf bifurcation are satisfied, giving the possibility of the birth of limit cycles around the equilibrium. ■

If we consider larger values of $m$ then (2.58) leads to higher order equations and therefore the stability analysis becomes far more complicated, and would usually require the use of computational methods.

However we can show that if $r > \frac{-1}{N-1}$, then all roots of (2.58) have negative real parts, so the equilibrium is asymptotically stable. On the contrary assume that $Re\lambda \geq 0$. Then

$$|\lambda + a| \geq a \quad \text{and} \quad \left|1 + \frac{\lambda T}{p}\right| \geq 1,$$

so that

$$\left|(\lambda + a)\left(\frac{T\lambda}{p} + 1\right)^{m+1}\right| \geq a > -ar(N-1) = |ar(N-1)|,$$

hence $\lambda$ cannot be root of (2.58).

So far we have made the simplifying assumption that $S = 0$ in equation (2.57). In order to illustrate a case when there are lags in the information on both the rivals' and own outputs consider (2.57) with positive $S$ and $T$ and with $m = l = 0$. The cubic equation

$$ST\lambda^3 + \lambda^2(S+T) + \lambda\,(1 + aT - (N-1)arS) + (a - (N-1)ar) = 0$$

is then obtained. All coefficients are positive, and the Routh–Hurwitz criterion implies that the roots have negative real parts if and only if

$$(S+T)\,(1 + aT - (N-1)arS) > ST\,(a - (N-1)ar)\,,$$

which can be rewritten as

$$S + T + aT^2 - (N-1)arS^2 > 0.$$

This inequality always holds, since $r \leq 0$. Hence the equilibrium is always locally asymptotically stable.

Continuously distributed time lags were originally introduced and applied in mathematical biology (see for example, Cushing (1977)). They have been introduced into economic modeling by Invernizzi and Medio (1991). Chiarella and Khomin (1996) applied these techniques to Cournot oligopolies with some simulation results. Chiarella and Szidarovszky (2001*a*) gave a detailed discussion of the problem with a general solution. The fundamentals of bifurcation theory are presented in many books, for example, Guckenheimer and Holmes (1983), Jackson (1989), and Kubicek and Marek (1986) are all good sources for the most important results. In the models discussed in this section nonlinear Volterra-type integro-differential equations were considered, a topic on which Volterra (1931) and Miller (1972) offer useful additional material.

# Chapter 3
# General Oligopolies

In the previous chapter we analyzed concave oligopolies where the best response functions were monotonic and therefore the local and global analysis of the corresponding dynamic processes were relatively simple. The examples discussed there have allowed the reader to become familiar with the major concepts and methods that we shall use in the rest of the book. If we drop the simplifying assumptions of the previous chapter then more complex dynamics may arise. In this chapter we will present a collection of such models.

We initiate our discussion in Sect. 3.1 where we consider oligopolies with isoelastic price functions and dynamics in discrete time. We give a detailed analysis of local and global stability of some particular examples. In Sect. 3.2 we return to the issue of oligopolies with cost externalities, which may display multiple interior Nash equilibria. The global analysis of some specific examples indicates how the oligopoly may converge to particular equilibria.

## 3.1 Isoelastic Price Functions

In this section we assume that the price function is isoelastic, as in Example 1.5. As in the previous chapters let $N$ denote the number of firms, let $x_k$ be the output of firm $k$ ($k = 1, 2, \ldots, N$) and $Q = \sum_{k=1}^{N} x_k$ the total output of the industry. Then the price function is $f(Q) = A/Q$ with some positive constant $A$. If no externalities are assumed and $C_k(x_k)$ denotes the cost of firm $k$, then its profit is given as

$$\varphi_k(x_1, \ldots, x_N) = \begin{cases} -C_k(0), & \text{if} \quad x_k = 0, \\ \dfrac{Ax_k}{Q_k + x_k} - C_k(x_k), & \text{if} \quad x_k > 0, \end{cases}$$

where we use again the simplifying notation $Q_k = \sum_{l \neq k} x_l$ so that $Q = Q_k + x_k$. In the following discussion we will assume that for all $k$, $C_k$ is twice continuously differentiable, increasing and convex, so that for all feasible values of $x_k$,

(D) $C_k'(x_k) > 0$ and $C_k''(x_k) \geq 0$.

G.I. Bischi et al., *Nonlinear Oligopolies*, DOI 10.1007/978-3-642-02106-0_3, 

We can now calculate the best response of firm $k$. Assume first that $Q_k = 0$, so that the other firms do not produce. Then

$$\varphi_k(x_1, \ldots, x_N) = \begin{cases} -C_k(0), & \text{if} \quad x_k = 0, \\ A - C_k(x_k), & \text{if} \quad x_k > 0. \end{cases}$$

In this case firm $k$ has no best choice, however it is in its interest to select a positive value of $x_k$ that is as small as possible. In other words, firm $k$ does not have a maximum profit for $Q_k = 0$, its profit has only a supremum at $x_k = 0$. If $Q_k > 0$, so that the other firms produce, then

$$\frac{\partial}{\partial x_k}\varphi_k(x_1, \ldots, x_N) = \frac{AQ_k}{(Q_k + x_k)^2} - C_k'(x_k), \tag{3.1}$$

and

$$\frac{\partial^2}{\partial x_k^2}\varphi_k(x_1, \ldots, x_N) = -\frac{2AQ_k}{(Q_k + x_k)^3} - C_k''(x_k) < 0,$$

showing that $\varphi_k$ is strictly concave in $x_k$ with fixed positive values of $Q_k$. If we assume again that each firm has a finite capacity limit, $L_k$, then the best response exists and is unique for each firm and is given by

$$R_k(Q_k) = \begin{cases} 0, & \text{if} \quad \dfrac{A}{Q_k} - C_k'(0) \le 0, \\ L_k, & \text{if} \quad \dfrac{AQ_k}{(Q_k + L_k)^2} - C_k'(L_k) \ge 0, \\ z_k^*, & \text{otherwise}, \end{cases}$$

where $z_k^*$ is the unique solution of the strictly monotonic equation

$$\frac{AQ_k}{(Q_k + z_k)^2} - C_k'(z_k) = 0 \tag{3.2}$$

in the interval $(0, L_k)$. The derivative of the best response function is obtained by implicit differentiation of the equivalent equation

$$AQ_k - C_k'(z_k)(Q_k + z_k)^2 = 0,$$

from which we have

$$A - C_k'' R_k'(Q_k + z_k)^2 - 2C_k'(Q_k + z_k)(1 + R_k') = 0$$

implying that

$$R_k'(Q_k) = \frac{A - 2C_k' Q}{C_k'' Q^2 + 2C_k' Q}. \tag{3.3}$$

Here the denominator is always positive but the sign of the numerator is indeterminate. Hence, $R_k(Q_k)$ is not necessarily monotonic, which stands in contrast to the concave case discussed in the previous chapter. If we express the best response functions in terms of the total output of the industry, then the resulting modified best response function $\widetilde{R}_k(Q)$ will not be monotonic either. Therefore the existence and uniqueness of the equilibrium cannot be examined in the same way as was done for concave oligopolies. However, by using a different approach, the existence of a unique equilibrium is proved in Szidarovszky and Okuguchi (1997), and this result is also presented with further details in Okuguchi and Szidarovszky (1999).

Consider now an interior equilibrium, then from (3.2),

$$A\bar{Q}_k - C_k'(\bar{x}_k)\bar{Q}^2 = 0$$

for all $k$. The numerator of (3.3) at the equilibrium becomes

$$A - \frac{2A\bar{Q}_k}{\bar{Q}} = \frac{A}{\bar{Q}}(\bar{Q} - 2\bar{Q}_k),$$

so $R_k'(\bar{Q}_k) \le 0$ if and only if $\bar{Q} \le 2\bar{Q}_k$.

Notice in addition that

$$R_k'(Q_k) > \frac{-C_k''Q^2 - 2C_k'Q}{C_k''Q^2 + 2C_k'Q} = -1. \tag{3.4}$$

It is interesting to note that this is exactly the same lower bound as in the concave case. If $N = 2$, then at a symmetric equilibrium $R_k' = 0$ for $k = 1, 2$. If the equilibrium is asymmetric, then $R_k'$ is positive for one firm and is negative for the other, so $R_1' R_2' < 0$. Assume next that $N \ge 3$, and for all firms, $x_k \le Q_k$. This condition means that there is no large firm dominating the rest of the industry. In this case $Q \le 2Q_k$ for all $k$, so $-1 < R_k' \le 0$ which is similar to the concave case. Notice that in the general case the condition $Q \le 2Q_k$ at the equilibrium can be violated by at most one firm, so there is at most one firm with positive derivative $R_k'$ at the equilibrium.

*Example 3.1.* In Example 1.5 we have already considered the isoelastic case with $p = f(Q) = A/Q$ and linear cost functions $C_k(x_k) = d_k + c_k x_k$. There we derived the equilibrium quantities of the firms which are given by

$$\bar{x}_k = \frac{(N-1)A}{\sum_l c_l} - \frac{(N-1)^2 A c_k}{(\sum_l c_l)^2},$$

for $k = 1, 2, \ldots, N$, and the total industry output

$$\bar{Q} = \frac{(N-1)A}{\sum_l c_l}.$$

Hence, we obtain

$$\bar{Q}_k = \bar{Q} - \bar{x}_k = \frac{(N-1)^2 A c_k}{(\sum_l c_l)^2}.$$

In order to guarantee that $\bar{x}_k \geq 0$ we have to assume that

$$c_k \leq \frac{\sum_l c_l}{N-1} \quad \text{or} \quad c_k \leq \frac{\sum_{l \neq k} c_l}{N-2}. \tag{3.5}$$

We can also find conditions such that $\bar{Q} \leq 2\bar{Q}_k$ for all $k$ implying that $-1 < R_k' \leq 0$ at the equilibrium, so the local asymptotic properties of the equilibrium become the same as in the concave case. This condition has the special form

$$\frac{(N-1)A}{\sum_l c_l} \leq \frac{2(N-1)^2 A c_k}{(\sum_l c_l)^2},$$

which can be rewritten as

$$c_k \geq \frac{\sum_l c_l}{2(N-1)}.$$

Notice that this lower bound is the half of the upper bound given in (3.5). The upper bound guarantees the non-negativity of the equilibrium outputs and the lower bound guarantees that the derivatives of the best responses at the equilibrium are between $-1$ and 0 as in the concave case. If $N = 2$, then this is true if $c_1 = c_2$, otherwise it holds for one firm and does not hold for the other. If $N \geq 3$, then this condition is certainly satisfied if none of the firms has very low marginal costs compared to its competitors.

### 3.1.1 Discrete Time Models and Local Stability

The local asymptotic behavior of the best reply dynamics with adaptive expectations and partial adjustment towards the best response with naive expectations (1.28)–(1.30) are equivalent to each other as has been shown earlier. So similar to the concave case we will discuss only system (1.30). The Jacobian of this dynamic system was derived in (2.20), where we did not use any special form of the best response functions, therefore the nonzero eigenvalues of the Jacobian of the isoelastic case are also the eigenvalues of the matrix $\bar{\boldsymbol{H}}$. Its characteristic equation is also given by (2.23), or equivalently by (2.24).

In the case when all $r_k = R_k'(Q_k)$ values are non-positive, all local stability results remain the same as demonstrated for the concave case. However in the general case the local asymptotic behavior of the equilibrium becomes more complicated.

Assume now that for a firm $k_0$, $r_{k_0} > 0$. Then $\bar{Q} > 2\bar{Q}_{k_0}$ or equivalently, $\bar{x}_{k_0} > \bar{Q}_{k_0}$. This condition means that firm $k_0$ produces more than the total output of the rest of the industry at the equilibrium, therefore $r_k > 0$ is possible for at most one firm. Similarly to the concave case we assume that $a_k = \alpha_k'(0) > 0$ for all $k$. Number the firms in such a way that the different $a_k(1 + r_k)$ values are

$$a_1(1 + r_1) > a_2(1 + r_2) > \cdots > a_s(1 + r_s),$$

and these values are repeated $m_1, m_2, \ldots, m_s$ times, respectively, among the $N$ firms. By adding the terms with identical denominators in the bracketed factor of (2.23) we obtain (2.24), where at most one $\theta_j$ can be positive. If all $\theta_j$ values are non-positive, then the problem remains the same as in the concave case with the same stability results. Therefore assume now that there is a $j_0$ such that $\theta_{j_0} > 0$. If $\theta_j \neq 0$ and $m_j = 1$, then $1 - a_j(1 + r_j)$ is not an eigenvalue of the Jacobian. Otherwise it is, and the other eigenvalues are the roots of the equation

$$1 + \sum_{j=1}^{s} \frac{\theta_j}{1 - a_j(1 + r_j) - \lambda} = 0,$$

where we assume that all $\theta_j \neq 0$.

Let $g(\lambda)$ denote again the left hand side of the last equation. Then clearly

$$\lim_{\lambda \to \pm\infty} g(\lambda) = 1,$$

$$\lim_{\lambda \to 1 - a_j(1+r_j) \pm 0} g(\lambda) = \begin{cases} \mp\infty & \text{if } j = j_0, \\ \pm\infty & \text{if } j \neq j_0, \end{cases}$$

however in contrast to the concave case, $g'(\lambda)$ has no definite sign, that is, $g$ is not necessarily monotonic. All poles are less than unity. Depending on the value of $j_0$ we have the following cases:-

Case 1. $j_0 = 1$.
The graph of $g(\lambda)$ for this case is shown in Fig. 3.1. There are $s - 2$ real roots between each pair of poles $1 - a_j(1 + r_j)$ and $1 - a_{j+1}(1 + r_{j+1})$ for $j = 2, \ldots, s - 1$. If the other two roots are real and they are between $1 - a_1(1 + r_1)$ and $1 - a_s(1 + r_s)$, then the equilibrium is locally asymptotically stable if $1 - a_1(1 + r_1) > -1$.

Case 2. $j_0 = s$.
The graph of $g(\lambda)$ in this case is shown in Fig. 3.2. All roots are real, one is before the smallest pole, one after the largest pole, and one between each pair of poles $1 - a_j(1 + r_j)$ and $1 - a_{j+1}(1 + r_{j+1})$ for $j = 1, \ldots, s - 2$. All roots are between -1 and 1 if $1 - a_1(1 + r_1) > -1$ and $g(-1) > 0$ and $g(1) > 0$.

Case 3. $1 < j_0 < s$.
The graph of $g(\lambda)$ is shown in Fig. 3.3. There are $s - 2$ real roots. If we assume that the remaining two roots are real and are between $1 - a_1(1 + r_1)$ and $1 - a_s(1 + r_s)$, then all roots are between $-1$ and $1$ if $1 - a_1(1 + r_1) > -1$ and $g(-1) > 0$.

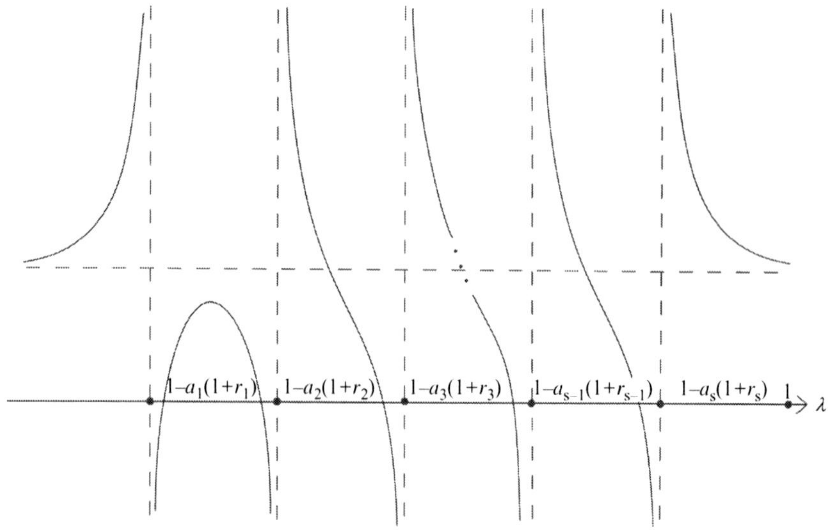


**Fig. 3.1** The oligopoly with isoelastic price function and convex cost functions with partial adjustment towards the best response with naive expectations. The graphical determination of the eigenvalues in the case $j_0 = 1$

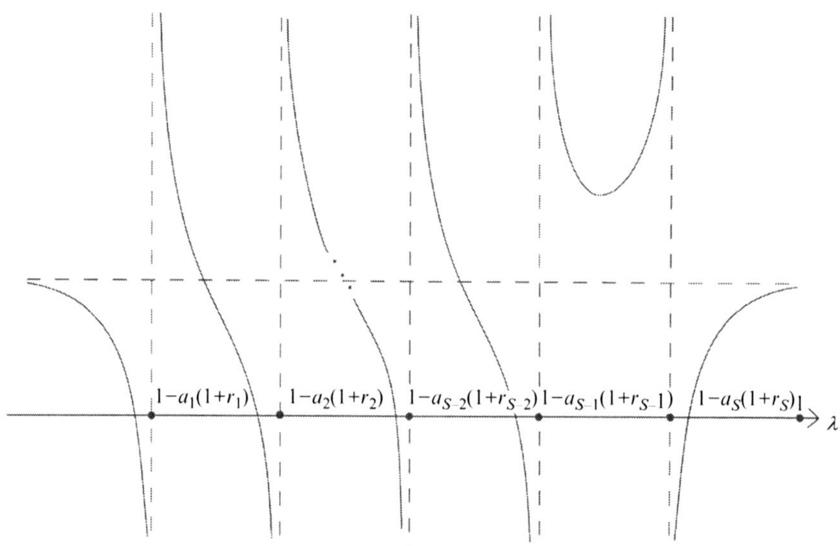


**Fig. 3.2** The oligopoly with isoelastic price function and convex cost functions with partial adjustment towards the best response with naive expectations. The graphical determination of the eigenvalues in the case $j_0 = s$

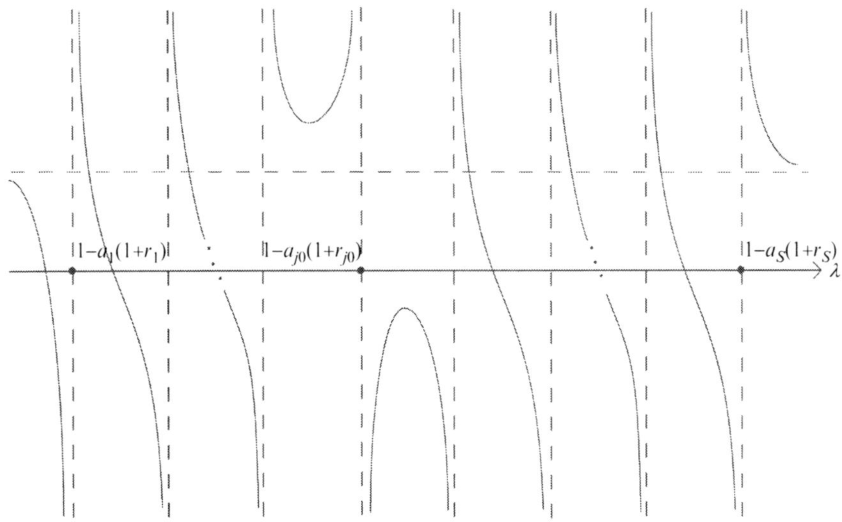


**Fig. 3.3** The oligopoly with isoelastic price function and convex cost functions with partial adjustment towards the best response with naive expectations. The graphical determination of the eigenvalues in the case $1 < j_0 < s$

Notice that conditions $g(-1) > 0$ and $g(1) > 0$ can be written as (2.22) and

$$\sum_{k=1}^{N} \frac{r_k}{1+r_k} < 1,$$

respectively.
In the case of complex roots, no similar stability condition can be given. The possibility of complex roots will be shown later in Example 3.2.

The assumption that $C_k$ is a convex function in its entire domain guarantees the existence of a Nash equilibrium. However if this condition is not satisfied everywhere and there is an interior equilibrium, then we have to assume that $C_k' > 0$ and $C_k'' \geq 0$ in its neighborhood in order to assure local asymptotic stability of that equilibrium. As an illustration consider a duopoly with linear cost functions and isoelastic price function.

*Example 3.2.* In this example we consider the duopoly case ($N = 2$). By using the notation of Example 3.1 we assume that the cost function of firm $k$ is $C_k(x_k) = d_k + c_k x_k$ ($k = 1, 2$), the price function is $f(Q) = A/Q$ with some positive constant $A$ and the capacity limits are sufficiently large. The equilibrium is positive, since condition (3.5), that is $c_k \leq c_1 + c_2$, is satisfied for both firms. Furthermore at the equilibrium

$$\bar{Q} = \frac{A}{c_1 + c_2} \qquad \text{and} \qquad \bar{Q}_k = \frac{Ac_k}{(c_1 + c_2)^2}.$$

From Example 1.5 we know that

$$R_k(Q_k) = \sqrt{\frac{AQ_k}{c_k}} - Q_k,$$

and so

$$R_k'(\bar{Q}_k) = \frac{c_1 + c_2}{2c_k} - 1,$$

therefore

$$r_1 = \frac{c_2 - c_1}{2c_1} \qquad \text{and} \qquad r_2 = \frac{c_1 - c_2}{2c_2}.$$

Assume that $c_1 \neq c_2$, and that the firms select identical adjustments, that is, $a_1 = a_2 \equiv a$. The characteristic equation of the Jacobian of the dynamic process with partial adjustment towards the best response is given in general by (2.23), which simplifies to

$$\prod_{k=1}^{2}(1 - a(1 + r_k) - \lambda)\left[1 + \frac{r_1 a}{1 - a(1 + r_1) - \lambda} + \frac{r_2 a}{1 - a(1 + r_2) - \lambda}\right] = 0.$$

This equation reduces to the quadratic

$$\lambda^2 + \lambda(2a - 2) + (1 - 2a + a^2 - a^2 r_1 r_2) = 0,$$

with roots

$$\lambda_{1,2} = (1 - a) \pm ia\sqrt{-r_1 r_2},$$

since

$$r_1 r_2 = -\frac{(c_1 - c_2)^2}{4c_1 c_2} < 0.$$

By an appropriate choice of the parameters $c_1$ and $c_2$ the quantity $r_1 r_2$ can take any negative value. Clearly if $c_1 \neq c_2$, then both roots are complex, and since

$$|\lambda_{1,2}|^2 = 1 - 2a + a^2(1 - r_1 r_2),$$

the roots can be both inside and outside the unit circle. The equilibrium is locally asymptotically stable if

$$a(1 - r_1 r_2) < 2,$$

and unstable if this condition is violated with strict inequality. An analogous condition for the stability of the equilibrium in the duopoly case with constant adjustment speeds has been derived by Puu (2003, Chap. 7). With fixed $r_1$ and $r_2$, stability occurs if the value of $a$ is sufficiently small. With a fixed value of $a \in (0, 1]$ we have stability if the product $|r_1 r_2|$ is sufficiently small, which holds if $c_1$ and $c_2$ are sufficiently close to each other.

*Example 3.3.* Next we examine an $N$-firm semi-symmetric oligopoly with linear cost functions, so we assume that firms $2, 3, \ldots, N$ have identical marginal costs, $c_k = c_2$ $(k = 2, 3, \ldots, N)$, identical capacity limits and common linear adjustment functions, and their initial outputs are also the same, so that $x_2(0) = \ldots = x_N(0)$. Given these assumptions the entire output trajectories of these firms are the same. Therefore we get a two-dimensional system with state variables $x_1$ and $x_2$ where $x_k = x_2$ for $k \geq 2$. In this case $Q_1 = (N-1)x_2$ and $Q_2 = x_1 + (N-2)x_2$. Assuming that the capacity limits $L_k$ are sufficiently large the general expressions for the equilibrium quantities given in Example 3.1 imply for the semi-symmetric case that

$$\begin{aligned}
\bar{x}_1 &= \frac{(N-1)A}{c_1 + (N-1)c_2}\left(1 - \frac{(N-1)c_1}{c_1 + (N-1)c_2}\right) \\
&= \frac{(N-1)A}{c_1 + (N-1)c_2}\left(\frac{(N-1)c_2 - (N-2)c_1}{c_1 + (N-1)c_2}\right) \\
\bar{x}_2 = \ldots = \bar{x}_N &= \frac{(N-1)A}{c_1 + (N-1)c_2}\left(1 - \frac{(N-1)c_2}{c_1 + (N-1)c_2}\right) \\
&= \frac{(N-1)A}{c_1 + (N-1)c_2}\left(\frac{c_1}{c_1 + (N-1)c_2}\right).
\end{aligned}$$

For the total industry output in equilibrium we obtain

$$\bar{Q} = \bar{x}_1 + (N-1)\bar{x}_2 = \frac{(N-1)A}{c_1 + (N-1)c_2}.$$

The derivatives of the best replies are obtained from (3.3) as

$$r_1 = R_1'(\bar{Q}_1) = \frac{A - 2c_1\bar{Q}}{2c_1\bar{Q}} = \frac{(N-1)c_2 + (3-2N)c_1}{2(N-1)c_1},$$

and

$$r_2 = R_2'(\bar{Q}_2) = \frac{A - 2c_2\bar{Q}}{2c_2\bar{Q}} = \frac{c_1 - (N-1)c_2}{2(N-1)c_2}.$$

Conditions (3.5) for $k = 1$ and $k = 2$ are of the form

$$c_1 \leq \frac{c_1 + (N-1)c_2}{N-1}, \quad c_2 \leq \frac{c_1 + (N-1)c_2}{N-1},$$

where the second inequality always holds and the first one can be written as

$$\kappa := \frac{c_2}{c_1} \geq \frac{N-2}{N-1},$$

where $\kappa$ denotes the cost ratio between the firms. In addition,

$$r_1 = \frac{(N-1)\kappa + (3-2N)}{2(N-1)} \quad \text{and} \quad r_2 = \frac{1-(N-1)\kappa}{2(N-1)\kappa}.$$

The dynamic process can be written as

$$\begin{aligned} x_1(t+1) &= (1-a_1)x_1(t) + a_1 R_1((N-1)x_2(t)), \\ x_2(t+1) &= (1-a_2)x_2(t) + a_2 R_2(x_1(t) + (N-2)x_2(t)), \end{aligned}$$

so the Jacobian has the special form

$$\begin{pmatrix} 1-a_1 & a_1 r_1(N-1) \\ a_2 r_2 & 1-a_2+a_2 r_2(N-2) \end{pmatrix}$$

where $r_1 = R_1'$ and $r_2 = R_2'$ at the equilibrium. The characteristic equation of this matrix can be written as

$$(1-a_1-\lambda)(1-a_2+a_2 r_2(N-2)-\lambda) - a_1 a_2 r_1 r_2 (N-1) = 0,$$

which can be simplified to

$$\begin{aligned} &\lambda^2 + \lambda(-2+a_1+a_2+(2-N)a_2 r_2) + (1-a_1-a_2+(N-2)a_2 r_2 \\ &\quad + a_1 a_2(1+(2-N)r_2+(1-N)r_1 r_2)) = 0. \end{aligned}$$

Using results from Appendix F we know that the roots are inside the unit circle if and only if

$$-a_1 + a_2((N-2)r_2 - 1) + a_1 a_2(1+(2-N)r_2+(1-N)r_1 r_2) < 0, \quad (3.6)$$

$$1+(2-N)r_2+(1-N)r_1 r_2 > 0, \quad (3.7)$$

$$4-2a_1+a_2(-2+(2N-4)r_2)+a_1 a_2(1+(2-N)r_2+(1-N)r_1 r_2) > 0. \quad (3.8)$$

The form of the stability region for $(a_1, a_2)$ depends on the number of firms and the actual values of the derivatives $r_1$ and $r_2$. Inserting the expressions for the derivatives $r_1$ and $r_2$ given above, the stability conditions can be written in terms of the cost ratio $\kappa = c_2/c_1$, the number of firms $N$, and the adjustment coefficients $a_1$ and $a_2$ as

$$-4a_1\kappa(N-1) + a_1 a_2(1+\kappa(N-1))^2 + 2a_2(-2+N(1+\kappa-\kappa N)) < 0, \quad (3.9)$$

$$(1+\kappa(N-1))^2 > 0, \quad (3.10)$$

$$-8(-2+a_1)\kappa(N-1) + a_1 a_2(1+\kappa(N-1))^2 + 4a_2(-2+N(1+\kappa-\kappa N)) > 0. \quad (3.11)$$

It is clear that the second inequality is always fulfilled. The properties of the stability region for $(a_1, a_2)$ depend on the number of firms and the ratio of the firms' unit costs.

Instead of giving a complete analysis in general we reconsider the duopoly case of Example 3.2, where $a_1 = a_2 = a$. In this special case the conditions (3.6)–(3.8) further simplify to

$$-2a + a^2(1 - r_1 r_2) < 0,$$

$$1 - r_1 r_2 > 0,$$

and

$$4 - 4a + a^2(1 - r_1 r_2) > 0.$$

The second and third inequalities are always satisfied, since in Example 3.2 we have shown that $r_1 r_2 < 0$. The first relation holds if and only if

$$a(1 - r_1 r_2) < 2.$$

This condition is the same as the one that was obtained earlier in Example 3.2.

The case of linear cost functions is examined in detail in the book of Okuguchi and Szidarovszky (1999) and Puu (2003).

### 3.1.2 Global Dynamics of Discrete Time Models

As we have seen in the discussion in Chap. 2 on concave oligopolies, the conditions for global asymptotic stability are very restrictive. In most cases of isoelastic price functions this is true as well.

Under condition (D) of Sect. 3.1, for at most one firm $r_k > 0$, and for all other firms, $-1 < r_k \leqslant 0$. If all $r_k$ values are non-positive, then the global stability conditions are still given by (2.31). However if one $r_k$ is positive, this condition can no longer be used, it has to be modified accordingly.

We also notice that the global stability condition given in Theorem B.3 cannot be applied either. At $Q_k = 0$, firm $k$ has no best response, which is clear from its definition given in Example 1.5 and in the first part of Sect. 3.1. Therefore the set where the dynamical system

$$x_k(t+1) = x_k(t) + \alpha_k(R_k(Q_k(t)) - x_k(t)), \quad (k = 1, 2, \ldots, N),$$

is defined is not closed, so the contraction mapping theorem (upon which the proof of Theorem B.3 relies) cannot be used. If we consider the continuous extension by defining $R_k(0) = 0$, then in addition to the Nash equilibrium the zero output vector also becomes a steady state of the above dynamical system, so the presence of multiple steady states excludes the possibility of global asymptotic stability.

In this subsection we start to investigate the kinds of dynamic behavior that we can observe when the restrictive conditions for global stability are not satisfied. A characterization of the global dynamics is not trivial, since we are dealing with an $N$-dimensional piecewise differentiable dynamical system. Therefore, our study is based on a combination of analytical, geometrical and numerical arguments. As has been demonstrated in previous chapters, qualitative changes of the dynamics are often caused by contacts between singularities known as critical sets (see Appendix C), lines of non-differentiability, and basin boundaries. In general such contacts can only be revealed numerically, since the equations of the curves which are involved in such contacts cannot be analytically expressed in terms of elementary functions. Hence, an analysis of global bifurcations is, in general, carried out by using both theoretical and numerical methods. The occurrence of such bifurcations is shown by computer-assisted proofs, and is based on the knowledge of the properties of the singularities involved and their graphical representation (see Mira et al. (1996) for many examples and see also Brock and Hommes (1997)). This "modus operandi" is quite common in the study of the global properties of nonlinear two-dimensional discrete dynamical systems. However an extension of such methods to higher-dimensional dynamical systems is obviously limited. A practical problem which arises is that the visualization of objects in a phase space of dimension greater than two and the detection of contacts between surfaces may become very difficult. Consequently, in the examples that follow we will (again) restrict ourselves to the case of duopoly or the semi-symmetric case of an oligopoly. It should be mentioned that in the case of isoelastic demand, the non-negativity of prices is always guaranteed. So, in contrast to the oligopolies with for example linear or quadratic price functions as considered before, we do not need to ensure this property by selecting the values of the model parameters carefully. On the other hand, we still need to look at the profits along the sequence of quantity decisions in order to see if the long-run dynamics are viable from an economic point of view. Although the problem of negative profits is regularly neglected in the literature on complex dynamics in oligopolies, it is a crucial element of the analysis of an adjustment type model. The dynamical system just represents the firms' individual production decisions, but does not directly tell us if the firms are profitable as a result of the collective outcome.

*Example 3.4.* We consider again the reaction functions in the model with isoelastic demand and linear cost functions derived at the beginning of this chapter, which in the current example becomes

$$R_k(Q_k) = \begin{cases} 0 & \text{if } z_k^* \leqslant 0, \text{i.e., } Q_k \geqslant \frac{A}{c_k}, \\ L_k & \text{if } z_k^* \geqslant L_k, \text{i.e., } Q_k^2 + \left(2L_k - \frac{A}{c_k}\right) Q_k + L_k^2 \leqslant 0, \\ z_k^* = \sqrt{\frac{AQ_k}{c_k}} - Q_k & \text{otherwise}, \end{cases} \tag{3.12}$$

where $k = 1, \ldots, N$. Notice that the constraint $z_k^* = L_k$ is ineffective if $L_k \geqslant A/(4c_k)$, otherwise we have $R_k = L_k$ for

$$Q_k \in \left[ \left( \frac{A}{2c_k} - L_k \right) - \frac{1}{2c_k} \sqrt{A\,(A - 4c_k L_k)}, \left( \frac{A}{2c_k} - L_k \right) + \frac{1}{2c_k} \sqrt{A\,(A - 4c_k L_k)} \right]$$

(see Fig. 1.9). In the duopoly case, $N = 2$, already considered in Example 3.2, partial adjustment towards the best response is governed by the discrete time dynamical system

$$\begin{aligned} x_1(t+1) &= (1 - a_1)x_1(t) + a_1 R_1(x_2), \\ x_2(t+1) &= (1 - a_2)x_2(t) + a_2 R_2(x_1), \end{aligned} \tag{3.13}$$

and the unique Nash equilibrium is given by

$$\bar{\boldsymbol{x}} = (\bar{x}_1; \bar{x}_2) = \left( \frac{Ac_2}{(c_1 + c_2)^2}; \frac{Ac_1}{(c_1 + c_2)^2} \right). \tag{3.14}$$

The local stability properties of $\bar{\boldsymbol{x}}$ in the duopoly case have already been derived in Example 3.2. For identical adjustment coefficients, $a_1 = a_2 = a$, the equilibrium is locally asymptotically stable if $a(1 - r_1 r_2) < 2$, where $r_k = R_k'(Q_k) = (c_1 + c_2 - 2c_k)/(2c_k)$. Inserting these expressions for the derivatives of the best replies allows us to express the stability condition in terms of the cost ratio $\kappa = c_2/c_1$ (cf. also Example 3.3 for the semi-symmetric case). Hence, in this case local asymptotic stability of the equilibrium given in (3.14) is ensured if

$$\frac{a(1+\kappa)^2}{4\kappa} < 2.$$

Consequently, for any given $a \in (0, 1]$, as long as

$$\kappa \in \left( \frac{4 - a - 2\sqrt{4 - 2a}}{a}, \frac{4 - a + 2\sqrt{4 - 2a}}{a} \right)$$

holds, the equilibrium is stable. Note that since $\kappa = 1$ is always inside this interval for all adjustment coefficients $a \in (0, 1]$, the equilibrium is always stable if firms have identical marginal costs. It is also worth pointing out that the cost difference between the firms has to be quite strong in order to render the equilibrium unstable. To demonstrate this, we look at a particular case of the best reply dynamics, namely $a_1 = a_2 = 1$. Here the Nash equilibrium (3.14) is stable if and only if the cost ratio $\kappa = c_2/c_1 \in (3 - 2\sqrt{2}, 3 + 2\sqrt{2}) \simeq (0.17, 5.83)$ (see also Puu (1991, 2003)). If, for example, $c_1 = 1$, this result shows that the unit cost of firm 2 has to be either at least almost 6 times higher than firm 1's unit cost or less than about $1/6$ of it in order that instability occurs. If the cost ratio $c_2/c_1$ exits this interval, then the Nash equilibrium

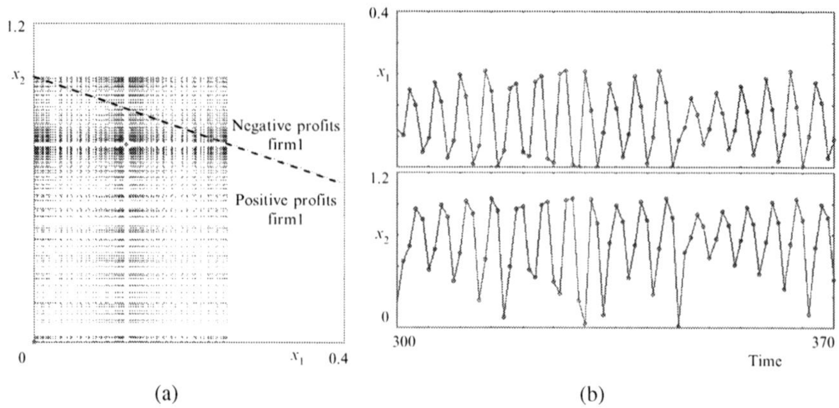


**Fig. 3.4** Example 3.4; discrete time oligopoly with isoelastic demand and linear cost functions – the duopoly case. The cost ratio $\kappa = c_2/c_1 = 0.16$. (**a**) The chaotic attractor in the $(x_1, x_2)$ plane and the dividing line between regions of negative and positive profits for firm 1. (**b**) Time series of a portion of the chaotic attractor

loses stability via a period doubling bifurcation. For values of the cost ratio outside the interval $(3-2\sqrt{2}, 3+2\sqrt{2})$ the asymptotic dynamics may converge to periodic cycles or even exhibit chaotic motion around the Nash equilibrium. A numerically computed chaotic trajectory is shown in Fig. 3.4, obtained for $A = 1, a_1 = a_2 = 1$, $c_1 = 1$, $c_2 = 0.16$. It can be noticed that the chaotic area is quite large, hence we expect no correlations between $x_1(t)$ and $x_2(t)$, in the sense that high values of $x_1(t)$ are associated either with high or with low values of $x_2(t)$ in the same time period; see Fig. 3.4b, where a portion of the chaotic trajectory of Fig. 3.4a is represented for the time periods $t \in [300, 370]$. Note that the profits for the firms are non-negative only if $x_1 + x_2 \leq A/c_k, k = 1, 2$. In Fig. 3.4a we depict the line of zero profits for firm 1, which is represented by the equation $x_1 + x_2 = A/c_1 = 1$. Notice that the zero profit line of firm 2, $x_1 + x_2 = 6.25$, is outside the area shown in the figure. This indicates that the profits for the low-cost firm 2 are always positive, whereas firm 1 makes a loss in some periods along any trajectory which describes the long-run dynamics. The latter point becomes even more obvious if we consider other kinds of long-run dynamics for the duopoly with best reply dynamics. For example, let $c_2 = 0.161$, with all other parameter values as before. The cost ratio is now outside the stability region, and the disequilibrium dynamics in this case are described by a 4-cyclic chaotic attractor[1] (see Fig. 3.5[2]). Of course, even if in this case chaotic dynamics are observed, the time series are much more regular, since they are characterized by a quasi-cyclic behavior (Fig. 3.5b). Furthermore, the zero

[1] An $n$-cyclic chaotic attractor consists of $n$ separate pieces that are visited cyclically in a given order.

[2] The particular "rectangular shape" of the attractors shown in Figs. 3.4 and 3.5 is related to the particular structure of the map in the case of best reply dynamics, see for example, Bischi et al. (2000b) and Agliari et al. (2002a)

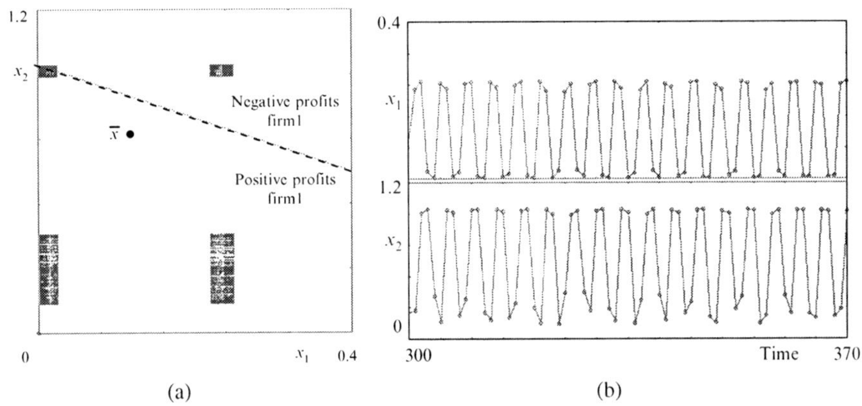


**Fig. 3.5** Example 3.4; discrete time oligopoly with isoelastic demand and linear cost functions – the duopoly case. The cost ratio $\kappa = c_2/c_1 = 0.161$. (**a**) A 4-cyclic chaotic attractor in the $(x_1, x_2)$ plane and dividing line between regions of positive and negative profits for firm 1. (**b**) Time series of a portion of the chaotic attractor. Note how they are more regular than those in Fig. 3.4b

profit line for firm 1 depicted in Fig. 3.5a indicates that the high-cost firm 1 would make a loss after every fourth period with certainty, and potentially also makes a loss after every third period. Consequently, given the regularity of the trajectories in this situation and the possibility of losses following a regular pattern, it seems that the assumption of naive expectations would be more plausible in the former case, where the chaotic attractor extends over a larger portion of the phase space.

Let us now turn to the *semi-symmetric* case obtained by assuming $c_2 = \cdots = c_N$, $a_2 = \cdots = a_N$, $L_2 = \cdots = L_N$ and $x_2(0) = \cdots = x_N(0)$. This particular situation, which has been already studied in Example 3.3, allows us to get some insight into the effects of increasing the number of competitors. As we have seen already in the previous chapters, if the firms partially adjust their production quantities towards the best replies, then the decisions made by firm 1 and the identical firms $2, \ldots, N$ are captured by the two-dimensional dynamical system

$$T : \begin{cases} x_1(t+1) = (1-a_1)x_1(t) + a_1 R_1((N-1)x_2), \\ x_2(t+1) = (1-a_2)x_2(t) + a_2 R_2(x_1 + (N-2)x_2). \end{cases}$$

Assuming an interior equilibrium, it is given by

$$\bar{x}_1 = \frac{(N-1)A}{c_1 + (N-1)c_2}\left(\frac{(N-1)c_2 - (N-2)c_1}{c_1 + (N-1)c_2}\right),$$

$$\bar{x}_2 = \ldots = \bar{x}_N = \frac{(N-1)A}{c_1 + (N-1)c_2}\left(\frac{c_1}{c_1 + (N-1)c_2}\right)$$

and it is locally asymptotically stable if

$$
\begin{aligned}
&-4a_1\kappa(N-1)+a_1a_2(1+\kappa(N-1))^2+2a_2(-2+N(1+\kappa-\kappa N))<0\\
&-8(-2+a_1)\kappa(N-1)+a_1a_2(1+\kappa(N-1))^2+4a_2(-2+N(1+\kappa-\kappa N))>0,
\end{aligned}
\tag{3.15}
$$

where $\kappa = c_2/c_1$ denotes the cost ratio between firms (see Example 3.3; recall that the other stability condition derived there is always fulfilled). For given adjustment coefficients and unit costs, these conditions tell us for which number of firms the equilibrium becomes unstable. Consider for example $A = 16$, $a_1 = 0.4$, $a_2 = 0.3$, $c_1 = 5$, $c_2 = 6$, $L_1 = L_2 = 2$. Then it is easy to see that the first condition holds always for $N > 2$, so we do not consider it in the following analysis. The second inequality becomes $-88N^2 + 1246(N-1) > 0$, and it holds as long as the number of firms $N \leq 13$. So in these cases the equilibrium is stable. For $N = 14$ this inequality is violated, showing that the equilibrium becomes unstable. Figure 3.6 shows a bifurcation diagram for $N$ in the range $[2, 30]$. As expected, the Nash equilibrium $\bar{x}$ is stable as long as the number of competitors is less than 13, and then it loses stability through a period doubling bifurcation. For even higher values of $N$ other bifurcations occur leading to more complicated kinds of asymptotic behavior. Since more detailed results can be easily derived on the basis of a standard local stability analysis, we now turn to the more interesting investigation of the global properties of our model.

In order to explain what kind of bifurcations and global dynamic properties are involved in the qualitative changes of the dynamics observed in Fig. 3.6, we study the properties of the piecewise smooth map $T$. We first divide the strategy space

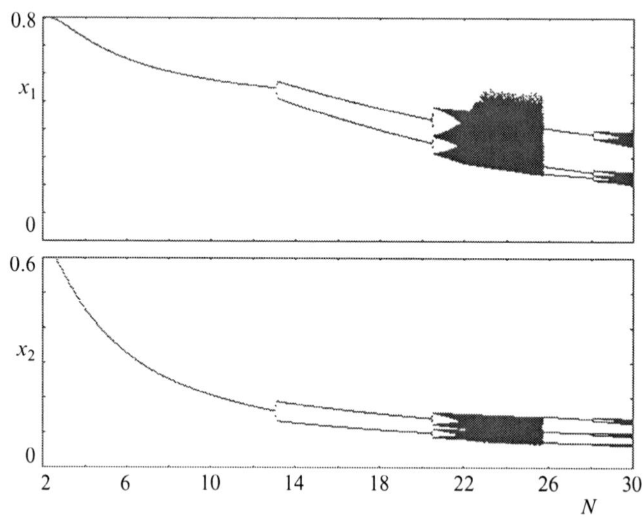


**Fig. 3.6** Example 3.4; discrete time oligopoly with isoelastic demand and linear cost functions – the semi-symmetric case. Bifurcation diagrams of outputs $x_1, x_2$ with respect to the number of firms

$\mathbb{D} = [0, L_1] \times [0, L_2]$ into regions $\mathbb{D}^{(k)}$ where the map $T$ has different expressions. As observed in Chap. 2, the curves that divide these regions are curves of non-differentiability, and these curves may play the role of folding curves (or critical curves, following the terminology used in Mira et al. (1996)). In order to write the expression of the map $T$ in the different regions $\mathbb{D}^{(k)}$, notice that for the set of parameters considered, in the expression of the reaction curve (3.12) of firm 1 we have

$$z_1^* < 0 \quad \text{for} \quad x_2 > \frac{16}{5\,(N-1)},$$

whereas the constraint $z_1^* > L_1$ is ineffective since $L_1 > A/(4c_1)$. Likewise, for the reaction function $R_2$ of firms $2, \ldots, N$, we have

$$z_2^* < 0 \quad \text{for} \; x_1 + (N-2)\,x_2 > 8/3$$

and the constraint $z_2^* > L_2$ is ineffective since $L_2 > A/(4c_2)$. The lines $x_2 = \frac{16}{5(N-1)}$ and $x_2 = \frac{8-3x_1}{3(N-2)}$ divide the strategy space $\mathbb{D}$ into 4 regions. In region $\mathbb{D}^{(1)}$, where $x_2 < \frac{16}{5(N-1)}$ and $x_2 < \frac{8-3x_1}{3(N-2)}$, we have

$$T|_{\mathbb{D}^{(1)}} : \begin{cases} x_1(t+1) = (1-a_1)x_1(t) + a_1\left[\sqrt{\frac{16(N-1)x_2(t)}{5}} - (N-1)x_2(t)\right], \\ x_2(t+1) = (1-a_2)x_2(t) + a_2\Big[\sqrt{\frac{16(x_1(t)+(N-2)x_2(t))}{6}} - x_1(t) \\ \qquad\qquad -(N-2)x_2(t)\Big]. \end{cases}$$

In region $\mathbb{D}^{(2)}$, where $x_2 < \frac{16}{5(N-1)}$ and $x_2 > \frac{8-3x_1}{3(N-2)}$, the map is

$$T|_{\mathbb{D}^{(2)}} : \begin{cases} x_1(t+1) = (1-a_1)x_1(t) + a_1\left[\sqrt{\frac{16(N-1)x_2(t)}{5}} - (N-1)x_2(t)\right], \\ x_2(t+1) = (1-a_2)x_2(t). \end{cases}$$

In region $\mathbb{D}^{(3)}$, where $x_2 > \frac{16}{5(N-1)}$ and $x_2 > \frac{8-3x_1}{3(N-2)}$, the map is

$$T|_{\mathbb{D}^{(3)}} : \begin{cases} x_1(t+1) = (1-a_1)x_1(t), \\ x_2(t+1) = (1-a_2)x_2(t). \end{cases}$$

In region $\mathbb{D}^{(4)}$, where $x_2 > \frac{16}{5(N-1)}$ and $x_2 < \frac{8-3x_1}{3(N-2)}$, we have

$$T|_{\mathbb{D}^{(4)}} : \begin{cases} x_1(t+1) = (1-a_1)x_1(t), \\ x_2(t+1) = (1-a_2)x_2(t) + a_2\Big[\sqrt{\frac{16(x_1(t)+(N-2)x_2(t))}{6}} - x_1(t) \\ \qquad\qquad -(N-2)x_2(t)\Big]. \end{cases}$$

The positive equilibrium

$$\bar{x} = \left( \frac{16\,(N-1)\,(N+4)}{(6N-1)^2}, \frac{80\,(N-1)}{(6N-1)^2} \right)$$

is in region $\mathbb{D}^{(1)}$, whereas no equilibria exist in regions $\mathbb{D}^{(k)}$, $k = 2, 3, 4$. In order to study the local stability of the positive fixed point $\bar{x}$, we consider the Jacobian matrix

$$J^{(1)} = \begin{pmatrix} 1 - a_1 & a_1 \left[ \frac{2(N-1)}{\sqrt{5(N-1)x_2}} - (N-1) \right] \\ a_2 \left[ \frac{\sqrt{2}}{\sqrt{3[x_1 + (N-2)x_2]}} - 1 \right] & 1 - a_2 + (N-2)a_2 \left[ \frac{\sqrt{2}}{\sqrt{3[x_1+(N-2)x_2]}} - 1 \right] \end{pmatrix} \tag{3.16}$$

computed at $\bar{x}$. Using the characteristic equation, the stability condition $-88N^2 + 1246(N-1) > 0$ given before follows after some calculation. As noticed above, the equilibrium $\bar{x}$ undergoes a flip (or period doubling) bifurcation for increasing $N$. After the first flip bifurcation, occurring at $N \simeq 13$, further period doublings occur and a route towards chaotic behavior is observed for increasing values of $N$. However, it is obvious from the stability conditions in (3.15) that the values of the two speeds of adjustment also play an important role. Stability of the positive equilibrium is always ensured for appropriately selected low values of the adjustment speed $a_2$. This can also be confirmed by numerical simulations. In Fig. 3.7 we show a bifurcation diagram obtained with $N = 23$, where all the other parameters are chosen as in Fig. 3.6 and with the bifurcation parameter $a_2$ spanning the whole range $(0, 1]$. For low values of $a_2$ the equilibrium is stable. For increasing values of $a_2$ several sudden transitions between chaotic and periodic behavior characterize the asymptotic dynamics. Many of these bifurcations are different from the common bifurcations observed for smooth dynamical systems as the reader might notice. The reason is that the bifurcations observed here are strongly influenced by the presence of the lines of non-differentiability. As already stressed in Chap. 2, these can be often classified as border collision bifurcations, occurring when an equilibrium point (or a periodic point) of a piecewise differentiable dynamical system crosses a curve of non-differentiability. Such a contact may produce many kinds of effects (transition to another cycle of any period or a sudden transition to chaos) depending on the eigenvalues of the two Jacobian matrices on the two adjacent sides of the curve of non-differentiability involved in the contact (see for example, Banerjee et al. (2000b)). Moreover, as we have shown in Chap. 2 (see also Appendix C) the lines of non-differentiability may represent "folding lines," and consequently they have a role similar to that of the critical curves, where the latter are defined as sets of points where the Jacobian determinant vanishes. In other words, candidates for the "folding curves" $F^{(i)}$ in the particular example we are considering are:

1. The curves of non-differentiability, that is the lines $x_2 = \frac{16}{5(N-1)}$ and $x_2 = \frac{8-3x_1}{3(N-2)}$;
2. The curves of vanishing Jacobian, where the Jacobian matrices in the regions $\mathbb{D}^{(k)}$, $k = 1, \ldots, 4$, are respectively $J^{(1)}$, given in (3.16),

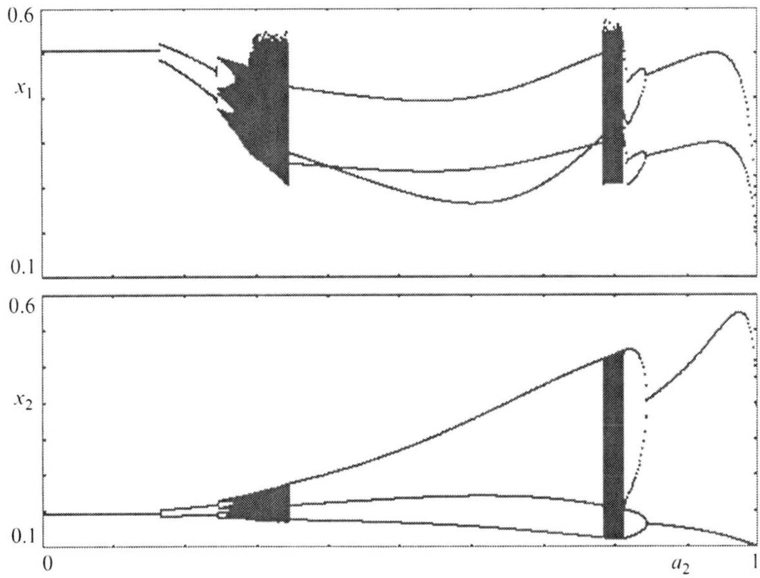


**Fig. 3.7** Example 3.4; discrete time oligopoly with isoelastic demand and linear cost functions – the semi-symmetric case. Bifurcation diagrams of outputs $x_1, x_2$ with respect to $a_2$ with the number of firms held fixed at $N = 23$. The parameters are otherwise as in Fig. 3.6

$$J^{(2)} = \begin{pmatrix} 1 - a_1 & a_1 \left( \frac{2(N-1)}{\sqrt{5(N-1)x_2}} - (N-1) \right) \\ 0 & 1 - a_2 \end{pmatrix},$$

$$J^{(3)} = \begin{pmatrix} 1 - a_1 & 0 \\ 0 & 1 - a_2 \end{pmatrix},$$

and

$$J^{(4)} = \begin{pmatrix} 1 - a_1 & 0 \\ a_2 \left( \frac{\sqrt{2}}{\sqrt{3[x_1 + (N-2)x_2]}} - 1 \right) & 1 - a_2 + (N-2)\, a_2 \left( \frac{\sqrt{2}}{\sqrt{3[x_1 + (N-2)x_2]}} - 1 \right) \end{pmatrix}.$$

Notice that only in regions $\mathbb{D}^{(1)}$ and $\mathbb{D}^{(4)}$ may we have points at which the Jacobian determinant vanishes.

After the foregoing preparations, we are now in a position to describe some border collision bifurcations as well as some methods to bound chaotic attractors that involve the lines of non-differentiability for a specific numerical example. Let us start from the set of parameters used to obtain the bifurcation diagram Fig. 3.7, that is $N = 23$, $A = 16$, $a_1 = 0.4$, $c_1 = 5$, $c_2 = 6$, $L_1 = L_2 = 2$. From the second stability condition in (3.15) we can deduce that at $a_2 = \frac{21120}{127781} \simeq 0.165$ the Nash equilibrium $\bar{x}$ loses stability through a flip bifurcation, at which it becomes a saddle point, and a stable cycle of period 2 is created around it. Just after this bifurcation,

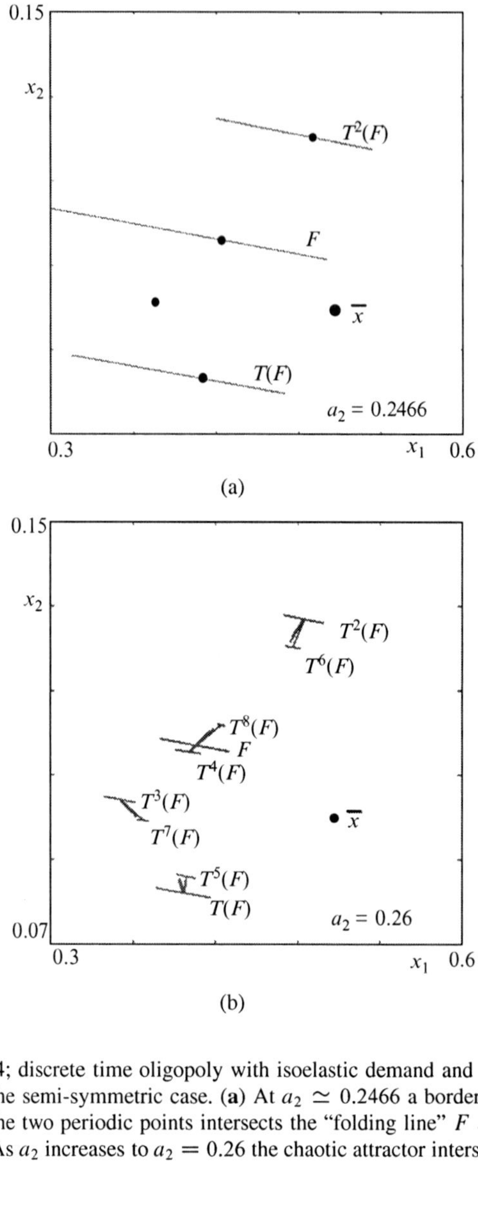


**Fig. 3.8** Example 3.4; discrete time oligopoly with isoelastic demand and linear cost functions. Global dynamics in the semi-symmetric case. (**a**) At $a_2 \simeq 0.2466$ a border collision bifurcation occurs when one of the two periodic points intersects the "folding line" $F$ and a 4-piece chaotic attractor is born. (**b**) As $a_2$ increases to $a_2 = 0.26$ the chaotic attractor intersects a "folding line"

the two periodic points are close to the saddle point $\bar{x}$, hence they belong to region $\mathbb{D}^{(1)}$. As the parameter $a_2$ is further increased, the two periodic points move away from the fixed point, and one of them intersects the boundary of region $\mathbb{D}^{(1)}$, denoted as "folding line" $F$ in Fig. 3.8. This first border crossing may produce many kinds of effects. However, in this case there are no evident effects: if one of the periodic points moves into region $\mathbb{D}^{(2)}$ (while the other remains in region $\mathbb{D}^{(1)}$), the 2-cycle remains attracting. This is an example of a border collision without any change in the qualitative dynamics. At $a_2 \simeq 0.2462$ the 2-cycle undergoes a flip bifurcation and a stable cycle of period 4 appears. As before, just after the bifurcation the four periodic points are close to the 2-cycle saddle, and far from the lines of non-differentiability. However, as the parameter $a_2$ is further increased, one of the periodic points moves towards the folding line $F$, and at $a_2 \simeq 0.2466$, a periodic point intersects the boundary of region $\mathbb{D}^{(1)}$, that is the "folding line" $F$ (see Fig. 3.8a). This marks the occurrence of a true border collision bifurcation, with the effect of a transition to a 4-piece chaotic attractor (see Fig. 3.8b with $a_2 = 0.26$).

As can be seen, the chaotic attractor crosses the folding line $F$. Hence, it is bounded by the images of this line, denoted as $T^{(i)}(F)$, $i = 1 \ldots, 8$, in Fig. 3.8b. This suggests that when a chaotic attractor intersects a folding line $F$, the boundary of the chaotic area includes points belonging to images of increasing rank of $F$. This is a well-known property of the critical lines of smooth noninvertible maps (see Appendix C), which is here extended to the lines of non-differentiability of a piecewise differentiable map (see Mira et al. (1996)). As $a_2$ is further increased, the 4-cyclic chaotic attractor becomes wider (see Fig. 3.9a) until the merging of the pieces occurs. This merging leads to a 2-cyclic chaotic attractor (this occurs at $a_2 \simeq 0.2765$) and then a unique large chaotic attractor emerges (see Fig. 3.9b), obtained for $a_2 \simeq 0.2965$). Also in this case, the boundary of the chaotic area is given by the images of a suitable portion of the folding line $F$. Finally, we once again point out that in the two cases shown in Fig. 3.9, the upper portion of the chaotic attractors is included in the region with negative profits, that is above the lines representing the equation $x_1 + (N-1)x_2 = A/c_k$, $k = 1, 2$. This means that along the chaotic trajectories that describe the long run time evolution of the production decisions of the firms, some periods with negative profits are involved.

### 3.1.3 Continuous Time Models and Local Stability

In this section model (1.31), describing the continuous time dynamics of partial adjustment towards the best response with naive expectations, is examined in the isoelastic case. The Jacobian of the system again has the form (2.46), and its characteristic equation has the special form of (2.47). We assume again that $a_k = \alpha_k'(0) > 0$ for all $k$. Here either all $r_k$ values are in the interval $(-1, 0]$, or exactly one $r_k$ value is positive. If none of the $r_k$ values is positive, then the local asymptotic behavior of the equilibrium is the same as in the concave case. By adding up the terms with identical denominators in the bracketed factor of (2.47) we obtain (2.48), where at most one $\theta_j > 0$. If all $\theta_j \leq 0$, then the problem is the same as in the concave case, so the equilibrium is always locally asymptotically stable. Therefore we may assume that $\theta_{j_0} > 0$ for some $j_0$. If $\theta_j \neq 0$ and $m_j = 1$, then $-a_j(1 + r_j)$ is not an eigenvalue of the Jacobian. Otherwise it is, and the other eigenvalues are the roots of the equation

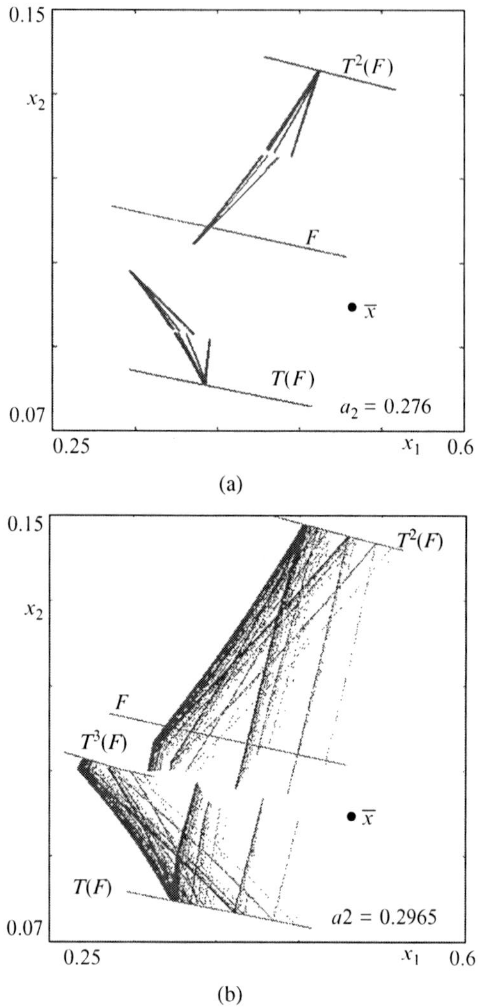


**Fig. 3.9** Example 3.4; discrete time oligopoly with isoelastic demand and linear cost functions. Global dynamics in the semi-symmetric case. Parameters are the same as in Fig. 3.8. (**a**) As $a_2$ increases further to $a_2 \simeq 0.2765$ the pieces of the chaotic attractor merge into a 2-cyclic chaotic attractor. (**b**) At $a_2 \simeq 0.2965$ a unique large chaotic attractor emerges

$$1 - \sum_{j=1}^{s} \frac{\theta_j}{a_j(1+r_j)+\lambda} = 0,$$

where we assume again that $\theta_j \neq 0$ for all $j$, $a_k > 0$ for all firms, and

$$a_1(1+r_1) > a_2(1+r_2) > \ldots > a_s(1+r_s).$$

If $g(\lambda)$ denotes again the left hand side of the above equation, then

$$\lim_{\lambda \to \pm\infty} g(\lambda) = 1,$$

$$\lim_{\lambda \to -a_j(1+r_j)\pm 0} g(\lambda) = \begin{cases} \mp\infty & \text{if } j = j_0, \\ \pm\infty & \text{if } j \neq j_0. \end{cases}$$

However similarly to the discrete time case, $g'(\lambda)$ has no definite sign. The graph of $g(\lambda)$ is the same as shown earlier in Figs. 3.1–3.3 with the only difference being that the poles are all negative and given by $-a_1(1+r_1), \ldots, -a_s(1+r_s)$. Therefore we have again three cases.

Case 1. If $j_0 = 1$, then there are $s-2$ real roots between each pair of poles $-a_j(1+r_j)$ and $-a_{j+1}(1+r_{j+1})$ for $j = 2, \ldots, s-1$. If the other two roots are real and are between $-a_1(1+r_1)$ and $-a_s(1+r_s)$, then the equilibrium is locally asymptotically stable.

Case 2. If $j_0 = s$, then all roots are real and are negative if $g(0) > 0$. This condition can be rewritten as

$$\sum_{k=1}^{N} \frac{r_k}{1+r_k} < 1.$$

Case 3. If $1 < j_0 < s$, then there are $s-2$ real roots, one before $-a_1(1+r_1)$, and one in between each pair of poles $-a_j(1+r_j)$ and $-a_{j+1}(1+r_{j+1})$ for $j = 1, \ldots, j_0-2, j_0+1, \ldots, s-1$. If we assume that the remaining two roots are real and between $-a_1(1+r_1)$ and $-a_s(1+r_s)$, then all roots are negative.

The possibility of complex roots will be shown later in Example 3.6. If there are complex roots, then no simple stability conditions can be given. We will next return to the case of Example 3.3, but under the assumption of continuous time dynamics.

*Example 3.5.* Consider again the $N$-person semi-symmetric oligopoly of Example 3.3, now under the assumption of continuous time adjustment of the outputs of the firms of the oligopoly. Assume again that $c_2 = \ldots = c_N$. Then $Q_1 = (N-1)x_2$ and $Q_2 = x_1 + (N-2)x_2$ by assuming that firms $2, \ldots, N$ select identical linear adjustment function and initial outputs. From Example 3.3 we know that at the interior equilibrium

$$\bar{Q} = \frac{(N-1)A}{c_1 + (N-1)c_2},$$

$$r_1 = R_1'(\bar{Q}_1) = \frac{(N-1)c_2 + (3-2N)c_1}{2(N-1)c_1},$$

$$r_2 = R_2'(\bar{Q}_2) = \frac{c_1 - (N-1)c_2}{2(N-1)c_2}.$$

Condition (3.5) for $k = 1$ and $k = 2$ is

$$c_1 \leq \frac{c_1 + (N-1)c_2}{N-1}, \quad c_2 \leq \frac{c_1 + (N-1)c_2}{N-1}.$$

The second inequality always holds, the first can be rewritten as

$$\frac{c_2}{c_1} \geq \frac{N-2}{N-1}. \tag{3.17}$$

By introducing again the notation $\kappa = c_2/c_1$ we have

$$\kappa \geq \frac{N-2}{N-1},$$
$$r_1 = \frac{(N-1)\kappa + (3-2N)}{2(N-1)} \quad \text{and} \quad r_2 = \frac{1-(N-1)\kappa}{2(N-1)\kappa}.$$

The two-dimensional system for the adjustment of firms' outputs has the form

$$\begin{aligned}\dot{x}_1 &= a_1(R_1((N-1)x_2) - x_1),\\ \dot{x}_2 &= a_2(R_2(x_1 + (N-2)x_2) - x_2),\end{aligned}$$

with Jacobian matrix

$$\begin{pmatrix} -a_1 & a_1 r_1(N-1) \\ a_2 r_2 & a_2(r_2(N-2)-1) \end{pmatrix}.$$

The characteristic equation can be written as

$$(-a_1 - \lambda)(a_2(r_2(N-2)-1) - \lambda) - a_1 a_2 r_1 r_2 (N-1) = 0$$

or

$$\lambda^2 + \lambda[a_1 + a_2(1 + r_2(2-N))] + a_1 a_2[1 + (2-N)r_2 - (N-1)r_1 r_2] = 0. \tag{3.18}$$

Clearly,

$$r_2 \leq \frac{1-(N-1)\frac{N-2}{N-1}}{2(N-1)\kappa} = \frac{3-N}{2(N-1)\kappa}.$$

Notice first that the linear coefficient of (3.18) is always positive since $r_2 \leq 0$. With the new variable $\mathcal{K} = (N-1)\kappa$, the multiplier of $a_1 a_2$ in the constant term of (3.18) has the form

$$\begin{aligned}&1 + (2-N)\frac{(1-\mathcal{K})}{2\mathcal{K}} - (N-1)\frac{(\mathcal{K} + (3-2N))}{2(N-1)}\frac{1-\mathcal{K}}{2\mathcal{K}}\\ &= \frac{1}{4\mathcal{K}}[4\mathcal{K} + (4-2N)(1-\mathcal{K}) - (1-\mathcal{K})(\mathcal{K}+3-2N)] = \frac{(\mathcal{K}+1)^2}{4\mathcal{K}} > 0.\end{aligned}$$

Then Lemma F.2 implies that the equilibrium is always locally asymptotically stable.

*Example 3.6.* Assume in the previous example that $N = 2, a_1 = a_2 = a$. Then (3.18) simplifies to

$$\lambda^2 + 2a\lambda + a^2(1 - r_1 r_2) = 0.$$

From Example 3.2 we know that $r_1 r_2 < 0$ if $c_1 \neq c_2$. In this case both the linear and constant coefficients are positive (as in the general case of the previous example), and the discriminant is

$$4a^2 - 4a^2(1 - r_1 r_2) = 4a^2 r_1 r_2 < 0.$$

So both roots are complex, showing that there is no guarantee that the eigenvalues are real, contrary to the case of concave oligopolies discussed in Sect. 2.5.

The book by Okuguchi and Szidarovszky (1999) contains some stability results in the case of linear cost functions. A detailed stability analysis is presented by Chiarella and Szidarovszky (2002) for the general nonlinear case. Models with continuously distributed time lags are identical to the concave case, so the derivations and the similar results are not duplicated here.

## 3.2 Cost Externalities and Multiple Interior Nash Equilibria

In Chap. 2 we demonstrated that under some standard assumptions on the demand function and on the cost functions of the oligopolists, the reaction functions of the firms are decreasing. However, there are several situations where the microeconomic fundamentals of an oligopoly model lead to reaction functions which are non-monotonic. For example, in the previous subsection we have shown that with isoelastic price functions the reaction functions are increasing over the range where the expected aggregate quantity of the other players is small, otherwise it is decreasing (see also Example 1.5 and Bulow et al. (1985*b*)). Using non-monotonic reaction functions, several authors have considered the best response dynamics and the partial adjustment towards the best response and have demonstrated that such adjustment processes may lead to non-convergence with complicated, but bounded fluctuations of the production sequences (for example, Rand (1978), Dana and Montrucchio (1986), Witteloostuijn and Lier (1990) and Puu (1991)). The focus of these contributions has been mainly towards questions of local stability of the Nash equilibria and the creation of complex attractors if convergence to an equilibrium fails. The emphasis of the analysis is, in this case, on the delineation of a trapping region in the space of production quantities, where the asymptotic dynamics of the oligopoly game are ultimately bounded.

In the present subsection we will turn our attention to externalities in the cost functions, which might also give rise to non-monotonic reaction functions (see Example 1.6, Kopel (1996), Puhakka and Wissink (1995), Bischi and Lamantia

(2002) and Furth (1986, 2009)). We will consider a duopoly market and we will show that in a simple model with cost externalities we obtain several coexisting equilibria. Since an equilibrium point can be considered as a convention that arises among firms interacting repeatedly, stability arguments are often used to solve this coordination problem. See for example, Van Huyck and Battalio (1998) and Van Huyck et al. (1984, 1997). If a stability argument selects a single equilibrium, this point can be considered as the solution of the oligopoly game. However, as we will see, in the model with cost externalities multiple equilibria survive this type of refinement and several (locally) stable equilibria coexist. Each of these equilibria has its own basin of attraction and, consequently, the dynamic process becomes path dependent. The long run outcome of the players' myopic output decisions crucially depends on the initial production quantity. Hence, in such a situation it is not sufficient to analyze the local stability properties. In order to be able to give some insight into the long run market outcome, it is important to gain some knowledge about the boundaries that separate the basins of attraction of the various coexisting equilibria, and to study the role of these boundaries in the occurrence of global bifurcations that drastically change the topological structure of the basins.

Recall from Example 1.6 that if the inverse demand function is linear, $p = f(Q) = A - BQ$, and the cost functions of the oligopolists are characterized by interfirm externalities, that is $C_k(x_k, Q_k) = x_k M_k(Q_k)$ with $M_k(Q_k) = A - B(1 + 2\mu_k)Q_k - 2B\mu_k Q_k^2$, then the best response of firm $k$ is given by

$$R_k(Q_k) = \begin{cases} 0 & \text{if} \quad \mu_k Q_k(1 - Q_k) \leq 0, \\ L_k & \text{if} \quad \mu_k Q_k(1 - Q_k) \geq L_k, \\ z_k^* & \text{otherwise,} \end{cases}$$

where $z_k^* = \mu_k Q_k(1 - Q_k)$ and $L_k$ denotes the capacity of firm $k$. The parameters $\mu_k$ measure the intensity of the interfirm cost externality (see Kopel (1996)). In what follows we consider a duopoly market ($N = 2$), so that $Q_1 = x_2$ and $Q_2 = x_1$. We let $\mu_k \in (1, 4]$ and for simplicity we assume that $L_k = 1$. Under these assumptions the reaction functions reduce to

$$R_1(x_2) = \mu_1 x_2(1 - x_2), \quad R_2(x_1) = \mu_2 x_1(1 - x_1). \tag{3.19}$$

The Nash equilibria of this duopoly are located at the intersections of the two reaction curves $x_1 = R_1(x_2)$ and $x_2 = R_2(x_1)$. The reaction functions are shown in Fig. 3.10, where the two panels illustrate that beside the trivial Nash equilibrium $O = (0, 0)$, multiple interior Nash equilibria can exist depending on the level of the cost externalities. For example, for $\mu_1 = 3, \mu_2 = 3.5$ there is just one interior Nash equilibrium $E_S$ (part (a)), whereas for $\mu_1 = 3.7, \mu_2 = 3.5$ there are two additional interior Nash equilibria $E_1$ and $E_2$ (part (b)). Analytically, the interior equilibria are obtained as the real solutions of the fourth degree algebraic system

$$x_1 = \mu_1 x_2(1 - x_2), \quad x_2 = \mu_2 x_1(1 - x_1),$$

and this system can have up to four solutions.

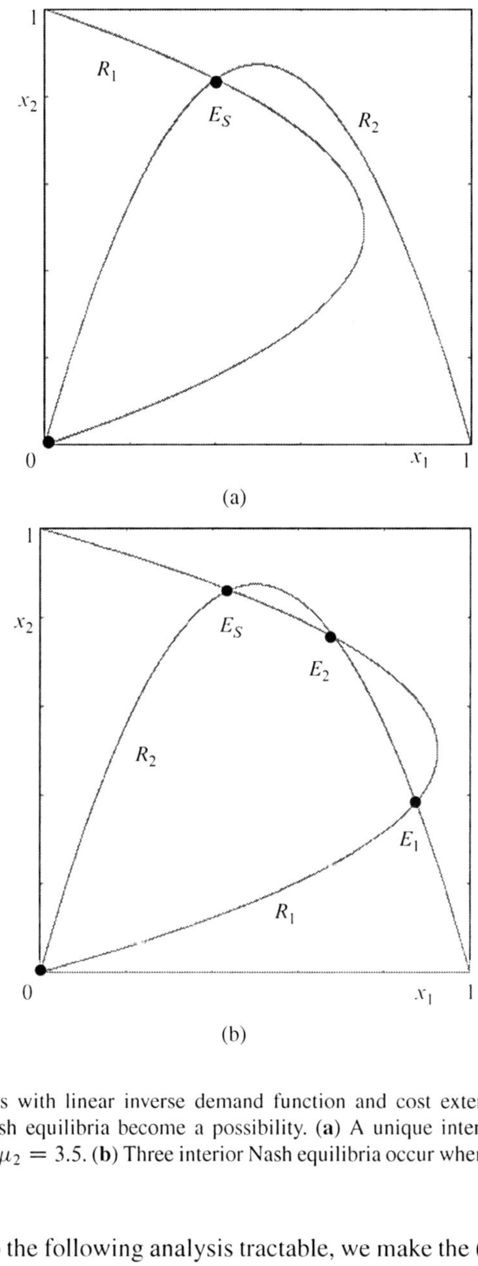


**Fig. 3.10** Oligopolies with linear inverse demand function and cost externalities. The case of duopoly, multiple Nash equilibria become a possibility. (**a**) A unique interior Nash equilibrium occurs when $\mu_1 = 3, \mu_2 = 3.5$. (**b**) Three interior Nash equilibria occur when $\mu_1 = 3.7, \mu_2 = 3.5$

In order to keep the following analysis tractable, we make the (rather reasonable) assumption that the influence of each firm's action on the marginal costs of the competitor is identical for both firms, that is

$$\mu_1 = \mu_2 = \mu. \tag{3.20}$$

In the case of $\mu > 1$ there is always an interior Nash equilibrium $E_S$ which belongs to the diagonal $\Delta = \{(x, x), x \in \mathbb{R}\}$. Its coordinates are given by

$$E_S = \left(1 - \frac{1}{\mu}, 1 - \frac{1}{\mu}\right),$$

and it is characterized by identical production quantities of the two firms. At $\mu > 3$ two further Nash equilibria exist. They are given by

$$\begin{aligned} E_1 &= \left(\frac{\mu + 1 + \sqrt{(\mu + 1)(\mu - 3)}}{2\mu}, \frac{\mu + 1 - \sqrt{(\mu + 1)(\mu - 3)}}{2\mu}\right), \\ E_2 &= \left(\frac{\mu + 1 - \sqrt{(\mu + 1)(\mu - 3)}}{2\mu}, \frac{\mu + 1 + \sqrt{(\mu + 1)(\mu - 3)}}{2\mu}\right), \end{aligned} \tag{3.21}$$

and they are located in symmetric positions with respect to the diagonal $\Delta$. Notice that for $\mu = 3$, $E_1$, $E_2$ and $E_S$ coincide. These Nash equilibria are characterized by different production quantities of the two players. It is easy to see that the market share of firm 1 (firm 2) is larger in $E_1$ ($E_2$). Obviously, in a situation where multiple Nash equilibria coexist, a coordination problem for the two firms arises. It is not clear which of the Nash equilibria the firms can agree upon as an outcome of the game. One possibility to discriminate among the equilibria is to assume that players start with quantity pairs out of equilibrium and adjust their production decision to evolving changes in their environment, for example, using their best replies or estimates of the gradient of the profit functions. Then we can use local stability, global dynamics, or for example, the extent of the basins of attraction in the case of multiple locally stable equilibria to obtain insights into the question about which of the equilibria is more likely to be a long run outcome of the game (see Kopel (2009) and Cox and Walker (1998)).

We will assume that in order to update their production decisions, the duopolists use partial adjustment towards the best response with naive expectations. Recall, however, that in Chap. 1 we have shown that in the duopoly case the best reply dynamics with adaptive expectations is identical to the dynamical system obtained by partial adjustment towards the best response with naive expectations (see (1.20) and (1.21)). Consequently, for our duopoly model with symmetric cost externalities, in either case the dynamical systems which generates the sequences of (expected) production quantities is given by

$$\begin{aligned} x_1(t+1) &= (1-a_1)x_1(t)+a_1 R_1(x_2(t)) = (1 - a_1) x_1(t) + a_1 \mu x_2(t) (1 - x_2(t)), \\ x_2(t+1) &= (1 - a_2)x_2(t)+a_2 R_2(x_1(t)) = (1-a_2)x_2(t) + a_2 \mu x_1(t) (1-x_1(t)). \end{aligned} \tag{3.22}$$

### 3.2.1 Identical Speeds of Adjustment

We first assume that the speeds of adjustment are identical for the two firms, that is

$$a_1 = a_2 = a.$$

Under this assumption, in contrast to the previous examples, the singularities that are involved in global bifurcations can be given in closed form. Moreover, the exact values for the parameters at which global bifurcations occur can be explicitly determined (see Bischi and Kopel (2001) for further details).

In this case, it is obvious that the steady states of this system correspond to the Nash equilibria of the game and are independent of the adjustment speed $a$. A proper study of the two-dimensional map $T : (x_1, x_2) \to (x_1', x_2')$ defined by

$$T : \begin{cases} x_1' = (1-a)\,x_1 + a\mu x_2\,(1-x_2), \\ x_2' = (1-a)\,x_2 + a\mu x_1\,(1-x_1), \end{cases} \tag{3.23}$$

should provide some answers to the questions stated above. Since we restrict ourselves to $\mu \in (1,4]$, the strategy space $\mathcal{S} = \{[0,1]\times[0,1]\}$ is trapping for each value of $a \in (0,1]$ and for each initial value of production quantities in $\mathcal{S}$.[3] In other words, any sequence of production quantities which starts inside $\mathcal{S}$ remains feasible for all $t \geq 0$.

We first turn to the question of local stability of the interior Nash equilibria and provide a characterization of the corresponding stability regions (see also Fig. 3.11).

**Proposition 3.1.** *Let* $\Omega = \left\{(\mu,a) \in \mathbb{R}^2 | 1 < \mu \leq 4, 0 < a \leq 1\right\}$ *denote the appropriate region in the parameter space. Then the following holds.*

*(i) The symmetric Nash equilibrium* $E_S = \{1-1/\mu, 1-1/\mu\}$ *exists for all* $(\mu,a) \in \Omega$. *It is locally asymptotically stable for* $(\mu,a) \in \Omega$, *if* $1 < \mu < 3$.

*(ii) The Nash equilibria* $E_i$, $i = 1,2$, *given in (3.21) exist for* $\mu > 3$. *They are locally asymptotically stable for* $(\mu,a) \in \Omega$, *if* $a < a_h(\mu) = 2/(\mu^2 - 2\mu - 3)$.

*(iii) In the set*

$$\Omega^s(E_i,C_2) = \left\{ (\mu,a) \in \Omega \mid \mu > 3, a_h(\mu) > a > a_p(\mu) = \frac{6-\sqrt{12\mu(\mu-2)}}{3+2\mu-\mu^2} \right\}, \tag{3.24}$$

*the two stable Nash equilibria* $E_i$, $i = 1,2$, *given in (3.21) coexist with a stable cycle of period two*

[3] This is so since the maxima of the reaction functions $R_k$ occur at $\mu_k/4$, and here we have $\mu_1 = \mu_2 = \mu$ with $0 < \mu \leq 4$.

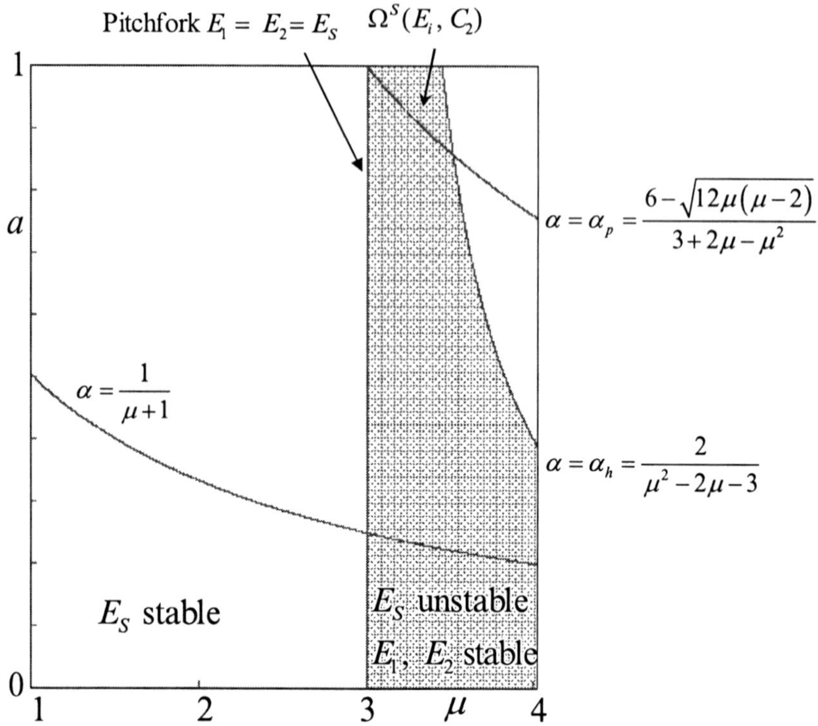


**Fig. 3.11** Oligopolies with linear inverse demand function and cost externalities. The case of duopoly with identical speeds of adjustment. Multiple Nash equilibria in the $(\mu, a)$ plane. Note that $E_S$ is unique and stable for $\mu < 3$. For $\mu > 3$ $E_S$ becomes unstable and two stable equilibria $E_1, E_2$ occur

$$C_2 = \{(p_1, p_1), (p_2, p_2)\} \in \Delta, \tag{3.25}$$

*with coordinates*

$$p_1 = \frac{a(\mu-1)+2-\sqrt{a^2(\mu-1)^2-4}}{2a\mu},$$

*and*

$$p_2 = \frac{a(\mu-1)+2+\sqrt{a^2(\mu-1)^2-4}}{2a\mu}.$$

For the interested reader it should be mentioned that for $(\mu, a) \in \Omega$ with $a > a_h(\mu)$, more complicated dynamics might be observed. The proof of this proposition is based on a standard analysis of the eigenvalues of the Jacobian matrix and is given in detail in Bischi and Kopel (2001).

The results given in this proposition show that for a large set of values of the cost externality $\mu$ and the adjustment speed $a$, multiple stable Nash equilibria are obtained (see the shaded area in Fig. 3.11). Additionally, for sufficiently high values of the adjustment coefficient $a$ in this area, namely for $a > a_p(\mu)$, a stable 2-cycle $C_2$ coexists with the two stable equilibria $E_1$ and $E_2$. This latter point seems to be important for the following reason. If the adjustment process converges to the equilibria only if initial conditions are chosen from a certain subset of $S$ and otherwise it cannot be observed, it becomes crucial to obtain information on the relative size of the set of initial conditions from which players can eventually coordinate their actions (see Mailath (1998), Fudenberg and Levine (1998)).

We will now turn to the analysis of the global dynamics of the model. Since we are not able to discriminate among the equilibria $E_1$ and $E_2$ on the basis of the local stability properties, to obtain further information on the stability properties of the Nash equilibria we will study their basins of attraction. Figure 3.12 depicts the basins of the locally stable equilibria $E_1$ and $E_2$ for two quite distinct situations. In Fig. 3.12a, obtained with $\mu = 3.4$ and $a = 0.2 < 1/(1+\mu) = 0.2273$, the basins have a quite simple structure. For initial production quantities in $S$ with $x_1(0) > x_2(0)$ the adjustment process (3.23) converges to the equilibrium $E_1$. On the other hand, if the reverse inequality holds, then the process converges to the equilibrium $E_2$. Therefore, if firm 1 (firm 2) initially dominates the market in terms of market share, this property prevails throughout and the equilibrium $E_1$ (equilibrium $E_2$) is eventually selected. In contrast to this, the situation shown in Fig. 3.12b, is quite different. It is obtained with the same value of the cost externality $\mu$, but with higher values of the adjustment coefficients, namely $a = 0.5 > 1/(1+\mu) = 0.2273$. In

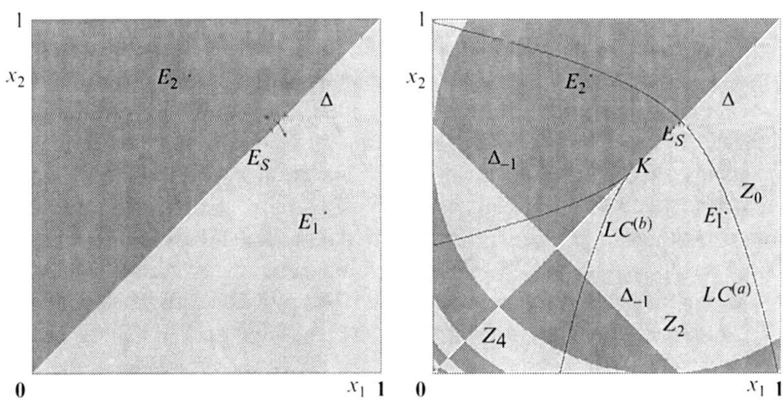


**Fig. 3.12** Oligopolies with linear inverse demand function and cost externalities. The case of duopoly with identical speeds of adjustment. Basins of attraction of the multiple Nash equilibria (**a**) Simple structure for $\mu = 3.4$ and $a = 0.2$. Convergence to either $E_1$ or $E_2$ depending on which firm dominates initially. (**b**) Non-connected basins for $\mu = 3.4$ and $a = 0.5$, now convergence to $E$ or $E_2$ cannot be determined on the basis of which firm dominates initially

this case the basins are no longer simply connected sets, and portions of each basin are present both in the region above and below the diagonal $\Delta$. The basins are now disconnected sets, and the adjustment process starting from initial conditions below or above the diagonal may lead to convergence to either $E_1$ or $E_2$.

The transition from simply connected basins to disconnected basins is caused by a global bifurcation. We will now describe the mechanism which causes this bifurcation in more detail. The argument begins by noticing that the map $T$ defined in (3.23) is noninvertible. Given a point $(x_1', x_2') \in \mathcal{S}$, its preimages are computed by solving with respect to $x_1$ and $x_2$ the algebraic system

$$\begin{cases} (1-a)x_1 + a\mu x_2(1-x_2) = x_1', \\ \\ (1-a)x_2 + a\mu x_1(1-x_1) = x_2'. \end{cases} \tag{3.26}$$

As noticed before, this is a fourth degree algebraic system, which may have four or two real solutions, or no real solution at all. Hence, the strategy set $\mathcal{S}$ can be subdivided into the regions $Z_4$, $Z_2$, and $Z_0$, separated by branches of the critical curve $LC$. For the differentiable map (3.23) the curve $LC_{-1}$ coincides with the set of points at which the determinant of the Jacobian matrix vanishes (see Appendix C) so that

$$\left(x_1 - \frac{1}{2}\right)\left(x_2 - \frac{1}{2}\right) = \frac{(1-a)^2}{4a^2\mu^2}. \tag{3.27}$$

Equation (3.27) represents an equilateral hyperbola. The curve $LC_{-1}$ is formed by the union of two disjoint branches, say $LC_{-1} = LC_{-1}^{(a)} \cup LC_{-1}^{(b)}$, which are depicted in Fig. 3.13a. Also its image $LC = T(LC_{-1})$ is the union of two branches, $LC^{(a)} = T(LC_{-1}^{(a)})$ and $LC^{(b)} = T(LC_{-1}^{(b)})$. This is shown in Fig. 3.13b. The branch $LC^{(a)}$ separates the region $Z_0$, whose points have no preimages, from the region $Z_2$, whose points have two distinct rank-1 preimages. The other branch $LC^{(b)}$ separates the region $Z_2$ from the region $Z_4$, whose points have four distinct preimages.[4] In order to give a geometrical interpretation of the "unfolding action" of the multivalued inverse $T^{-1}$, it is useful to consider a region $Z_k$ as the superposition of $k$ sheets, each associated with a different inverse. Such a representation is known as *Riemann foliation* of the plane (see for example, Mira et al. (1996)). Different sheets are connected by folds joining two sheets, and the projections of such folds on the phase plane are arcs of $LC$. The foliation associated with the map (3.23) is qualitatively represented in Fig. 3.13c. It can be noticed that the cusp point of $LC^{(b)}$ denoted by $K$ is characterized by three merging preimages at the junction of two folds.

This cusp point $K$ of $LC^{(b)}$ plays a crucial role in the analysis, since when $K$ enters the strategy set $\mathcal{S}$ (for $a(\mu+1) > 1$, see below), suddenly points of $\mathcal{S}$ have

---

[4] Following the terminology of Mira et al. (1996), we say that the map (3.23) is a noninvertible map of $Z_4 > Z_2 - Z_0$ type, where the symbol "$>$ " denotes the presence of a cusp point in the branch $LC^{(b)}$.

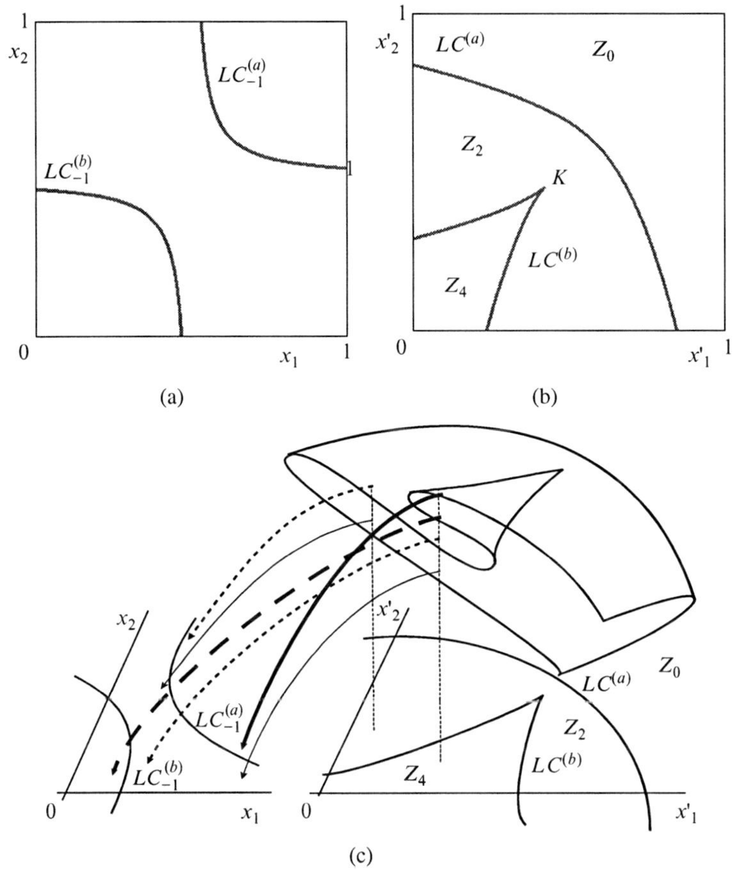


**Fig. 3.13** Oligopolies with linear inverse demand function and cost externalities, the case of duopoly with identical speeds of adjustment. (**a**) The two disjoint branches, $LC_{-1}^{(a)}$ and $LC_{-1}^{(b)}$ of the curve $LC_{-1}$. (**b**) The critical curves $LC = T(LC_{-1})$. Note the cusp at $K$. (**c**) Illustrating the Riemann foliation of the $(x_1, x_2)$ plane

a higher number of preimages then before. The unfolding process of the inverse of the map $T$ then causes the creation of disconnected components of the basins. The bifurcation occurring at $a(\mu + 1) = 1$ is a *global (or contact) bifurcation*, which is characterized by a contact between the stable set of $E_S$ along the diagonal $\Delta$ and a critical curve $LC$. The coordinates of the cusp point of $LC^{(b)}$ can be easily computed in our case. Using (3.27) it is easy to see that the intersection of $LC_{-1}^{(b)}$ with the diagonal $\Delta$ occurs at

$$K_{-1} = LC_{-1}^{(b)} \cap \Delta = (k_{-1}, k_{-1}) \quad \text{with} \quad k_{-1} = \frac{a(\mu + 1) - 1}{2a\mu}.$$

Then the coordinates of the cusp point of the curve $LC^{(b)} = T(LC_{-1}^{(b)})$ are given by

$$K = LC^{(b)} \cap \Delta = (k, k) \quad \text{with} \quad k = f(k_{-1}) = \frac{(a(\mu+1)-1)(a\mu + 3(1-a))}{4a\mu},$$

where the one-dimensional map $f(x) = (1 + a(\mu - 1))x - a\mu x^2$ is the restriction of the map $T$ to the diagonal. It now becomes obvious that at $a(\mu + 1) = 1$ the cusp point $K$ enters the strategy set $\mathcal{S}$ and that after this bifurcation there are points in the strategy set that have a higher number of preimages.

To elaborate a little further on the workings of the mechanism which transforms the basins from simply connected sets to disconnected sets, consider the origin $O = (0, 0)$. If $0 < a < 1/(\mu + 1)$, then $O \in Z_2$ and there are just two rank-1 preimages of $O$. Both belong to the diagonal $\Delta$, with one preimage is $O$ itself (since $O$ is a fixed point), and the other preimage is

$$O_{-1}^{(1)} = \left( \frac{1 + a(\mu - 1)}{a\mu}, \frac{1 + a(\mu - 1)}{a\mu} \right).$$

This can be easily seen by using the restriction of the map $T$ to the diagonal. The situation is depicted in Fig. 3.14a, where for the sake of mathematical exposition we show the whole extent of the basins of attraction of the locally stable equilibria $E_1$ and $E_2$ (and not just the region belonging to the strategy space $\mathcal{S}$ as in Fig. 3.12). Observe that as long as the cusp point is outside the basins of attraction, the basins are simple and connected sets. If however $a > 1/(\mu + 1)$, then the origin $O \in Z_4$ since the cusp point has entered $\mathcal{S}$, and two more rank-1 preimages of $O$ exist. These

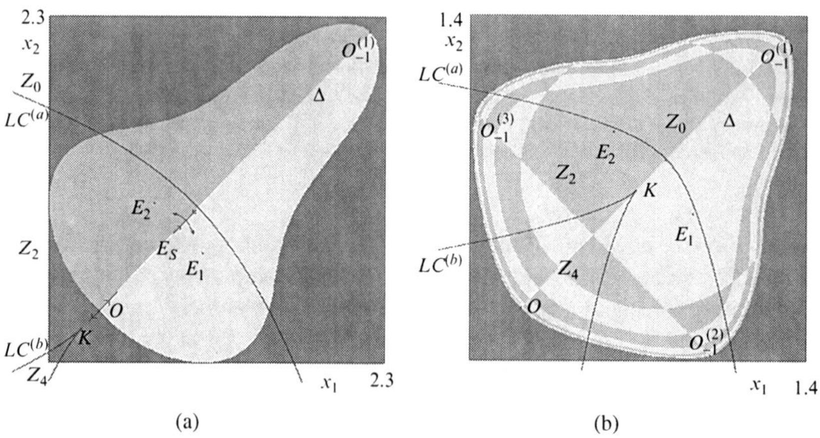


**Fig. 3.14** Linear inverse demand function and cost externalities. The case of duopoly with identical speeds of adjustment - basins of attraction of the two equilibria $E_1$ and $E_2$. (**a**) Here $\mu = 3.4$, $a = 0.2 < 1/(\mu+1)$, and the basins of attraction are simple and connected sets. (**b**) Here $\mu = 3.4$, $a = 0.5 < 1/(\mu - 1)$, and the basins of attraction become disconnected

two further preimages, $O_{-1}^{(2)}$ and $O_{-1}^{(3)}$, are located on the line $\Delta_{-1}$ of the equation[5]

$$x_1 + x_2 = 1 + \frac{1}{\mu}\left(1 - \frac{1}{a}\right).$$

in symmetric positions with respect to $\Delta$ (see Fig. 3.14b). Hence

$$O_{-1}^{(2)} = \left(\frac{a(\mu+1) - 1 + \sqrt{a^2\mu^2 + 2a\mu(1-a) - 3(a^2+1) + 6a}}{2a\mu}, \right.$$
$$\left. \frac{a(\mu+1) - 1 - \sqrt{a^2\mu^2 + 2a\mu(1-a) - 3(a^2+1) + 6a}}{2a\mu}\right) \qquad (3.28)$$

and the symmetric point $O_{-1}^{(3)}$ is obtained from $O_{-1}^{(2)}$ by swapping the two coordinates.

To conclude this subsection, we would like to reflect on several issues. First, the occurrence of the bifurcation which transforms the basins from simply connected to disconnected sets causes a loss of predictability concerning the long-run outcome of the adjustment process. The presence of many disjoint components of both basins causes a sensitivity with respect to the initial production quantities, in the sense that a small perturbation may lead to a crossing of the boundary which separates the two basins and, consequently, the trajectory may converge to a different Nash equilibrium. Second, for increasing values of the adjustment coefficient $a$, as the line $\Delta_{-1}$ in Fig. 3.14b moves upwards, certain connected parts of the basins of the equilibria come closer to the corresponding other equilibrium. That is, initial production quantities which eventually lead to convergence to $E_i$ are located close to the equilibrium $E_j, i \neq j$, and vice versa. In contrast to a global analysis, a study based only on the local properties of the process around the equilibria would not have been able to provide us with information on the size of the neighborhood from which convergence to the corresponding equilibrium is achieved. Finally, our global analysis also reveals that for $(\mu, a) \in \Omega^s(E_i, C_2)$ three coexisting attractors are present[6]. Hence the outcome of the oligopoly game is highly path dependent and could end up at any of the attractors depending on the initial conditions.

### 3.2.2 *Non-Identical Speeds of Adjustment*

We now turn to the case of different speeds of adjustment. In contrast to the previous situation, a rigorous mathematical analysis cannot be provided. However, guided by

[5] This can be seen by setting $x_1' = x_2'$ in (3.26) and adding or subtracting the two symmetric equations.

[6] We remind the reader that the stability region of $E_1$, $E_2$ and $C_2$ is defined in Proposition 3.1.

the knowledge of the critical curves, we can still analyze the structure of the basins of the two coexisting stable Nash equilibria and we can characterize the bifurcations that cause their qualitative changes using numerical and graphical procedures.

As in the case of identical speeds of adjustment, there exists a rather large set of parameter values for $\mu, a_1$, and $a_2$ for which two stable equilibria exist. Moreover, it is easy to realize that small differences between the two adjustment coefficients do not cause significant changes in the local stability properties, that is in the modulus of the eigenvalues. On the other hand, as will be demonstrated below, such small differences may cause drastic effects with regard to the structure of the basins. Many of the arguments given in the previous section for the study of the boundaries of the basins and their global bifurcations continue to hold for non-identical adjustment speeds. However, there are some important differences.

- The main difference is that the diagonal $\Delta$ is no longer invariant. Even if the fixed points remain the same, the basins are no longer symmetric with respect to $\Delta$.
- The preimages of the unstable fixed point $O$ belong to the boundary of the set of points which generate bounded trajectories, but a simple analytical expression of the preimages of $O$ cannot be obtained. Since they are solutions of a fourth degree algebraic equation, they can be computed by standard numerical routines.
- For increasing values of $\mu$ or $a_i$ the point $O$ enters the region $Z_4$. However the exact values of the parameters at which this occurs cannot be computed analytically.
- Although the boundary which separates the basins of $E_1$ and $E_2$ is still formed by the whole stable set of $E_S$, in the case of $a_1 \neq a_2$ the local stable set of $E_S$ is not along the diagonal $\Delta$. The contact between the stable set of $E_S$ and the critical curve $LC^{(b)}$, which causes the transition from simple to complex basins, does not occur at the fixed point $O$ (since now the origin $O$ does not belong to the stable set of $E_S$) and no longer involves the cusp point of $LC^{(b)}$. Again, the parameter values at which such contact bifurcations occur cannot be computed analytically. However, the bifurcation is always caused by a contact between $LC$ and a basin boundary.

We will finally demonstrate that the occurrence of these bifurcations can be detected by computer-assisted proofs, based on the knowledge of the properties of the critical curves and their graphical representation. As mentioned before, this "modus operandi" is typical in the study of the global bifurcations of nonlinear two-dimensional maps. Figure 3.15a shows the situation obtained for $\mu = 3.6$ and $a_1 = 0.55, a_2 = 0.7$. The stable set of $E_S$ forms the boundary of the basin of $E_1$. On the one hand, the effect of such a small asymmetry in the adjustment speeds on the local stability properties is negligible. The eigenvalues of the two fixed points are exactly the same and are very close to the eigenvalues obtained for identical adjustment speeds with the same value of $\mu$ and with, for example, $a = (a_1 + a_2)/2$. On the other hand, as far as the global dynamics is concerned, non-identical adjustment speeds have a strong effect on the structure of the basins of attraction of the Nash equilibria $E_1$ and $E_2$. Our numerical simulations show that in general the Nash equi-

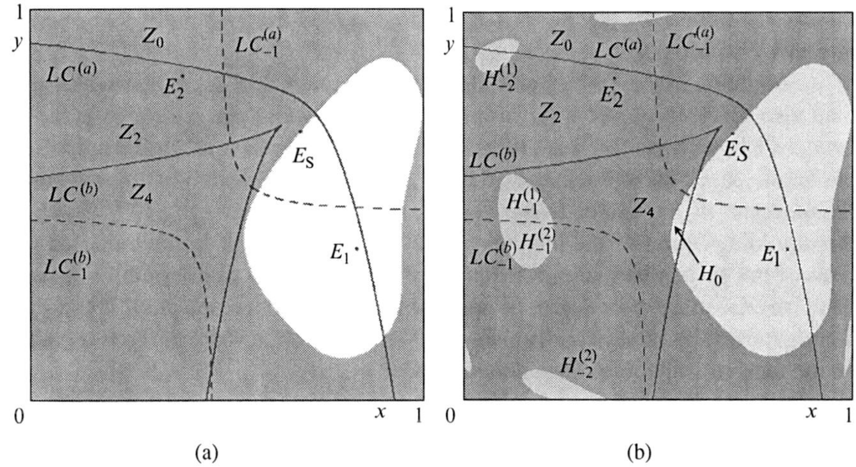


**Fig. 3.15** Linear inverse demand function and cost externalities. The case of duopoly with different speeds of adjustment. (**a**) Here $\mu = 3.6$, $a_1 = 0.55$, $a_2 = 0.7$ - the basin of $E_1$ forms an island inside the basin of $E_2$. (**b**) Here $\mu = 3.6$, $a_1 = 0.59$, $a_2 = 0.7$ - a contact bifurcation has occurred and the basin of $E_2$ becomes a set of disjoint islands inside the basin $E_2$

librium $E_i$ dominates $E_j$ in terms of the size of the basin if $a_i > a_j$. Figure 3.15a shows that although the basin of $E_1$ is a simply connected set, the basin of $E_2$ is now multiply connected. The basin of $E_1$ forms a big "hole" (or "island," to use the term of Mira et al. (1996)) inside the basin of $E_2$. The stable set of $E_S$, that is the boundary which separates the two basins, is entirely included inside the regions $Z_2$ and $Z_0$. Note, however, that the stable set of $E_S$ is close to the critical curve $LC$, which is a signal for the occurrence of a global bifurcation. If a change in parameters causes a contact between the stable set of $E_S$ (a basin boundary) and $LC$, then this contact marks a bifurcation which normally causes a qualitative change in the structure of the basins.

This is demonstrated in Fig. 3.15b, where $\mu = 3.6$ and $a_1 = 0.59$, $a_2 = 0.7$. Such a small change in the adjustment speed of player 2 causes a portion of the basin of $E_1$ to enter the region $Z_4$ (denoted by $H_0$ in the figure). Consequently, new rank-1 preimages of that portion will appear near $LC_{-1}^{(b)}$, and such preimages must belong to the basin of $E_1$. These rank-1 preimages, denoted by $H_{-1}^{(1)}$ and $H_{-1}^{(2)}$, are located at opposite sides with respect to $LC_{-1}^{(b)}$ and merge onto it. Obviously, the set $H_{-1} = H_{-1}^{(1)} \cup H_{-1}^{(2)}$ constitutes a disconnected portion of the basin of $E_1$. Moreover, since $H_{-1}$ belongs to the region $Z_4$, it also has four rank-1 preimages. Two of them are located in the strategy space $S$ and are denoted by $H_{-2}^{(j)}$, $j = 1, 2$. Points belonging to these "islands" are mapped into $H_0$ in two iterations of the map $T$. Indeed, infinitely many higher rank preimages of $H_0$ exist, even if only some of them are inside the strategy space $S = [0, 1] \times [0, 1]$, thus giving smaller disjoint "islands" of the basin of $E_1$. Hence, at the contact between the stable set of $E_S$ and the critical curve $LC$, the basin of $E_1$ is transformed from a simply connected set into a disconnected set.

In summary, in the case of non-identical adjustment speeds, parameter changes may also result in global bifurcations. Such bifurcations are related to a contact between a basin boundary and critical curves and change the qualitative structure of the basins. Since the whole basin of $E_1$ is given by the union of the infinitely many preimages of its immediate basin $\mathcal{B}_0(E_1)$, that is $\mathcal{B}(E_1) = \bigcup_{k \geq 0} T^{-k}(\mathcal{B}_0(E_1))$, the unfolding action of the inverses of the map $T$ can result in disconnected portions of the basin which are quite far away from the Nash equilibrium. In a sense, this gives rise to a higher degree of uncertainty with respect to the possibility of predicting the effects of any small change in the initial market share of the competitors on the long-run outcome of the duopoly game.

# Chapter 4
# Modified and Extended Oligopolies

The previous chapters have introduced and analyzed the classical Cournot model under a number of assumptions. In this chapter we discuss some important modifications and extensions. We first introduce market share attraction games where the dynamics are driven by a generalization of the gradient adjustment process introduced in Chaps. 1 and 2. We carry out both a local and global analysis of the stability of these games. In Sect. 4.2 we consider labor-managed oligopolies with best response dynamics. We give a detailed discussion of the local stability in the discrete time case and via an example show the type of global dynamical behavior that is possible in this model type. The section concludes with a brief discussion of the local stability of a continuous time version of the labor-managed oligopoly. In Sect. 4.3 we introduce intertemporal demand interaction effects, brought about for example by habit formation, into dynamic oligopolies with best response dynamics. We give a local and global stability analysis of the model in discrete time. For the continuous time version we study the local stability of the dynamics, including also the case when there are information lags. In Sect. 4.4 we analyze oligopolies with production adjustment costs. For the case of best reply dynamics in discrete time we give local stability conditions. In the final section we consider oligopolies where there is partial cooperation amongst the firms of the industry. We show various properties of the best response function, give local stability for best reply dynamics in continuous time, and analyze the global dynamics of a particular example under discrete time best response dynamics.

## 4.1 Market Share Attraction Games

Market share attraction models have been used in a variety of contexts to describe the behavior of competitors in a market. Not only have they been employed frequently in empirical applications, they are also prevalent in the economics, game theory and operations research literature. In the marketing literature, market share attraction models are often used to describe the competition between several brands of a product in the market (see for example, Hanssens et al. (1990) and Cooper and Nakanishi (1988)). The models are then sometimes referred to as brand competition

G.I. Bischi et al., *Nonlinear Oligopolies*, DOI 10.1007/978-3-642-02106-0_4, 

models. A typical model of this type specifies that the market share of a competitor is equal to the attraction of its product, divided by the total attraction of all the competitors' products in the market. Each competitor's attraction is given in terms of its competitive effort allocations. To provide an example, let us consider the case of two competitors, who compete against each other in the market on the basis of marketing efforts expended. If $x_1 > 0$ denotes the marketing effort of competitor 1 and $x_2 > 0$ the marketing effort of competitor 2, then $\alpha_1 x_1^{\beta_1}$ and $\alpha_2 x_2^{\beta_2}$ represent the attractions of customers to the products of competitors 1 and 2, respectively. The positive parameters $\alpha_1$ and $\alpha_2$ in this context denote the relative effectiveness of efforts and the parameters $\beta_1 > 0$ and $\beta_2 > 0$ are the elasticities of the products' attractions with respect to the marketing efforts. The competitors' market shares are then given by

$$s_1 = \frac{\alpha_1 x_1^{\beta_1}}{\alpha_1 x_1^{\beta_1} + \alpha_2 x_2^{\beta_2}}, \qquad s_2 = \frac{\alpha_2 x_2^{\beta_2}}{\alpha_1 x_1^{\beta_1} + \alpha_2 x_2^{\beta_2}}. \tag{4.1}$$

Such a specification has the theoretically appealing property that it is logically consistent in the sense that it yields market shares that are between zero and one, and the market shares sum to one across all the competitors in the market. If $A > 0$ denotes the sales potential of the market (in monetary terms) and $c_i$ the marginal cost of effort of firm $i$, then the one-period profits of firm 1 and 2 are

$$\varphi_1 = As_1 - c_1 x_1, \qquad \varphi_2 = As_2 - c_2 x_2. \tag{4.2}$$

The reader should notice that by introducing the new decision variables $z_1 = \alpha_1 x_1^{\beta_1}$, $z_2 = \alpha_2 x_2^{\beta_2}$ and cost functions $C_1(z_1) = c_1 \left(\frac{z_1}{\alpha_1}\right)^{1/\beta_1}$ and $C_2(z_2) = c_2 \left(\frac{z_2}{\alpha_2}\right)^{1/\beta_2}$, the market share attraction game is identical to an oligopoly game with isoelastic market demand function which we have discussed in the previous chapter. Therefore, the results obtained there are valid for market share attraction games as well.

Recall that $\beta_1$ and $\beta_2$ are the elasticities of the products' attractions with respect to the marketing efforts. Hence, we typically have $\beta_i \in (0, 1)$, or $1/\beta_i > 1$, so the functions $C_1$ and $C_2$ are strictly convex. Consequently, in applications a unique Nash equilibrium is obtained. In the general case, a closed-form solution for the Nash equilibrium cannot be given. However, for the symmetric case, that is for identical elasticities $\beta_1 = \beta_2 = \beta$, identical marginal costs of effort $c_1 = c_2 = c$, and identical effectiveness parameters $\alpha_1 = \alpha_2$, the Nash equilibrium can be easily calculated. It is characterized by identical efforts of the two competitors,

$$E^* = \left(\frac{A\beta}{4c}, \frac{A\beta}{4c}\right). \tag{4.3}$$

In the existing literature, market share attraction models are predominantly used in a *static* framework. In this literature, similarly to that on oligopolistic competition, the emphasis of the investigation lies on demonstrating the existence and uniqueness of Nash equilibria (see Friedman (1958), Mills (1961) and Schmalensee (1976)) and on studying the properties of these Nash equilibria (see Monahan (1987) and Karnani (1985)). Only a few papers have addressed the problem of local stability of these equilibria (Schmalensee (1976) and Balch (1971)), but issues concerning the global dynamics of these types of models have been completely disregarded. This is quite surprising, since Schmalensee (1976) remarked: "Ideally, analysis of the dynamic behavior of a model of this sort away from equilibrium can perform two services. First, if it turns out that additional parameter restrictions are needed to ensure global stability, more comparative static information may be obtained. Second, such analysis can provide a further test of the model's plausibility, since systems that go to equilibrium only if they begin life in a neighborhood thereof are unattractive." (p. 502). One reason for a lack of understanding of the global properties of dynamic models is apparently that there has been a lack of appropriate methods to carry out such an analysis. In this subsection we will introduce a dynamic version of a market share attraction model. We will assume that competitive effort allocations for the two brands are adaptively adjusted over time, and characterize the global properties of this model. Our main concern here is to provide a rigorous description of the set of initial effort allocations which leads to convergence to the Nash equilibrium, and the changes of this set if parameters of the model are varied.

A dynamic version of a market share attraction model can, for example, be obtained on the basis of marginal profits. We assume that at time $t$ the marketing efforts of the next period, $x_1(t+1)$ and $x_2(t+1)$, are determined according to the adjustment process

$$\begin{aligned} x_1(t+1) &= x_1(t) + \lambda_1(x_1(t)) \left[ \frac{\partial \varphi_1(x_1(t), x_2(t))}{\partial x_1} \right], \\ x_2(t+1) &= x_2(t) + \lambda_2(x_2(t)) \left[ \frac{\partial \varphi_2(x_1(t), x_2(t))}{\partial x_2} \right]. \end{aligned} \tag{4.4}$$

Notice that this dynamic process is a generalization of the gradient adjustment process, since in this case the constant speeds of adjustment of each firm are replaced by speeds of adjustment dependent on the marketing effort of the particular firm. In Sect. 1.3.3 a similar model was examined.

The expressions $\lambda_i(\cdot)$ determine by how much efforts can vary from period to period and they can be interpreted as the "speeds of reaction." Obviously, the steady states of the dynamical system (4.4) are given as solutions of the equations

$$\lambda_1(x_1)\frac{\partial \varphi_1}{\partial x_1} = 0, \qquad \lambda_2(x_2)\frac{\partial \varphi_2}{\partial x_2} = 0.$$

Any interior Nash equilibrium of the underlying market share attraction game is obtained as the positive solution of the first order conditions $\partial \varphi_1/\partial x_1 = 0$,

$\partial\varphi_2/\partial x_2 = 0$ assuming that the second order conditions are satisfied. Note, however, that fixed points which are not Nash equilibria may exist. Furthermore, it should be mentioned that the functional form of the speeds of reaction $\lambda_i(\cdot)$ are inconsequential for the computation of the Nash equilibrium.

To keep our analysis simple, we will assume that $\lambda_1(x_1) = v_1 x_1$ and $\lambda_2(x_2) = v_2 x_2$. In economic terms, the dynamical system then incorporates the idea that the *relative* change in marketing efforts is proportional to the marginal profits, where the positive parameters $v_1$ and $v_2$ are the proportionality factors. Using the expressions for the market shares $s_1$ and $s_2$ given in (4.1) and the profits in (4.2), the resulting dynamic market share attraction model (4.4) can be written as

$$T : \begin{cases} x_1(t+1) = (1 - v_1 c_1) x_1(t) + v_1 \beta_1 A k \dfrac{x_1(t)^{\beta_1} x_2(t)^{\beta_2}}{\left(x_1(t)^{\beta_1} + k x_2(t)^{\beta_2}\right)^2}, \\ x_2(t+1) = (1 - v_2 c_2) x_2(t) + v_2 \beta_2 A k \dfrac{x_1(t)^{\beta_1} x_2(t)^{\beta_2}}{\left(x_1(t)^{\beta_1} + k x_2(t)^{\beta_2}\right)^2}, \end{cases} \tag{4.5}$$

where $k = \alpha_2/\alpha_1$. The two-dimensional map

$$T : (x_1(t), x_2(t)) \to (x_1(t+1), x_2(t+1))$$

generates the sequences of marketing efforts resulting from the decisions of the two competitors. The corresponding market shares are then obtained via (4.1).

### 4.1.1 Local Stability

Although the Jacobian matrix for our dynamical system can be easily derived, the fact that the Nash equilibrium for the general case cannot be given in closed-form makes a standard stability analysis intractable. Here we have to rely on numerical methods. However, for the symmetric case, where $\beta_1 = \beta_2 = \beta$, $c_1 = c_2 = c$, $\alpha_1 = \alpha_2 = \alpha$, and $v_1 = v_2 = v$, an analytic characterization of the local stability properties of the symmetric equilibrium (4.3) is possible. In this case, the Jacobian matrix computed at the Nash equilibrium $E^*$ becomes $(1 - vc)\mathbf{I}$, where $\mathbf{I}$ is the identity matrix. Therefore, in the symmetric case the unique Nash equilibrium (4.3) is locally asymptotically stable if $0 < vc < 2$ (see Bischi and Kopel (2003*b*), for more details).

### 4.1.2 The Feasible Set and Global Stability

We now turn to the question as to whether the Nash equilibrium is globally stable and if so, under which conditions. Obviously, it only makes sense to consider situations where both firms expend positive efforts. That is, mathematically the map

(4.5) is defined only for positive values of the dynamic variables $x_1$ and $x_2$. Consequently, the first question that arises is, under which conditions does the sequences of efforts remain positive? Given initial efforts $x_1(0), x_2(0)$, we will say that a trajectory is *feasible* if $(x_1(t), x_2(t)) = T^t(x_1(0), x_2(0))$, $t = 0, 1, 2, \ldots$ is entirely contained in the positive orthant $\mathbb{R}^2_+ = \{(x_1, x_2) | x_1 > 0 \text{ and } x_2 > 0\}$. The *feasible set* is the subset of $\mathbb{R}^2_+$ whose points generate feasible trajectories.

For our dynamical system it is obvious that under the conditions $v_i c_i < 1$ for $i = 1, 2$, it follows that if the efforts in period $t$ are positive, then the efforts in the subsequent period are positive as well. That is, if $(x_1(t), x_2(t)) \in \mathbb{R}^2_+$ then $(x_1(t+1), x_2(t+1)) \in \mathbb{R}^2_+$. Furthermore, it is easy to check that any feasible trajectory of our dynamical system is bounded (Bischi and Kopel (2003*b*)). Hence, the conditions $v_1 c_1 < 1$ and $v_2 c_2 < 1$ are sufficient for the feasibility and boundedness of all points in $R^2_+$. It turns out that these conditions are also necessary for the feasibility of the whole region $\mathbb{R}^2_+$. If at least one of these two inequalities does not hold, then points of $\mathbb{R}^2_+$ exist that generate infeasible trajectories. In order to see this, note first that the coordinate axes are invariant: $x_i(t) = 0$ implies $x_i(t+1) = 0$. The dynamics along the invariant $x_i$-axis is governed by the one-dimensional linear map

$$x_i(t+1) = (1 - v_i c_i) x_i(t) . \tag{4.6}$$

For example, if $v_1 c_1 < 1$, then given a point $(x_1, 0)$, with $x_1 > 0$, the map (4.6) generates a sequence of points on the $x_1$-axis with $x_1 > 0$. By continuity, the same holds for points $(x_1, x_2)$ with arbitrarily small $x_2$. Hence, in this case the feasible region includes the $x_1$-axis. Instead, if $v_1 c_1 > 1$, then a point $(x_1, 0)$, with $x_1 > 0$, generates a negative point after the first iteration of (4.6). In this case the whole $x_1$-axis must belong to the set of infeasible points. Clearly, the same reasoning applies to the $x_2$-axis.

In order to obtain an exact delineation of the boundary of the feasible region, we consider the invariant coordinate axes and their preimages. The map $T$ is a noninvertible map. If we consider a generic point $(0, x_2')$, $x_2' > 0$, on the $x_2$-axis, then its preimages are the positive solutions of the system

$$(1 - v_1 c_1) x_1 \left( x_1^{\beta_1} + k x_2^{\beta_2} \right)^2 + v_1 \beta_1 A k x_1^{\beta_1} x_2^{\beta_2} = 0,$$

$$((1 - v_2 c_2) x_2 - x_2') \left( x_1^{\beta_1} + k x_2^{\beta_2} \right)^2 + v_2 \beta_2 A k x_1^{\beta_1} x_2^{\beta_2} = 0,$$

obtained from (4.5) with $x_i(t) = x_i$ as unknowns and $x_1(t+1) = 0, x_2(t+1) = x_2'$ taken as parameters. If $v_2 c_2 < 1$, then one solution always exists on the $x_2$-axis. It is given by $x_1 = 0$, $x_2 = \frac{1}{1 - v_2 c_2} x_2'$. Solutions with $x_1 > 0$ cannot exist if $v_1 c_1 < 1$, because in this case the first equation can never be satisfied. On the other hand, if $v_1 c_1 > 1$, two preimages with $x_1 > 0$ exist. They are located on the curves with equation

$$x_2 = \left\{ \frac{x_1^{\beta_1-1}}{2k(v_1c_1-1)} \left[ v_1\beta_1 A - 2(v_1c_1-1)x_1 \pm \sqrt{v_1^2\beta_1^2A^2 - 4v_1\beta_1 A(v_1c_1-1)x_1} \right] \right\}^{\frac{1}{\beta_2}} . \tag{4.7}$$

The same arguments, applied to the preimages of a generic point of the $x_1$-axis $(x_1', 0)$, can be used to prove that points of the invariant $x_1$-axis have preimages in the positive quadrant $\mathbb{R}_+^2$ only if $v_1c_1 > 1$. Such preimages are located on the curves with equation

$$x_1 = \left\{ \frac{kx_2^{\beta_2-1}}{2(v_2c_2-1)} \left[ v_2\beta_2 A - 2(v_2c_2-1)x_2 \pm \sqrt{v_2^2\beta_2^2A^2 - 4v_2\beta_2 A(v_2c_2-1)x_2} \right] \right\}^{\frac{1}{\beta_1}} . \tag{4.8}$$

These results on the preimages of the invariant axes are of crucial importance to determine the boundaries of the feasible region, as we will now demonstrate. Let us first look at the *symmetric* case, where all parameters of the competitors are identical. Recall that the Nash equilibrium $E^*$ given in (4.3) is locally asymptotically stable if $0 < vc < 2$. From the arguments above, we know that $0 < vc < 1$ is sufficient and necessary for any trajectory to be feasible and bounded, hence it is also a necessary condition for the global stability of $E^*$. Indeed, we numerically see that whenever $0 < vc < 1$, the basin of attraction of $E^*$ is given by the whole positive quadrant $\mathbb{R}_+^2$, so that the Nash equilibrium is globally stable. On the other hand, the results given above also show that this is no longer true if $1 < vc < 2$. In this case the basin $B(E^*)$ is a proper subset of the positive quadrant $\mathbb{R}_+^2$, and this subset is bounded by the preimages of the coordinate axes. Figure 4.1 illustrates the situation for $vc = 1.05 > 1$. The white region represents the basin of attraction of the Nash equilibrium $E^*$, and the black region indicates the infeasible set of marketing efforts. As the figure shows, the rank-1 preimages (denoted by $(X_1)_{-1}$ and $(X_2)_{-1}$) of the axes are curves starting at the origin, they are symmetric with respect to the diagonal, and join at the rank one preimage of the origin $O_{-1} = \left(\frac{vmB\beta}{4(vc-1)}, \frac{vmB\beta}{4(vc-1)}\right)$. Thus, for $vc > 1$, the length of the segment $OO_{-1}$ gives a rough idea of the extent of the feasible region. If $vc$ is decreased below 1, then $E^*$ becomes globally stable. For $vc = 1$ a global bifurcation occurs which causes the feasible set to be bounded. If $vc$ is further increased, with the other parameters held constant, the feasible set shrinks. If $vc$ is increased beyond the value $vc = 2$, then the Nash equilibrium loses its stability and becomes repelling, and we numerically see that the generic trajectory then becomes infeasible.

To conclude this subsection, we now briefly turn to the question of the robustness of the results derived for the case of identical competitors. That is, we are trying to see if the qualitative descriptions given above are still valid if we assume that the parameters which characterize the two competitors and their effort decisions are different. It turns out that the answer is yes. Also in this case, if $v_ic_i < 1$ for $i = 1, 2$, the feasible region coincides with the whole positive quadrant $\mathbb{R}_+^2$, because no preimages of the coordinate axes exist inside $\mathbb{R}_+^2$. Our numerical simulations show that the Nash equilibrium in this case is globally asymptotically stable. Every combination of initial marketing efforts in $\mathbb{R}_+^2$ generates a sequence of efforts which converges to

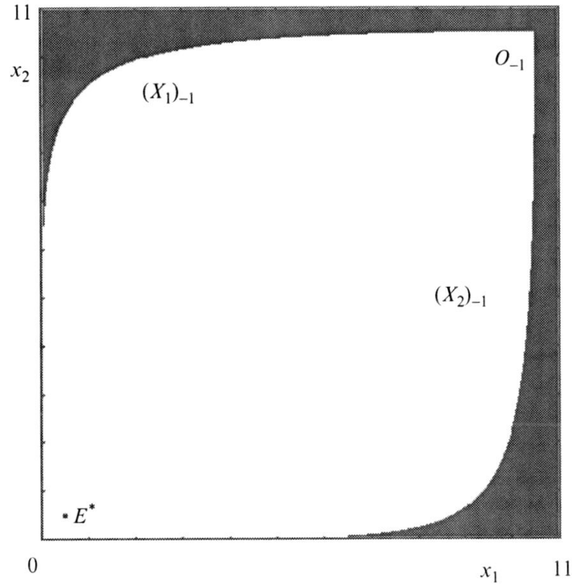


**Fig. 4.1** The market share attraction game – the symmetric case. The *white region* represents the basin of attraction of the stable Nash equilibrium $E^*$, the *black region* represents the infeasible set. This figure is obtained with parameters $A = 20$, $k = 1$, $v_1 = v_2 = 0.35$, $\beta_1 = \beta_2 = 0.3$, $c_1 = c_2 = 3$

the equilibrium of the market share game. Like in the symmetric case, a wide range of parameter values exist such that the Nash equilibrium is stable. If one (or both) of the above inequalities is (are) reversed, then the Nash equilibrium only attracts points of the feasible set, which no longer covers the whole area $\mathbb{R}^2_+$. The boundary of the feasible set can then be again determined using (4.8) and (4.7). This can be seen as follows. If $v_1c_1 < 1$ and $v_2c_2 > 1$, then the feasible set is an *unbounded region* (extending for arbitrarily large $x_1$) with the upper boundary formed by the rank-1 preimage of the $x_1$-axis, say $(X_1)_{-1}$ (see Fig. 4.2a). The equation of this preimage is given by (4.8) with the "+" sign. The curve $(X_1)_{-1}$ is tangent to the $x_1$-axis at the origin. Analogously, if $v_1c_1 > 1$ and $v_2c_2 < 1$, then the feasible set is an unbounded region (extending for arbitrarily large $x_2$) with the right boundary formed by the rank-1 preimage of the $x_2$-axis, say $(X_2)_{-1}$, whose equation is given by (4.7) with the "+" sign.

If both inequalities are reversed, so that $v_1c_1 > 1$ and $v_2c_2 > 1$, then the feasible set is a *bounded region*, whose boundary is formed by the curves $(X_1)_{-1}$ and $(X_2)_{-1}$, starting at the origin $O$ tangent to the axes and intersecting at the preimage of the origin $O_{-1}$ (see Fig. 4.1). Hence, the conditions $v_ic_i = 1$ and $v_jc_j > 1$, $i \neq j$, denote the occurrence of a global bifurcation, at which the feasible region is changed from unbounded to bounded. It should be noticed that other bifurcations that change the topological structure of the boundaries of the feasible region

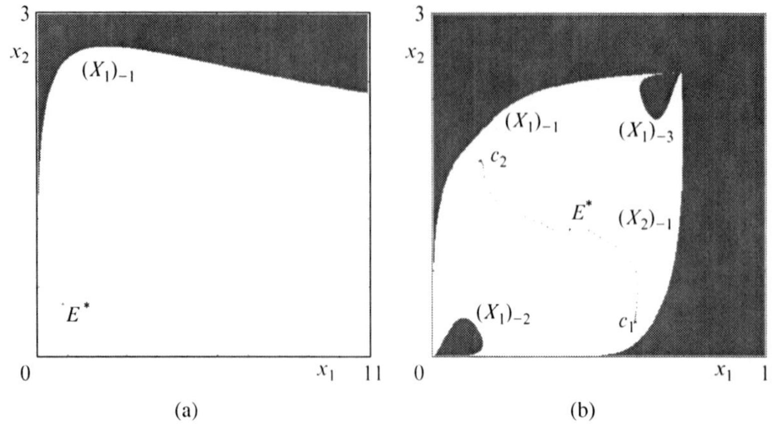


**Fig. 4.2** The market share attraction game – the asymmetric case. (*a*) The parameters are $A = 20$, $k = 1.2, v_1 = v_2 = 0.3, \beta_1 = 0.5, \beta_2 = 0.3, c_1 = 3, c_2 = 4$ so that $v_1c_1 < 1$ and $v_2c_2 > 1$. The basin of attraction of the Nash equilibrium $E^*$ is unbounded along the $x_1$-direction, since $v_1c_1 < 1$. (*b*) With $v_1 = 0.75, v_2 = 0.91, c_1 = 3, c_2 = 2$ so that now $v_1c_1 > 1$ and $v_2c_2 > 1$. The Nash equilibrium $E^*$ is unstable and a stable cycle of period two attracts the trajectories that start in the *white region*: one of these trajectories, starting from an initial condition close to the Nash equilibrium is represented by a sequence of *dots*

may occur. This is due to the fact that higher order preimages of the coordinate axes appear inside $\mathbb{R}^2_+$. In fact, in the situation depicted in Fig. 4.2b, a preimage of rank-$k$ of a coordinate axis bounds a region of the phase space whose points are infeasible, since points in this set are mapped into points with a negative coordinate after $k$ iterations. Two such regions are shown and they have the shape of small lobes starting from $O$ and $O_{-1}$. They are bounded by preimages of rank-2 and rank-3 of the $x_1$-axis, say $(X_1)_{-2}$ and $(X_1)_{-3}$.

The Nash equilibrium loses stability as one or both of the expressions $v_ic_i$ are increased even further. In contrast to the symmetric case, more complex bounded attractors (such as periodic cycles) may exist around the unstable Nash equilibrium. Hence, in the asymmetric case the long-run dynamics may be characterized by bounded periodic (or even aperiodic) oscillations around the Nash equilibrium. However, the occurrence of such local bifurcations, at which new bounded attracting sets appear inside the feasible region, is not related to the global bifurcations that change the shape of the boundaries of the feasible region. Further details on the global dynamics of market share attraction models can be found in Bischi and Kopel (2003*b*) and Bischi et al. (2000a).

## 4.2 Labor-Managed Oligopolies

Suppose that $N$ firms produce a single good, or offer identical services and the payoff function of each firm is the surplus per labor unit of the firm. If $f$ denotes the price function, $W$ the competitive wage rate, $d_k$ the fixed cost of firm $k$, and $h_k$

the number of labor units in firm $k$ as a function of its production level $x_k$, then the payoff of firm $k$ is given as

$$\varphi_k(x_1, \dots, x_N) = \frac{x_k f(Q) - W h_k(x_k) - d_k}{h_k(x_k)}, \tag{4.9}$$

with $Q = \sum_{l=1}^{N} x_l$ as before. Here the total production costs are given by

$$C_k(x_k) = W h_k(x_k) + d_k.$$

Notice that no externalities are included in this model.

The existence of the static Nash equilibrium has been proved by Okuguchi (1996) under realistic conditions. This result has been also discussed in detail in Okuguchi and Szidarovszky (1999).

In this section we assume that the price function $f$ and the functions $h_k$, for all $k$, are twice continuously differentiable. Furthermore, we assume that

(A) $f'(Q) < 0$,

(B) $x_k f''(Q) + f'(Q) < 0$,

(C) $h_k'(x_k) > 0$, and $h_k''(x_k) \geq 0$,

for all $k$ and all feasible values of $x_k$ and $Q$.

Condition (C) states that the functions $h_k$ are convex and increasing, which means that for additional outputs increasingly more labor units are required.

Consider an interior equilibrium. In its neighborhood the best response of firm $k$ is the solution of the single variable equation

$$\frac{\partial \varphi_k}{\partial x_k} = \frac{[f(x_k + Q_k) + x_k f'(x_k + Q_k)]h_k(x_k) - [x_k f(x_k + Q_k) - d_k]h_k'(x_k)}{h_k(x_k)^2} = 0,$$

which can be written as

$$[f(x_k + Q_k) + x_k f'(x_k + Q_k)]h_k(x_k) - [x_k f(x_k + Q_k) - d_k]h_k'(x_k) = 0. \tag{4.10}$$

Notice that the derivative of the left hand side with respect to $x_k$ is given by

$$(2f' + x_k f'')h_k - (x_k f - d_k)h_k''.$$

Under assumptions (A) and (B), the first term is negative. If we make the natural assumption that at the equilibrium the firms have non-negative payoffs, then the second term is non-positive, so the derivative of the left hand side of (4.10) is negative. Therefore in the neighborhood of the equilibrium the best response function is unique. By implicitly differentiating (4.10) with respect to $Q_k$ and noting that

$x_k = R_k(Q_k)$ we have

$$\begin{aligned}&\left[f'(1+R_k')+R_k'f'+x_kf''(1+R_k')\right]h_k+\left[f+x_kf'\right]h_k'R_k'\\&-\left[R_k'f+x_kf'(1+R_k')\right]h_k'-\left[x_kf-d_k\right]h_k''R_k'=0,\end{aligned}$$

implying that

$$R_k' = \frac{-(f'+x_kf'')h_k + x_kf'h_k'}{(2f'+x_kf'')h_k - (x_kf-d_k)h_k''}. \tag{4.11}$$

Notice that the denominator coincides with the derivative of the left hand side of (4.10). As we have just shown above, this expression is negative. The first term of the numerator is positive and the second term is negative. Therefore $R_k'$ does not have a definite sign, however it is easy to see that $R_k' > -1$ always holds.

*Example 4.1.* Consider linear price and labor functions, $f(Q) = A - BQ$, and $h_k(x_k) = q_k + p_kx_k$ with all coefficients being positive. In this case $f' = -B$, $f'' = 0$, $h_k' = p_k$ and $h_k'' = 0$ so that

$$R_k' = \frac{B(q_k+p_kx_k)+x_k(-B)p_k}{-2B(q_k+p_kx_k)} = -\frac{q_k}{2(q_k+p_kx_k)},$$

which lies between $-\frac{1}{2}$ and 0. Therefore the $r_k = R_k'(\bar{Q}_k)$ values satisfy the conditions that hold in the concave oligopoly case, so the asymptotic properties of this model are the same as those discussed for concave oligopolies in Chap. 2.

In the general case however the $R_k'(\bar{Q}_k)$ values can be positive. Contrary to the case of isoelastic price functions there is the possibility that more than one firm has positive $r_k$ values.

Labor-managed oligopolies were introduced and first discussed by Ward (1958). Hill and Waterson (1983) investigated profit maximizing and labor-managed models with identical cost functions. The non-symmetric case was examined by Neary (1984). The works of Okuguchi (1993) and Okuguchi (1996) contain the most general existence results.

### 4.2.1 Discrete Time Models and Local Stability

The dynamic models with discrete time scales have exactly the same general forms as the best response dynamics with adaptive expectations (1.28)–(1.29) and the partial adjustment towards the best response (1.30) in the case of concave oligopolies. Therefore the eigenvalue equation is also the same as given in (2.24), which we repeat here for the sake of convenience:

$$\prod_{j=1}^{s}\left(1-a_j(1+r_j)-\lambda\right)^{m_j}\left[1+\sum_{j=1}^{s}\frac{\theta_j}{1-a_j(1+r_j)-\lambda}\right]=0. \tag{4.12}$$

Here we assume that $a_k = \alpha_k'(0) > 0$ for all firms, the different $a_k(1 + r_k)$ values are

$$a_1(1 + r_1) > a_2(1 + r_2) > \cdots > a_s(1 + r_s),$$

and these values are repeated $m_1, m_2, \ldots, m_s$ times, respectively, among the $N$ firms. The value of $\theta_j$ is the sum of all products $r_k a_k$ such that $a_k(1+r_k) = a_j(1+r_j)$. If $\theta_j \neq 0$ and $m_j = 1$, then $1 - a_j(1 + r_j)$ is not an eigenvalue of the Jacobian, and if $\theta_j = 0$ or $m_j \geq 2$, then $1 - a_j(1 + r_j)$ is an eigenvalue. Since $r_j > -1$, these eigenvalues are inside the unit circle, if $1 - a_j(1 + r_j) > -1$, that is, when

$$a_j(1 + r_j) < 2.$$

Let $g(\lambda)$ denote the bracketed factor in (4.12). It is easy to see that

$$\lim_{\lambda \to \pm\infty} g(\lambda) = 1,$$

$$\lim_{\lambda \to 1 - a_j(1+r_j) \pm 0} g(\lambda) = \begin{cases} \pm\infty, & \text{if} \quad \theta_j < 0, \\ \mp\infty, & \text{if} \quad \theta_j > 0. \end{cases}$$

Since the derivative of $g$ has no definite sign, no monotonocity property of $g$ can be established. Notice in addition, that all poles $1 - a_j(1 + r_j)$ are less than 1. The possible presence of complex conjugate roots makes stability analysis intractable in the general case. In such cases computational methods can be used to find the roots and check stability conditions. However, if there is at most one sign change in sequence $\theta_1, \theta_2, \ldots, \theta_s$ and it is from "−" to "+", then we always have only real eigenvalues and we can derive simple stability conditions.

Case 1. All $\theta_j > 0$. The graph of $g(\lambda)$ is as shown in Fig. 4.3. Clearly all roots are real, and all are between $-1$ and $+1$ if all poles are larger than $-1$ and $g(1) > 0$.

Case 2. All $\theta_j < 0$. Then the graph of $g(\lambda)$ is as illustrated in Fig. 4.4. All roots are real and they are between $-1$ and $+1$ if all poles are larger than $-1$ and $g(-1) > 0$.

Case 3. There is a sign change in the sequence $\theta_1, \theta_2, \ldots, \theta_s$. The corresponding graph of $g$ is shown in Fig. 4.5. We have again $s$ real roots and they are between $-1$ and $+1$, if all poles are larger than $-1$ and both $g(-1)$ and $g(1)$ are positive.

We note that conditions $g(-1) > 0$ and $g(1) > 0$ can be rewritten as (2.22) and

$$\sum_{k=1}^{N} \frac{r_k}{1 + r_k} < 1,$$

respectively.

*Example 4.2.* Consider again the linear case examined in the previous example and assume that $h_k(0) = 0$ for all $k$, that is, $q_k = 0$. Then $R_k' = 0$ for all $k$, so $R_k$

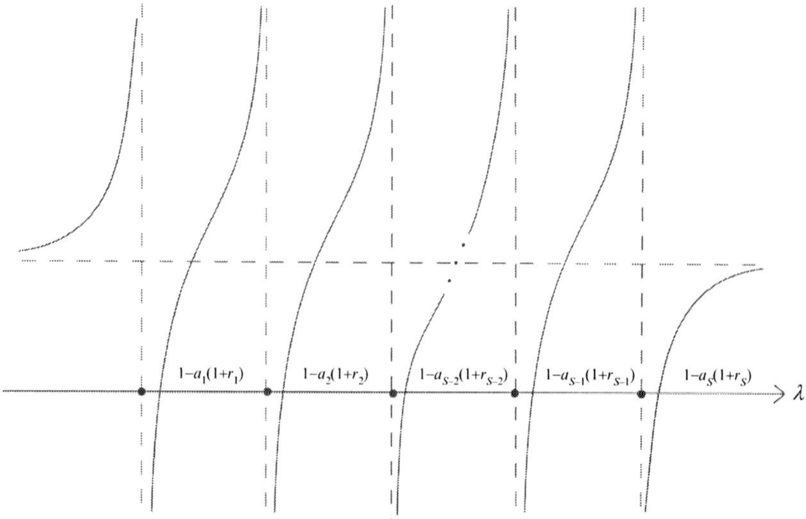


**Fig. 4.3** The discrete time model of labor-managed oligopolies. The determination of the eigenvalues that are roots of the graph of $g(\lambda)$, plotted here for $\theta_j > 0$ for all $j$

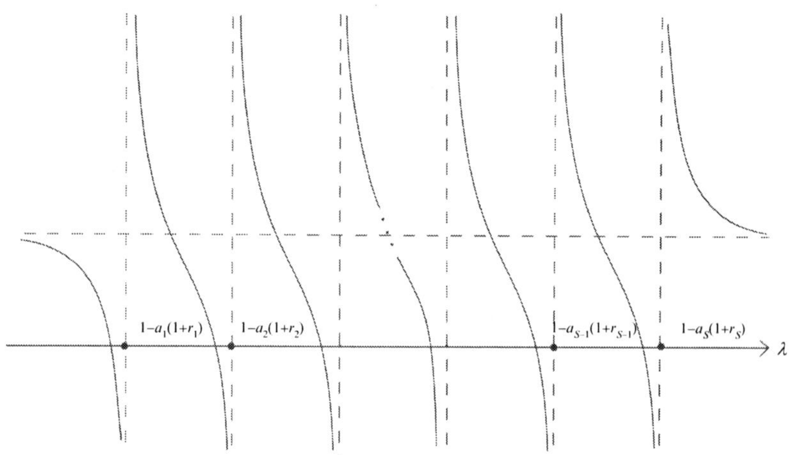


**Fig. 4.4** The discrete time model of labor-managed oligopolies. The determination of the eigenvalues that are roots of the graph of $g(\lambda)$, plotted here for $\theta_j < 0$ for all $j$

has to be constant. This property also follows directly from the special form of the payoff function

$$\varphi_k(x_1, \ldots, x_N) = \frac{x_k(A - Bx_k - BQ_k) - d_k}{p_k x_k} - W \tag{4.13}$$

$$= -\frac{d_k}{p_k x_k} + \left( \frac{A - Bx_k - BQ_k}{p_k} - W \right), \tag{4.14}$$

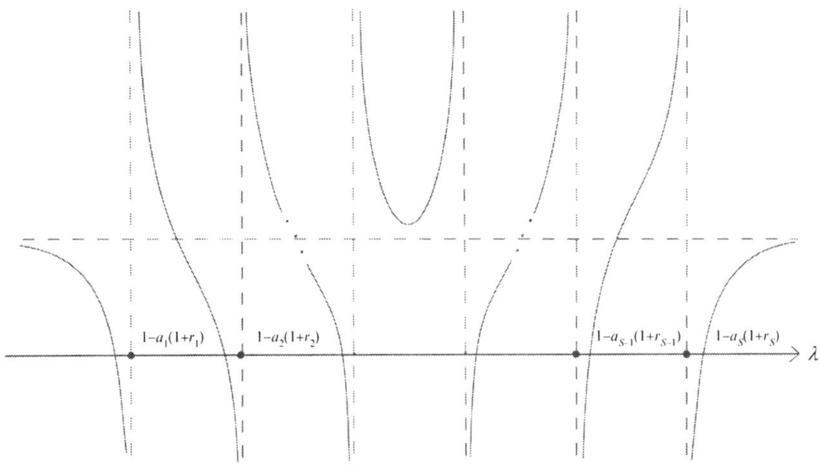


**Fig. 4.5** The discrete time model of labor-managed oligopolies. The determination of the eigenvalues that are roots of the graph of $g(\lambda)$, plotted here with a sign change in the $\theta_j$ values

which is clearly strictly concave. The first order condition

$$\frac{d_k}{p_k x_k^2} - \frac{B}{p_k} = 0$$

implies that

$$x_k = \sqrt{\frac{d_k}{B}},$$

and this is independent of the output selections of the competitors. Hence there is a unique equilibrium with $\bar{x}_k = \sqrt{d_k/B}$ for all $k$, and since $R_k' \equiv 0$, the matrix (2.20) becomes diagonal with diagonal elements $1 - a_1, \ldots, 1 - a_N$, which are the nonzero eigenvalues of the Jacobian. So the equilibrium is locally asymptotically stable if $a_j < 2$ for all $j$ and is unstable if for at least one $j$, $a_j > 2$.

*Example 4.3.* Let us modify the payoff function (4.13) of Example 4.2 by assuming an isoelastic price function and that the labor-independent cost is a linear function of the output $x_k$. In this case we have

$$f(Q) = \frac{A}{Q}, \quad h_k(x_k) = p_k x_k, \quad C_k(x_k) = d_k + c_k x_k,$$

so that

$$\begin{aligned}\varphi_k(x_1, \ldots, x_N) &= \frac{x_k \cdot \dfrac{A}{x_k + Q_k} - (d_k + c_k x_k)}{p_k x_k} - W \\ &= \frac{A}{p_k(x_k + Q_k)} - W - \frac{c_k}{p_k} - \frac{d_k}{p_k x_k}. \end{aligned} \tag{4.15}$$

Assuming an interior optimum in $x_k$, the first order condition is

$$-\frac{A}{p_k(x_k+Q_k)^2}+\frac{d_k}{p_k x_k^2}=0,$$

implying that

$$R_k(Q_k)=\frac{\sqrt{d_k}}{\sqrt{A}-\sqrt{d_k}}Q_k. \tag{4.16}$$

In order to ensure that $x_k>0$, we have to assume that $d_k<A$ for all $k$. Simple differentiation shows that the second order conditions are satisfied at the best response. Next we will show the existence of infinitely many equilibria under realistic conditions. From (4.16) and noting that $x_k=R_k(Q_k)$ we find that

$$Q=Q_k+x_k=\left(1+\frac{\sqrt{A}-\sqrt{d_k}}{\sqrt{d_k}}\right)x_k=x_k\frac{\sqrt{A}}{\sqrt{d_k}},$$

implying that

$$1=\frac{\sum_{k=1}^{N}x_k}{Q}=\frac{\sum_{k=1}^{N}\sqrt{d_k}}{\sqrt{A}}.$$

The payoff of firm $k$ at any equilibrium is

$$\frac{A}{p_k\bar{Q}}-W-\frac{c_k}{p_k}-\frac{d_k}{p_k\bar{x}_k}=\frac{A}{p_k\bar{Q}}\left(1-\frac{\sqrt{d_k}}{\sqrt{A}}\right)-W-\frac{c_k}{p_k},$$

which is positive for all $k$ if $\bar{Q}$ is sufficiently small, in particular if $\bar{Q}$ satisfies

$$\bar{Q}<\min_k\left\{\frac{\sqrt{A}(\sqrt{A}-\sqrt{d_k})}{p_k W+c_k}\right\}.$$

Hence we have shown that if $d_k<A$ for all $k$, then positive equilibria exist if and only if

$$\sum_{k=1}^{N}\sqrt{d_k}=\sqrt{A}. \tag{4.17}$$

In this case there are infinitely many equilibria, and the set of equilibria are all points on the ray

$$\bar{x}_k=\frac{\sqrt{d_k}}{\sqrt{A}}\bar{Q} \tag{4.18}$$

for any $\bar{Q}>0$. In addition, if $\bar{Q}$ is sufficiently small, then the profits of all firms are positive at the equilibrium.

From (4.16) we know that the derivatives of the best response functions are

$$r_k = R'_k(Q_k) = \frac{\sqrt{d_k}}{\sqrt{A} - \sqrt{d_k}}, \tag{4.19}$$

and simple substitution of $\lambda = 1$ into the eigenvalue equation (4.12) of the Jacobian shows that it is always an eigenvalue.[1] Therefore we cannot establish local asymptotic stability in the usual sense. This result clearly should be the case, since a small move away from any given equilibrium along the ray (4.18) would result in another equilibrium and since the state remains there for all future time, the trajectory of the state variable does not converge back to the original equilibrium.

*Example 4.4.* We will consider now a special $N$-firm labor-managed oligopoly. Assume that the firms have identical capacity limits, $L$, and the price function is $f(Q) = LN - Q$. Notice that the price is always non-negative. We also assume that the number of labor units is a quadratic function for each firm, $h_k(x_k) = p_k x_k^2$. Then the profit (4.9) per labor unit of firm $k$ is given by

$$\frac{x_k(LN - x_k - Q_k) - Wp_k x_k^2 - d_k}{p_k x_k^2} = \frac{LN - Q_k}{p_k x_k} - \frac{d_k}{p_k x_k^2} - (\frac{1}{p_k} + W).$$

Notice that the value of $p_k$ has no effect on the best response of firm $k$, it only affects the optimal profit. The derivative of this profit function can be written as

$$-\frac{LN - Q_k}{p_k x_k^2} + \frac{2d_k}{p_k x_k^3} = \frac{1}{p_k x_k^3}(2d_k - x_k(LN - Q_k)),$$

implying that the profit function is increasing for $x_k < 2d_k/(LN - Q_k)$ and decreasing if $x_k > 2d_k/(LN - Q_k)$. So the stationary point is

$$z_k^* = \frac{2d_k}{LN - Q_k}.$$

Since $z_k^*$ is necessarily positive, the best response of firm $k$ is

$$R_k(Q_k) = \begin{cases} z_k^*, & \text{if } z_k^* \leq L, \\ L, & \text{if } z_k^* > L, \end{cases}$$

which is illustrated in Fig. 4.6. If $LN - 2d_k/L \leq 0$, then $R_k(Q_k) = L$ for all $Q_k$.

We can also show that $0 < r_k < 1$ if the best response is interior. In this case

$$r_k = R'_k(Q_k) = \frac{2d_k}{(LN - Q_k)^2} = \frac{z_k^{*2}}{2d_k} > 0.$$

[1] Recall the definition of $\theta_j$ above (2.24), and make use of (4.17) and (4.19).

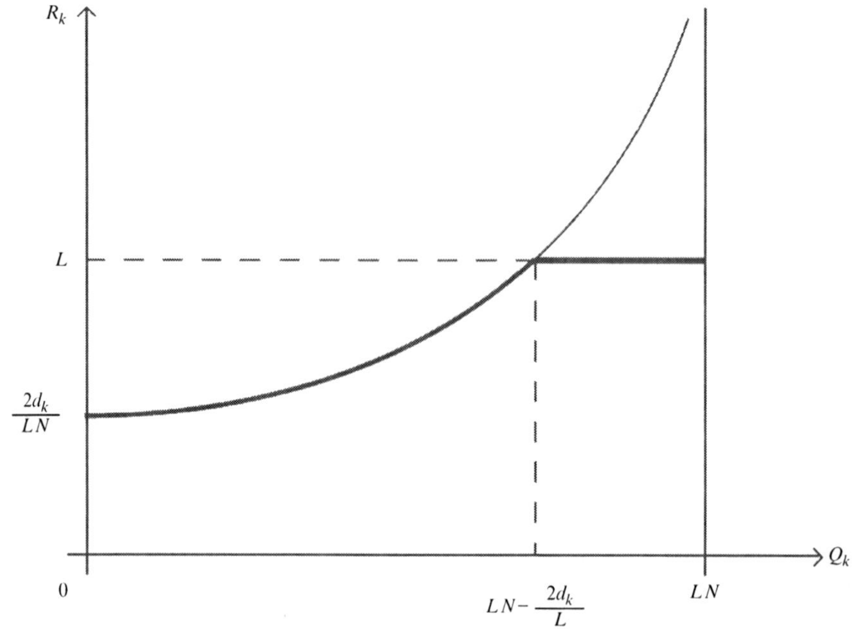


**Fig. 4.6** Example 4.4; the discrete time model of an $N$-firm labor-managed oligopoly. Linear price function and quadratic labor unit functions. Best response of firm $k$

From the first order condition we have

$$-z_k^*(LN - Q + z_k^*) + 2d_k = 0,$$

so that

$$z_k^{*2} = 2d_k - z_k^*(LN - Q),$$

implying that

$$r_k = \frac{2d_k - z_k^*(LN - Q)}{2d_k} < 1.$$

Hence Case 1 (shown in Fig. 4.3) occurs with all $\theta_j > 0$, so the equilibrium is locally asymptotically stable if for all $k$,

$$a_k(1 + r_k) < 2$$

and

$$\sum_{k=1}^{N} \frac{r_k}{1 + r_k} < 1.$$

Since

$$\frac{r_k}{1 + r_k} = \frac{1}{1 + \frac{1}{r_k}} < \frac{1}{2},$$

the local asymptotic stability of the equilibrium is guaranteed for $N = 2$ and $a_k \le 1$ for $k = 1, 2$. Since the best response functions are continuous, the $N$-dimensional best response mapping

$$\boldsymbol{R}(x_1, \ldots, x_N) = \left( R_1\left(\sum_{l \neq 1} x_l\right), \ldots, R_N\left(\sum_{l \neq N} x_l\right)\right)$$

maps the convex compact set $X_{k=1}^{N}[0, L_k]$ into itself, the Brouwer fixed point theorem guarantees the existence of at least one equilibrium. The uniqueness of the equilibrium however cannot be guaranteed as the following example shows.

*Example 4.5.* In Example 4.4 select $N = 3$, $L = 4$, $d_k = 8.5$ $(k = 1, 2, 3)$. We can easily show that both $\bar{x}_k^{(1)} = 3 - \sqrt{0.5}$ and $\bar{x}_k^{(2)} = 3 + \sqrt{0.5}$ are symmetric equilibria by verifying that both satisfy the best response relations. Clearly $0 \le \bar{x}_k^{(i)} \le L$ for all $k$ and $i$, furthermore

$$\begin{aligned} R_k(\bar{Q}_k^{(i)}) &= \frac{2d_k}{LN - \bar{Q}_k^{(i)}} = \frac{2d_k}{LN - 2\bar{x}_k^{(i)}} = \frac{17}{12 - 2(3 \mp \sqrt{0.5})} \\ &= \frac{8.5}{3 \pm \sqrt{0.5}} = \frac{8.5}{3 \pm \sqrt{0.5}} \cdot \frac{3 \mp \sqrt{0.5}}{3 \mp \sqrt{0.5}} = \frac{8.5(3 \mp \sqrt{0.5})}{9 - 0.5} \\ &= 3 \mp \sqrt{0.5} = \bar{x}_k^{(i)}. \end{aligned}$$

Okuguchi and Szidarovszky (1999) discussed the discrete time dynamic model of the linear case given in Example 4.1. The general nonlinear case with local stability analysis has been examined in Li and Szidarovszky (1999*b*). The equilibrium analysis of Example 4.3 has been presented in Li et al. (2003) but no stability analysis was given in the discrete time case.

### 4.2.2 Discrete Time Models and Global Dynamics

The global dynamic behavior in discrete-time labor-managed oligopolies is first illustrated with the following example.

*Example 4.6.* Consider again the situation of Example 4.4. We now consider a semi-symmetric oligopoly, that is, firms $k$, with $k \ge 2$, have identical fixed costs, $d_k = d_2$ $(k \ge 2)$, identical constant speeds of adjustment $a_k = a_2$ $(k \ge 2)$, as well as identical initial outputs. Then their entire trajectories are identical. In this case as before $Q_1 = (N-1)x_2$ and $Q_2 = x_1 + (N-2)x_2$. If we assume that the capacity limits of all firms are identical and equal to $L$, and the firms use partial adjustment towards the best response to update their quantity selections, then the adjustment process is represented by the two-dimensional dynamical system

$$x_1(t+1) = (1-a_1)x_1(t) + a_1 R_1((N-1)x_2(t)),$$
$$x_2(t+1) = (1-a_2)x_2(t) + a_2 R_2(x_1(t) + (N-2)x_2(t)),$$

with $0 < a_1, a_2 \leq 1$. From the expression of the best response $R_k(Q_k)$ of firm $k$ in Example 4.4, we obtain

$$x_1(t+1) = (1-a_1)x_1(t) + a_1\left[\min\left\{\frac{2d_1}{LN-(N-1)x_2}, L\right\}\right],$$

$$x_2(t+1) = (1-a_2)x_2(t) + a_2\left[\min\left\{\frac{2d_2}{LN-x_1(t)-(N-2)x_2(t)}, L\right\}\right].$$

The presence of capacity constraints makes the resulting dynamical system piecewise differentiable. The phase space $\mathbb{D} = [0, L]\times[0, L]$ can be divided into different subregions, denoted by $\mathbb{D}^{(i)}$ in Fig. 4.7, inside which the dynamical system is differentiable. These regions are separated by lines (or borders) of non-differentiability $b_1$ and $b_2$, where $b_1$ and $b_2$ are represented by the equations $x_2 = \frac{NL^2-2d_1}{(N-1)L}$ and $x_2 = -\frac{1}{N-2}x_1 + \frac{NL^2-2d_2}{(N-2)L}$ respectively. Of course, some of these subregions may be empty depending on the values of the model parameters. This subdivision is important for the computation of the equilibria. In fact, interior equilibria are located inside region $\mathbb{D}^{(1)}$, where the dynamical system assumes the form

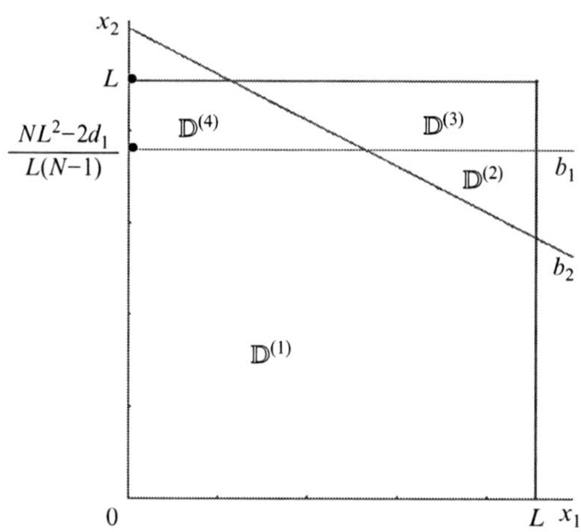


**Fig. 4.7** Example 4.6; the discrete time model of an $N$-firm labor-managed oligopoly in the semi-symmetric case. Linear price function and quadratic labor unit functions. The phase space structure in the plane of outputs

$$T|_{\mathbb{D}^{(1)}} : \begin{cases} x_1(t+1) = (1-a_1)x_1(t) + \dfrac{2a_1 d_1}{LN-(N-1)x_2}, \\ x_2(t+1) = (1-a_2)x_2(t) + \dfrac{2a_2 d_2}{LN - x_1(t) - (N-2)x_2(t)}, \end{cases}$$

and the equilibrium outputs are the solutions of the algebraic system

$$x_1 = \frac{2d_1}{LN-(N-1)x_2},$$
$$x_2 = \frac{2d_2}{LN - x_1(t) - (N-2)x_2(t)},$$

provided that these solutions are inside the region $\mathbb{D}^{(1)}$. The local asymptotic stability of any equilibrium point inside region $\mathbb{D}^{(1)}$ is determined by the study of the eigenvalues of the Jacobian matrix

$$J^{(1)} = \begin{pmatrix} 1-a_1 & \dfrac{2a_1 d_1 (N-1)}{(LN-(N-1)x_2)^2} \\ \dfrac{2a_2 d_2}{(LN-x_1-(N-2)x_2)^2} & 1-a_2+\dfrac{2a_2 d_2(N-2)}{(LN-x_1-(N-2)x_2)^2} \end{pmatrix}.$$

computed at the equilibrium. However, boundary equilibria can also exist, located in regions $\mathbb{D}^{(i)}$, $i = 2, 3, 4$. For example, in the region $\mathbb{D}^{(2)}$, where the map assumes the form

$$T|_{\mathbb{D}^{(2)}} : \begin{cases} x_1(t+1) = (1-a_1)x_1(t) + \dfrac{2a_1 d_1}{LN-(N-1)x_2}, \\ x_2(t+1) = (1-a_2)x_2(t) + a_2 L, \end{cases}$$

we can have a boundary equilibrium with coordinates $E = (\overline{x}_1, \overline{x}_2) = (2d_1/L, L) \in \mathbb{D}^{(2)}$, provided that $\overline{x}_2 < \frac{NL^2-2d_1}{(N-1)L}$ and $\overline{x}_2 > -\frac{1}{N-2}\overline{x}_1 + \frac{NL^2-2d_2}{(N-2)L}$. This holds if $\sqrt{2d_1} < L < \sqrt{d_1+d_2}$. It is clear that if the equilibrium $E$ exists, that is if the above inequalities are satisfied, then it is locally asymptotically stable, because inside the region $\mathbb{D}^{(2)}$ the Jacobian matrix is triangular with eigenvalues $(1-a_1)$ and $(1-a_2)$. Similar arguments apply to the region $\mathbb{D}^{(3)}$, where the map assumes the form

$$T|_{\mathbb{D}^{(3)}} : \begin{cases} x_1(t+1) = (1-a_1)x_1(t) + a_1 L, \\ x_2(t+1) = (1-a_2)x_2(t) + a_2 L, \end{cases}$$

and a boundary equilibrium of coordinates $E' = \left(\bar{x}_1', \bar{x}_2'\right) = (L, L) \in \mathbb{D}^{(3)}$ exists provided that $\bar{x}_2' > \frac{NL^2-2d_1}{(N-1)L}$ and $\bar{x}_2' > -\frac{1}{N-2}\bar{x}_1' + \frac{NL^2-2d_2}{(N-2)L}$, which imply that $L < \min\left(\sqrt{2d_1}, \sqrt{2d_2}\right)$. Whenever these inequalities are satisfied, then the boundary point $(L, L)$ is a locally asymptotically stable equilibrium, since the Jacobian

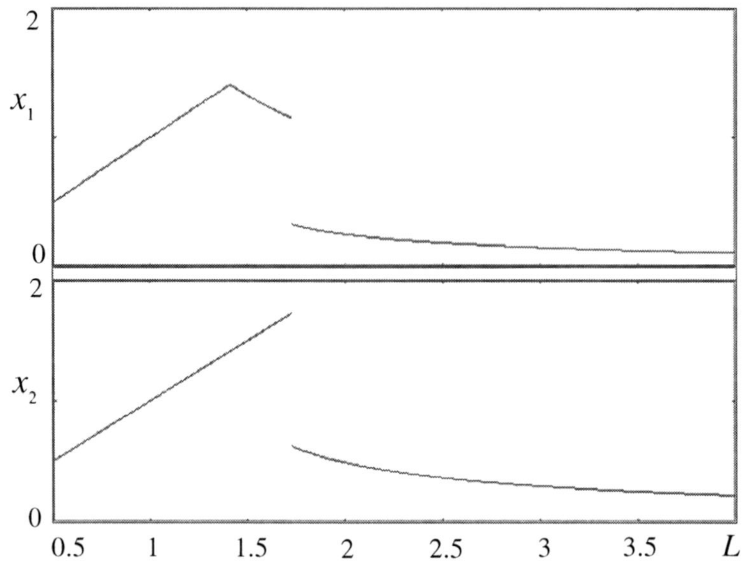


**Fig. 4.8** Example 4.6; the discrete time model of an $N$-firm labor-managed oligopoly in the semi-symmetric case. Linear price function and quadratic labor unit functions. Bifurcation diagrams of outputs with respect to the common capacity constraint $L$

matrix in the region $\mathbb{D}^{(3)}$ is diagonal with eigenvalues $(1 - a_1)$ and $(1 - a_2)$. These results stress the important role of the capacity constraints in the global dynamic properties of the discrete-time labor-managed oligopoly model. This can be also clearly seen by considering the bifurcation diagram in Fig. 4.8, obtained with the numerical values $N = 5$, $d_1 = 1$, $d_2 = 2$, $a_1 = 0.9$, $a_2 = 0.8$ and capacity constraint $L$ taken as a bifurcation parameter varying in the range $[0.5, 4]$. As long as $L < \sqrt{2d_1} = \sqrt{2}$, the only global equilibrium is $E' = (L, L)$. Numerically, this equilibrium appears to be globally asymptotically stable (see also Fig. 4.9(a) obtained for $L = 1$). For $\sqrt{2d_1} < L < \sqrt{d_1 + d_2}$, i.e., $\sqrt{2} < L < \sqrt{3}$ in our numerical example, the unique equilibrium is $E = (2d_1/L, L)$, and also in this case it numerically appears to be globally asymptotically stable (see for example Fig. 4.9(b) obtained for $L = 1.5$).

However, the existence of boundary equilibria does not exclude the coexistence of interior equilibria. Indeed, in our numerical example, as the capacity limit $L$ is further increased a saddle node bifurcation leads to the creation of two interior equilibria in the region $\mathbb{D}^{(1)}$. They are denoted by $E_1$ and $E_2$ in Fig. 4.10a, obtained with $L = 1.55$, where $E_1$ is a stable node and $E_2$ is a saddle point. As long as the inequality $\sqrt{2} < L < \sqrt{3}$ holds, the stable node $E_1$ coexists with the stable boundary equilibrium $E$, each with its own basin of attraction. In Fig. 4.10a the white portion of the strategy space $\mathbb{D} = [0, L] \times [0, L]$ represents the basin of the interior equilibrium $E_1$ and the grey shaded region represents the basin of the locally asymptotically stable boundary equilibrium $E$. The boundary that separates these

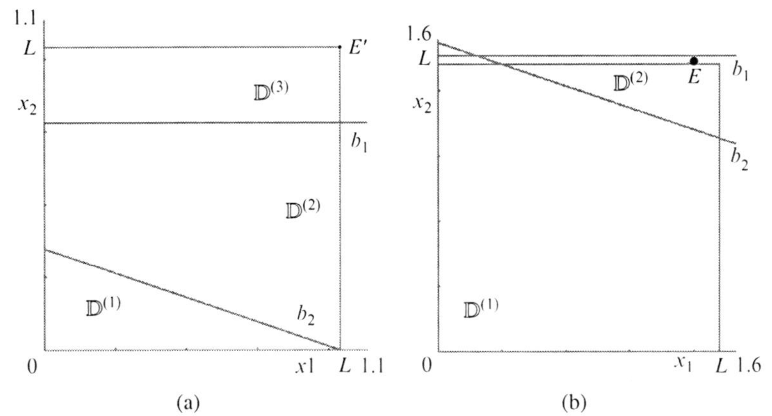


**Fig. 4.9** Example 4.6; the discrete time model of an $N$-firm labor-managed oligopoly in the semi-symmetric case. Linear price function and quadratic labor unit functions. (**a**) The phase space for $L = 1$. (**b**) The phase space for $L = 1.5$

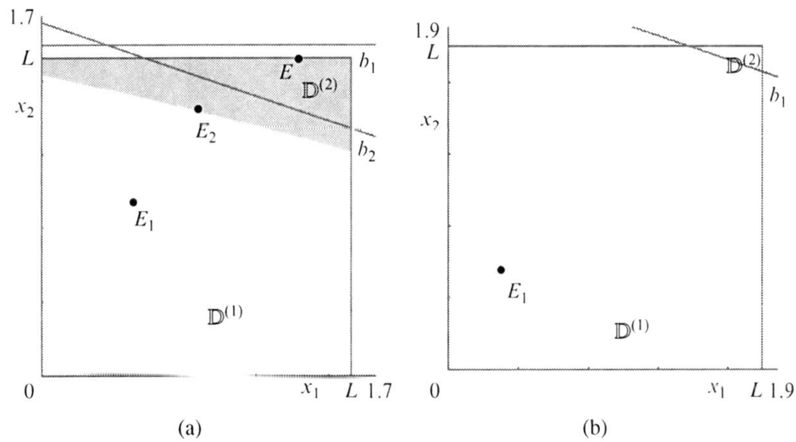


**Fig. 4.10** Example 4.6; the discrete time model of an $N$-firm labor-managed oligopoly in the semi-symmetric case. Linear price function and quadratic labor unit functions. (**a**) Co-existence of interior equilibria and their basins of attraction for $L = 1.55$, (**b**) At $L = 1.8$ the second co-existing equilibrium $E_2$ has disappeared and $E_1$ becomes the unique and globally stable attractor

two basins is given by the stable set of the saddle point $E_2$. A further increase of $L$ above the bifurcation value $\sqrt{d_1 + d_2} = \sqrt{3}$ marks a remarkable qualitative change in the global dynamic scenario: the boundary equilibrium $E$ and the saddle point $E_2$ disappear (they exit the respective regions $\mathbb{D}^{(1)}$ and $\mathbb{D}^{(2)}$ after merging along the boundary) and the interior equilibrium $E_1$ remains the unique (and globally stable) attractor (see Fig. 4.10b obtained for $L = 1.8$). This explains the sudden jump occurring at $L = \sqrt{3}$ in the bifurcation diagram of Fig. 4.8.

### 4.2.3 Continuous Time Models

The dynamic model with continuous time scales has the same form (1.31) as the one introduced in Sect. 1.2 and applied in Chaps. 2 and 3 for concave and isoelastic oligopolies. Therefore the Jacobian and its eigenvalue equation are also the same as the one given in (2.48), which we repeat here for the sake of convenience:

$$\prod_{j=1}^{s} \left(-a_j(1+r_j)-\lambda\right)^{m_j} \left[1-\sum_{j=1}^{s} \frac{\theta_j}{a_j(1+r_j)+\lambda}\right] = 0, \tag{4.20}$$

where $a_1(1+r_1) > a_2(1+r_2) > \cdots > a_s(1+r_s)$ are the different $a_k(1+r_k)$ values and $\theta_j$ is the sum of all $r_k a_k$ values such that $a_k(1+r_k) = a_j(1+r_j)$. We also assume that $a_k > 0$ for all firms. Since in general $r_k$ does not have a definite sign, the same holds for $\theta_j$. If $\theta_j = 0$ or $m_j \geq 2$, then $-a_j(1+r_j)$ is an eigenvalue of the Jacobian. Notice that they are all negative, since $r_k > -1$. All other eigenvalues are the solutions of the equation

$$1-\sum_{j=1}^{s} \frac{\theta_j}{a_j(1+r_j)+\lambda} = 0.$$

If $g(\lambda)$ denotes again the left hand side of the above equation, then similarly to the discrete time case we have

$$\lim_{\lambda\to\pm\infty} g(\lambda) = 1,$$

$$\lim_{\lambda\to -a_j(1+r_j)\pm 0} g(\lambda) = \begin{cases} \pm\infty, & \text{if} \quad \theta_j < 0, \\ \mp\infty, & \text{if} \quad \theta_j > 0. \end{cases}$$

Since $g'(\lambda)$ has no definite sign, no monotonicity property of $g$ can be established. Notice that all poles are negative. For the sake of mathematical simplicity assume again that there is at most one sign change in the sequence $\theta_1, \theta_2, \ldots, \theta_s$, and it is from "−" to "+". Under this condition we have the same three possibilities as in the discrete case (see Figs. 4.3–4.5), and the graph of function $g$ is the same as in the discrete case with the only difference being that all poles are now negative.

Case 1. All $\theta_j > 0$. Local asymptotic stability occurs if $g(0) > 0$.
Case 2. All $\theta_j < 0$. Then the equilibrium is always locally asymptotically stable.
Case 3. There is a sign change in the sequence $\theta_1, \theta_2, \ldots, \theta_s$. Local asymptotic stability occurs if $g(0) > 0$.

Notice that the condition $g(0) > 0$ can be rewritten as

$$\sum_{k=1}^{N} \frac{r_k}{1+r_k} < 1,$$

which does not depend on the adjustment scheme, it depends only on the derivatives of the best response functions.

*Example 4.7.* In the special case introduced earlier in Example 4.1 we have seen that $-\frac{1}{2} < R'_k < 0$, so $\theta_j < 0$ for all $j$. Therefore the equilibrium is always locally asymptotically stable.

*Example 4.8.* Consider next the case of Example 4.3, where we have infinitely many equilibria. By using the special form of $r_k$ given in (4.19) it can easily be proved that zero is always an eigenvalue, so we cannot establish local asymptotic stability of the equilibrium. Clearly this should be the case, since if the initial state is selected close to any given equilibrium on the ray (4.18), then the state will remain there for all future times and will not converge back to the original equilibrium. However Li et al. (2003) have proved that the ray (4.18) is a strongly attracting set, meaning that any point near the ray is attracted (that is, the trajectory starting at this point converges) to some particular point on the ray. The basin of attraction contains a cone which is centered at the ray. In order to prove this result the theory of differentiable manifolds was used (see for example, Hirsch et al. (1977)), a topic the discussion of which would take us beyond the scope of this book.

Models with continuously distributed time lags can be discussed in the same way as was done in Chap. 2. The only difference being that there is no sign restriction on the $r_k$ values.

*Example 4.9.* Consider again the symmetric case described by characteristic equation (2.58). The case $-1 < r < 0$ has been examined in Sect. 2.6. In the case of $r = 0$, the eigenvalues are $-a$ and $-(p/T)$, both of which are negative implying the local asymptotic stability of the equilibrium. That leaves us to consider the case $r > 0$.
If $T = 0$ then (2.58) becomes

$$\lambda + a - (N-1)ar = 0$$

with solution

$$\lambda = ((N-1)r - 1)\, a,$$

which is negative if $r < 1/(N-1)$ implying the local asymptotic stability of the equilibrium. If $r > 1/(N-1)$ then the equilibrium is unstable.
If $T > 0$ and $m = 0$, then (2.58) becomes the quadratic equation

$$\lambda^2 T + \lambda(1 + aT) + a(1 - (N-1)r) = 0.$$

Both roots have negative real parts if all coefficients are positive, which occurs if $1 - (N - 1)r > 0$. That is, if $r < 1/(N - 1)$, then the equilibrium is locally asymptotically stable, and if $r > 1/(N - 1)$, then it is unstable.

In the case when $m = 1$, the cubic equation (2.59) is obtained. All coefficients are positive if $r < 1/(N - 1)$, and the Routh–Hurwitz criterion shows that the eigenvalues have negative real parts if (2.61) is satisfied. Since all coefficients on the left hand side are positive, this inequality always holds implying the local asymptotic stability of the equilibrium. If $r > 1/(N - 1)$, then the equilibrium is unstable.

The special case of Example 4.1 has been examined in Okuguchi and Szidarovszky (1999) with continuous time scales. The further special case of Example 4.3 with infinitely many equilibria was investigated by Li et al. (2003), in which the theory of differentiable manifolds was used to prove that the equilibrium ray is a strongly attracting set.

## 4.3 Oligopolies with Intertemporal Demand Interaction

In this section we consider an $N$ firm oligopoly without externalities, but with intertemporal demand interaction. As in the earlier chapters, let $f$ denote the market price function and $C_k$ the cost function of firm $k$ $(1 \leq k \leq N)$. Intertemporal demand interaction is often a realistic assumption, since previous consumption might saturate the market, or might contribute to taste and habit formation for the consumers, to mention only some of the most common phenomena.

Okuguchi and Szidarovszky (2003), Szidarovszky and Zhao (2006) and Chiarella and Szidarovszky (2008b) introduced and analyzed various dynamic models that extend the classical oligopoly models to include intertemporal demand interaction. The special case of market saturation was examined by Szidarovszky et al. (2006).

Consider first discrete time scales, and let $S(t)$ represent the cumulative effect of the earlier consumptions up to time period $t$. If for example, market saturation is considered, then after each time period a certain proportion of goods already in use by the consumers remains in usable condition, while the rest has to be replaced. It is assumed that variable $S(t)$ follows the dynamic rule

$$S(t+1) = \beta_S S(t) + \sum_{k=1}^{N} x_k(t+1), \tag{4.21}$$

where $0 \leq \beta_S < 1$ is a given constant. This constant represents how past experience with the product affects current demand, and in the case of market saturation it shows the fraction of goods remaining in usable condition after each time period.

If we assume that the price depends on the current value of the variable $S$, then the profit of firm $k$ at time period $t + 1$ can be written as

$$x_k f\left(x_k + Q_k(t+1) + \beta_S S(t)\right) - C_k(x_k) \tag{4.22}$$

where $Q_k = \sum_{l \neq k} x_l$ as before. If $R_k\big(Q_k(t+1)\big)$ denotes the best response function of firm $k$ without intertemporal demand interaction, then $R_k\big(Q_k(t+1) + \beta_S S(t)\big)$ is the best response when it is taken into account. At time period $t+1$, when firm $k$ makes its decision on its production level, the simultaneous decisions of the competitors are not known, so instead of the true output $Q_k(t+1)$ of the rest of the industry, firm $k$ uses some expectation $Q_k^E(t+1)$ of this value. If we assume best reply dynamics with the adaptive expectations scheme (1.18), then the resulting dynamical system becomes

$$x_k(t+1) = R_k\left(Q_k^E(t) + \alpha_k\left(\sum_{l\neq k} x_l(t) - Q_k^E(t)\right) + \beta_S S(t)\right), \quad (1 \le k \le N) \tag{4.23}$$

$$Q_k^E(t+1) = Q_k^E(t) + \alpha_k\left(\sum_{l\neq k} x_l(t) - Q_k^E(t)\right), \quad (1 \le k \le N) \tag{4.24}$$

$$S(t+1) = \beta_S S(t) + \sum_{k=1}^{N} R_k\left(Q_k^E(t) + \alpha_k\left(\sum_{l\neq k} x_l(t) - Q_k^E(t)\right) + \beta_S S(t)\right), \tag{4.25}$$

where $\alpha_k$ is a sign-preserving function for all $k$.

Clearly $(\bar{x}_1, \ldots, \bar{x}_N, \bar{Q}_1^E, \ldots, \bar{Q}_N^E, \bar{S})$ is a steady state of the dynamical system (4.23)–(4.25) if and only if for all $k$,

$$\bar{Q}_k^E = \sum_{l\neq k} \bar{x}_l, \tag{4.26}$$

$$\bar{x}_k = R_k(\bar{Q}_k^E + \beta_S \bar{S}), \tag{4.27}$$

and

$$(1-\beta)\bar{S} = \sum_{k=1}^{N} \bar{x}_k. \tag{4.28}$$

Assume next continuous time scales. If we rewrite the discrete equation (4.21) as

$$S(t+1) - S(t) = \sum_{k=1}^{N} x_k(t+1) - (1-\beta_S)S(t),$$

we see that $S(t)$ in the continuous case is driven by the differential equation

$$\dot{S} = \sum_{k=1}^{N} x_k - \gamma_S S, \tag{4.29}$$

where $\gamma_S = 1 - \beta_S > 0$. For the market saturation example this equation can also be interpreted as expressing the fact that during each time period the value of $S(t)$

increases by the new sales, however a certain proportion of the new sales has to be used for replacement of goods which are not in usable condition anymore. Assume that the price decreases if either the total output of the industry or the value of $S$ increases. For the sake of simplicity we will assume that the unit price is a function of a linear combination of these two factors, so it is given as

$$f\left(\sum_{l=1}^{N} x_l + \beta_S^* S\right),$$

with some $\beta_S^* > 0$. Therefore the profit of firm $k$ can be written as

$$x_k f\left(\sum_{l=1}^{N} x_l + \beta_S^* S\right) - C_k(x_k). \tag{4.30}$$

If $R_k(Q_k)$ denotes the best response function of firm $k$ without intertemporal market interaction, then its best response becomes $R_k(Q_k + \beta_S^* S)$ when taking it into account, and so the dynamical system in continuous time becomes

$$\dot{x}_k = \alpha_k \left( R_k\left(\sum_{l \neq k} x_l + \beta_S^* S\right) - x_k \right), \quad (1 \leq k \leq N) \tag{4.31}$$

$$\dot{S} = \sum_{l=1}^{N} x_l - \gamma_S S, \tag{4.32}$$

where $\alpha_k$ is a sign-preserving function for all firms $k$. Clearly $(\bar{x}_1, \ldots, \bar{x}_N, \bar{S})$ is a steady state of this system if and only if

$$\bar{x}_k = R_k\left(\sum_{l \neq k} \bar{x}_l + \beta_S^* \bar{S}\right) \tag{4.33}$$

and

$$\sum_{k=1}^{N} \bar{x}_k = \gamma_S \bar{S}. \tag{4.34}$$

### 4.3.1 Discrete Time Models and Local Stability

The main result of this section on the stability of the discrete time intertemporal demand interaction dynamical system (4.23)–(4.25) is the following:

**Theorem 4.1.** *Assume that for all $k$, $a_k>0$, $-1<r_k \leq 0$ and $(a_k+\beta_S)(1+r_k)<1$, furthermore $\sum_{k=1}^{N} \frac{-r_k}{1-a_k(1+r_k)} \geq 1$ is satisfied. Then the equilibrium of the system (4.23)–(4.25) is locally asymptotically stable if*

$$\sum_{k=1}^{N} \frac{r_k(-a_k - 2\beta_S)}{2 - a_k(1 + r_k)} < \beta_S + 1,$$

*and is unstable if*

$$\sum_{k=1}^{N} \frac{r_k(-a_k - 2\beta_S)}{2 - a_k(1 + r_k)} > \beta_S + 1.$$

*Proof.* In analyzing the local asymptotic stability of system (4.23)–(4.25) we first have to determine its Jacobian matrix evaluated at the equilibrium, which turns out to have the form

$$\begin{pmatrix} \bar{J}_{11} & \bar{J}_{12} & \bar{J}_{13} \\ \bar{J}_{21} & \bar{J}_{22} & \bar{J}_{23} \\ \bar{J}_{31} & \bar{J}_{32} & \bar{J}_{33} \end{pmatrix}, \tag{4.35}$$

with

$$\bar{J}_{11} = \begin{pmatrix} 0 & r_1 a_1 & \dots & r_1 a_1 \\ r_2 a_2 & 0 & \dots & r_2 a_2 \\ \vdots & \vdots & & \vdots \\ r_N a_N & r_N a_N & \dots & 0 \end{pmatrix},$$

$$\bar{J}_{12} = \begin{pmatrix} r_1(1-a_1) & & & 0 \\ & r_2(1-a_2) & & \\ & & \ddots & \\ 0 & & & r_N(1-a_N) \end{pmatrix}, \quad \bar{J}_{13} = \begin{pmatrix} \beta_S r_1 \\ \beta_S r_2 \\ \vdots \\ \beta_S r_N \end{pmatrix},$$

$$\bar{J}_{21} = \begin{pmatrix} 0 & a_1 & \dots & a_1 \\ a_2 & 0 & \dots & a_2 \\ \vdots & \vdots & & \vdots \\ a_N & a_N & \dots & 0 \end{pmatrix}, \quad \bar{J}_{22} = \begin{pmatrix} 1 - a_1 & & & 0 \\ & 1-a_2 & & \\ & & \ddots & \\ 0 & & & 1-a_N \end{pmatrix}, \quad \bar{J}_{23} = \begin{pmatrix} 0 \\ 0 \\ \vdots \\ 0 \end{pmatrix},$$

$$\bar{J}_{31} = \left( \sum_{l \neq 1} r_l a_l, \dots, \sum_{l \neq N} r_l a_l \right),$$

$$\bar{J}_{32} = (r_1(1-a_1), \dots, r_N(1-a_N)), \quad \bar{J}_{33} = \left(1 + \sum_{l=1}^{N} r_l\right) \beta_S,$$

where $r_k = R'_k$ at the equilibrium and $a_k = \alpha'_k(0)$ as before. The eigenvalue equation of the Jacobian with eigenvalue $\lambda$ and eigenvector $(u_1, \ldots, u_N, v_1, \ldots, v_N, w)^\top$ thus has the special form

$$\boldsymbol{RAu} + \boldsymbol{RDv} + \boldsymbol{Rb}w = \lambda \boldsymbol{u}, \tag{4.36}$$

$$\boldsymbol{Au} + \boldsymbol{Dv} = \lambda \boldsymbol{v}, \tag{4.37}$$

$$\boldsymbol{1}^T \boldsymbol{RAu} + \boldsymbol{1}^T \boldsymbol{RDv} + (\beta_S + \boldsymbol{1}^T \boldsymbol{Rb})w = \lambda w, \tag{4.38}$$

where

$$\boldsymbol{R} = \begin{pmatrix} r_1 & & & 0 \\ & r_2 & & \\ & & \ddots & \\ 0 & & & r_N \end{pmatrix}, \quad \boldsymbol{A} = \begin{pmatrix} 0 & a_1 & \ldots & a_1 \\ a_2 & 0 & \ldots & a_2 \\ \vdots & \vdots & & \vdots \\ a_N & a_N & \ldots & 0 \end{pmatrix}, \quad \boldsymbol{b} = \begin{pmatrix} \beta_S \\ \beta_S \\ \vdots \\ \beta_S \end{pmatrix} = \beta_S \boldsymbol{1},$$

$$\boldsymbol{D} = \begin{pmatrix} 1 - a_1 & & 0 & \\ & 1 - a_2 & & \\ & & \vdots & \\ 0 & & & 1 - a_N \end{pmatrix},$$

$$\boldsymbol{1}^T = (1, 1, \ldots, 1), \quad \boldsymbol{u} = (u_1, \ldots, u_N)^T, \quad \boldsymbol{v} = (v_1, \ldots, v_N)^T.$$

Subtracting the $\boldsymbol{R}$-multiple of (4.37) from (4.36) we get

$$\boldsymbol{Rb}w = \lambda(\boldsymbol{u} - \boldsymbol{Rv}).$$

We can assume that $\lambda \neq 0$, since a zero eigenvalue cannot destroy local asymptotic stability. Then

$$\boldsymbol{v} = -\frac{1}{\lambda}\boldsymbol{b}w + \boldsymbol{R}^{-1}\boldsymbol{u}, \tag{4.39}$$

where we assume that $r_k \neq 0$ for all $k$. Multiply (4.36) by $\boldsymbol{1}^T$ and subtract the resulting equation from (4.38), to obtain

$$\beta_S w = \lambda(w - \boldsymbol{1}^T \boldsymbol{u}).$$

We also assume that $\lambda \neq \beta_S$, since $\beta_S \in [0, 1)$. Then,

$$w = \frac{\lambda}{\lambda - \beta_S} \boldsymbol{1}^T \boldsymbol{u}, \tag{4.40}$$

and from (4.39),

$$\boldsymbol{v} = \left( -\frac{1}{\lambda - \beta_S} \boldsymbol{b} \boldsymbol{1}^T + \boldsymbol{R}^{-1} \right) \boldsymbol{u}. \tag{4.41}$$

Combining (4.41) and (4.37) leads to a single equation for the vector $\boldsymbol{u}$, namely

$$\left[A + (D - \lambda I)\left(-\frac{1}{\lambda - \beta_S} b 1^T + R^{-1}\right)\right] u = \mathbf{0}. \qquad (4.42)$$

If $\boldsymbol{u} = \mathbf{0}$, then from (4.41), $\boldsymbol{v} = \mathbf{0}$, and from (4.40), $w = 0$. Since the eigenvector has to be nonzero, $\boldsymbol{u}$ must differ from zero, so the determinant of the coefficient matrix has to be zero.

We will next rewrite this matrix in the special form (E.2) introduced in Appendix E, so the characteristic polynomial of the system can be obtained in a simple form. A straightforward calculation shows that the coefficient matrix can take the form

$$a 1^T - (I - D) + (D - \lambda I) R^{-1} - \frac{1}{\lambda - \beta_S}(D - \lambda I) b 1^T,$$

where $\boldsymbol{a} = (a_1, \ldots, a_N)^T$. The determinant of this matrix can be factored as

$$det\big((D - \lambda I)R^{-1} - (I - D)\big) \cdot det\big(I + ((D - \lambda I)R^{-1} \\ -(I - D))^{-1}\left(a - \frac{1}{\lambda - \beta_S}(D - \lambda I)b\right) 1^T\big) = 0. \qquad (4.43)$$

The first factor of the last equation is zero if

$$\frac{1 - a_k - \lambda}{r_k} - a_k = 0,$$

which implies that

$$\lambda = 1 - a_k(1 + r_k). \qquad (4.44)$$

The second factor can be simplified by using identity (E.1), and the resulting equation is

$$1 + 1^T((D - \lambda I)R^{-1} - (I - D))^{-1}\left(a - \frac{1}{\lambda - \beta_S}(D - \lambda I)b\right) = 0,$$

that is,

$$1 + \sum_{k=1}^{N} \frac{a_k - \dfrac{\beta_S(1 - a_k - \lambda)}{\lambda - \beta_S}}{\dfrac{1 - a_k - \lambda}{r_k} - a_k} = 0,$$

which can be rewritten as

$$\sum_{k=1}^{N} \frac{r_k[\lambda(a_k + \beta_S) - \beta_S]}{1 - a_k(1 + r_k) - \lambda} = \beta_S - \lambda. \qquad (4.45)$$

Assume now that the price function $f$ and cost functions $C_k$ satisfy the conditions (A)–(C) of concave oligopolies given at the beginning of Sect. 2.1. Then $-1 < r_k \leq 0$ for all $k$. Let $g(\lambda)$ denote the left hand side of (4.45) and assume that all $a_k > 0$ and the $1 - a_k(1 + r_k)$ values are different, otherwise we can add the terms with identical denominators similarly to (2.24). Clearly,

$$\lim_{\lambda \to \pm\infty} g(\lambda) = -\sum_{k=1}^{N} r_k(a_k + \beta_S) \geq 0,$$

and it is positive unless all $r_k = 0$, which case is excluded from discussion. Furthermore

$$\lim_{\lambda \to 1-a_k(1+r_k)\pm 0} g(\lambda) = \pm\infty$$

and

$$g'(\lambda) = \sum_{k=1}^{N} \frac{a_k r_k (1 - (a_k + \beta_S)(1 + r_k))}{(1 - a_k(1 + r_k) - \lambda)^2} < 0$$

by assuming that for all $k$, $(a_k + \beta_S)(1 + r_k) < 1$. The graph of $g(\lambda)$ is shown in Fig. 4.11, and notice that under this assumption all poles of $g$ are positive and below 1. Notice also that

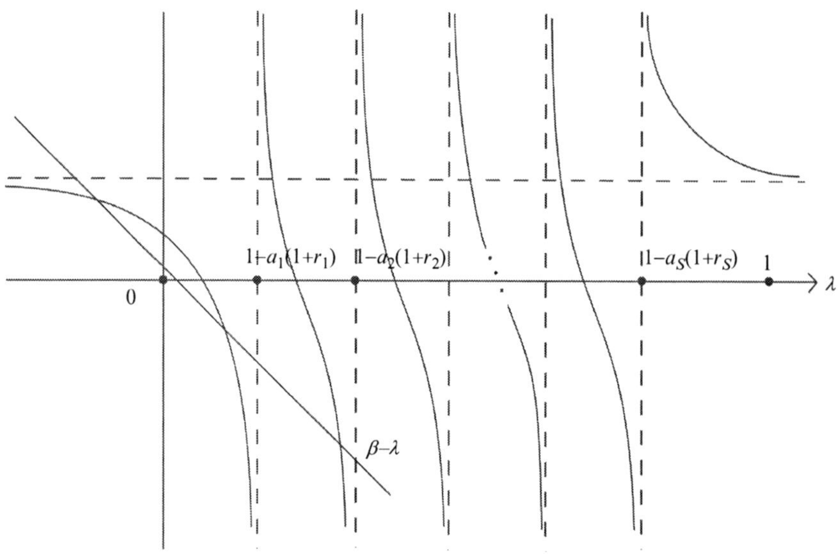


**Fig. 4.11** The oligopoly model with intertemporal demand interaction and best reply dynamics with adaptive expectations in the discrete time case. Graph of $g(\lambda)$ the roots of which are the eigenvalues of the Jacobian matrix

$$g(0) = \beta_S \sum_{k=1}^{N} \frac{-r_k}{1 - a_k(1 + r_k)},$$

and if we assume that

$$\sum_{k=1}^{N} \frac{-r_k}{1 - a_k(1 + r_k)} \geq 1, \tag{4.46}$$

then $g(0) \geq \beta_S$, so there is a root between each pair of consecutive poles and there is either a positive and a negative root, or zero and a positive or negative root, or zero with multiplicity two before the smallest pole. So all roots are real and they are between $-1$ and $+1$ if $g(-1) < \beta_S + 1$. ■

*Example 4.10.* As a special case consider linear price and cost functions i.e., $p = f(Q) = A - BQ$ and $C_k(x_k) = d_k + c_k x_k$ respectively. Then from Example 1.1 we know that $r_k = -1/2$ for all $k$. In this case

$$(a_k + \beta_S)(1 + r_k) = \frac{a_k + \beta_S}{2} < 1,$$

if $a_k + \beta_S$ is below 2, which always holds if both $a_k$ and $\beta_S$ are less than or equal to 1 and at least one of them is below one. Condition (4.46) also has the special form

$$\sum_{k=1}^{N} \frac{1}{2 - a_k} \geq 1,$$

which clearly holds if $N \geq 2$ and $0 \leq a_k \leq 1$. In this special case

$$g(-1) = \sum_{k=1}^{N} \frac{r_k(-a_k - 2\beta_S)}{2 - a_k(1 + r_k)} = \sum_{k=1}^{N} \frac{a_k + 2\beta_S}{4 - a_k},$$

so the equilibrium is locally asymptotically stable if

$$\sum_{k=1}^{N} \frac{a_k + 2\beta_S}{4 - a_k} < \beta_S + 1.$$

Notice that this relation can be rewritten as

$$\beta_S \left( \sum_{k=1}^{N} \frac{2}{4 - a_k} - 1 \right) < 1 - \sum_{k=1}^{N} \frac{a_k}{4 - a_k}.$$

If the firms select identical adjustment schemes, then $a_1 = \cdots = a_N = a$, so this relation simplifies to

$$\beta_S \left( \frac{2N}{4 - a} - 1 \right) < 1 - \frac{Na}{4 - a},$$

that is,

$$\beta_S(2N - 4 + a) < 4 - (N + 1)a.$$

Since the multiplier of $\beta_S$ is positive for $N \geq 2$ and $a > 0$, this relation can be rewritten as

$$\beta_S < \frac{4 - (N + 1)a}{2N - 4 + a}.$$

The right hand side is positive if

$$a < \frac{4}{N + 1},$$

so both $a$ and $\beta_S$ have to be sufficiently small. Notice that the above bound on $a$ decreases in $N$ and converges to zero as $N \to \infty$. So increasing the number of firms reduces the stability region.

Models with isoelastic price functions can be examined similarly, the details are not given here but are illustrated in the next subsection.

The stability of equilibria in multiproduct oligopolies with intertemporal demand interaction was first examined in Szidarovszky (1990). These results with some extensions are also reported in Okuguchi and Szidarovszky (1999). The model and results presented in this section are slight generalizations of those given in Szidarovszky and Zhao (2004).

### 4.3.2 Discrete Time Models and Global Stability

The global asymptotic stability of oligopolies with intertemporal demand interaction can be discussed in a similar fashion to the case of concave Cournot models in Chap. 2. In the following example we illustrate some global dynamic properties and complex asymptotic behavior by using the methods applied in earlier chapters.

*Example 4.11.* In this example we will consider $N$ firms, isoelastic price function, $f(Q, S) = A/(Q + \beta_S S)$, and linear cost functions, $C_k(x_k) = d_k + c_k x_k$ for $k = 1, 2, \ldots, N$. Then the profit (4.22) of firm $k$ becomes

$$\frac{Ax_k}{x_k + Q_k + \beta_S S} - (d_k + c_k x_k).$$

Assuming an interior optimum, the first order condition shows that at the optimum,

$$\frac{A(x_k + Q_k + \beta_S S) - Ax_k}{(x_k + Q_k + \beta_S S)^2} - c_k = 0,$$

implying that the solution is

$$z_k^* = \sqrt{\frac{A}{c_k}(Q_k + \beta_S S)} - (Q_k + \beta_S S).$$

Let $L_k$ denote the capacity limit of firm $k$, and since the payoff of firm $k$ is strictly concave in $x_k$, the best response of firm $k$ is

$$R_k(Q_k, S) = \begin{cases} 0 & \text{if} \quad z_k^* \leq 0, \\ L_k & \text{if} \quad z_k^* \geq L_k, \\ z_k^* & \text{otherwise.} \end{cases}$$

In the following discussion we consider the symmetric case of $N$ identical firms, so that $c_k = c$, $d_k = d$, $L_k = L$ for all $k$, and we assume adaptive output adjustments with identical speeds $a_k = a$. If all firms are assumed to start with the same initial output $x(0)$ then their outputs remain the same for all future periods, and therefore $Q_k = (N-1)x$ for all $k$. Due to the presence of the state variable $S$, by assuming partial adjustment towards the best response the dynamic model obtained is a two-dimensional discrete time dynamical system given by

$$\begin{aligned} x(t+1) =& (1-a)x(t) + aR\left((N-1)x(t), S(t)\right), \\ S(t+1) =& \beta_S S(t) + Nx(t+1) = \beta_S S(t) \\ &+ N\left[(1-a)x(t) + aR\left((N-1)x(t), S(t)\right)\right], \end{aligned} \tag{4.47}$$

where

$$R\left((N-1)x(t), S(t)\right) = \begin{cases} 0 \ \text{if} \ z^* < 0, \\ L \ \text{if} \ z^* > L, \\ z^* \ \text{otherwise,} \end{cases}$$

with

$$z^* = \sqrt{\frac{A}{c}\left((N-1)x + \beta_S S\right)} \quad (N-1)x - \beta_S S.$$

All parameters are non-negative, with the constraints $0 < a \leqslant 1$, $0 \leqslant \beta_S < 1$. Notice that in the limiting case $\beta_S = 0$ the best response coincides with the best response in Example 1.5 and in Example 3.4 for the one-dimensional symmetric case of $N$ identical firms. In the following we are mainly interested in the role of the parameter $\beta_S$, which measures the inertia of the effects of past sales, on the global dynamical properties of the model. Again, the presence of non-negativity and capacity constraints makes the dynamical system piece-wise differentiable, and the phase space $\mathbb{D} = [0, L] \times [0, +\infty]$ can be divided into subregions. In each of these subregions, the dynamical system is differentiable and these regions are separated by lines (or borders) of non-differentiability:

$\mathbb{D}^{(2)} = \{(x, S) : (N-1)\,x + \beta_S S > A/c\}$ where $z^*$ is negative,

$\mathbb{D}^{(3)} = \{(x, S) : z_1 < (N-1)\,x + \beta_S S < z_2\}$

with $z_{1,2} = \frac{A/c - 2L \pm \sqrt{A/c(A/c - 4L)}}{2}$, where $z^* > L$,

$\mathbb{D}^{(1)} = \mathbb{D} \setminus \left(\mathbb{D}^{(2)} \cup \mathbb{D}^{(3)}\right)$.

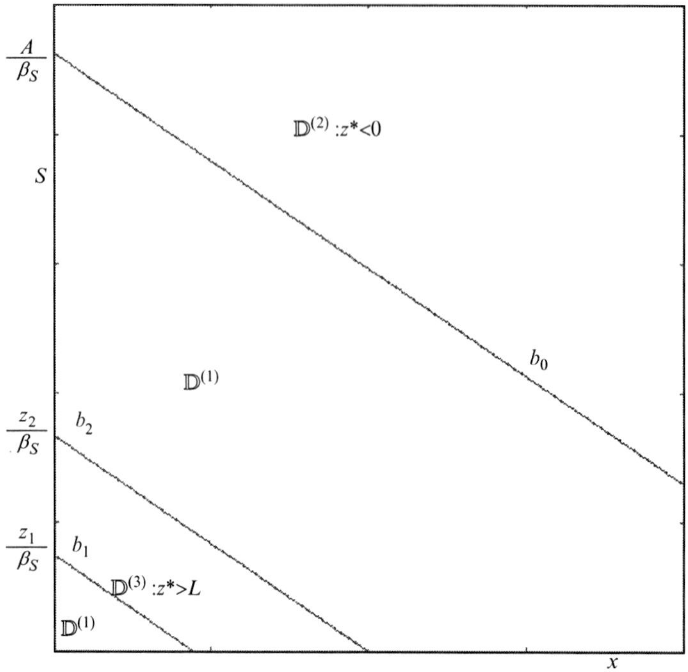


**Fig. 4.12** Example 4.11; the oligopoly model with intertemporal demand interaction and adaptive adjustment in the discrete time case. The phase space for the $N$-firm symmetric model with isoelastic price function and linear cost functions.

Notice that the region $\mathbb{D}^{(3)}$ is empty if $L > A/(4c)$. In this case the capacity constraint is ineffective since it is larger than the maximum value attained by $z^*$ (see also Fig. 1.9). In Fig. 4.12 these regions are shown for the case $L < A/(4c)$.

This kind of subdivision is important for the computation of the equilibrium points. In fact, the map in the different regions is given by the expressions

$$T|_{\mathbb{D}^{(1)}} : \begin{cases} x' = (1-a)\,x + a\left(\sqrt{\frac{A}{c}\left((N-1)\,x + \beta_S S\right)} - \left((N-1)\,x + \beta_S S\right)\right), \\ S' = \beta_S S + N\,[(1-a)\,x \\ \qquad + a\left(\sqrt{\frac{A}{c}\left((N-1)\,x + \beta_S S\right)} - \left((N-1)\,x + \beta_S S\right)\right)], \end{cases}$$

$$T|_{\mathbb{D}^{(2)}} : \begin{cases} x' = (1-a)\,x, \\ S' = \beta_S S + N\,(1-a)\,x, \end{cases}$$

$$T|_{\mathbb{D}^{(3)}} : \begin{cases} x' = (1-a)\,x + aL, \\ S' = \beta_S S + N\,[(1-a)\,x + aL]\,. \end{cases}$$

It is straightforward to calculate that $T|_{\mathbb{D}^{(3)}}$ has the unique fixed point

$$E_3 = \left( L, \frac{NL}{1 - \beta_S} \right),$$

which is an equilibrium for the dynamical system (4.47) provided that $E_3 \in \mathbb{D}^{(3)}$, that is if $z^*(E_3) > L$. This condition is equivalent to

$$L < \frac{A(1 - \beta_S)(N - 1 + \beta_S)}{cN^2}.$$

$T|_{\mathbb{D}^{(2)}}$ has the unique fixed point $O = (0, 0)$ which is not inside $\mathbb{D}^{(2)}$, so it is not an equilibrium of the dynamic process.

The fixed points of $T|_{\mathbb{D}^{(1)}}$ are the solutions of the algebraic system

$$\begin{aligned} x &= \sqrt{\tfrac{A}{c}((N-1)x + \beta_S S)} - ((N-1)x + \beta_S S), \\ (1 - \beta_S)\, S &= N\,[(1 - a)\,x + ax]\,. \end{aligned}$$

From the second equation we get $x = (1 - \beta_S)\, S/N$ and after substituting this expression into the first equation we get

$$\frac{A}{Nc}((N-1)(1 - \beta_S)\, S + N\beta_S S) = S^2.$$

Using these relations it is easy to see that $T|_{\mathbb{D}^{(1)}}$ has the unique fixed point

$$E_1 = (\bar{x}_1, \bar{S}_1) \text{ with } x_1 = \frac{A\,(1 - \beta_S)\,(N - 1 + \beta_S)}{N^2 c}, \quad \bar{S}_1 = \frac{A\,(N - 1 + \beta_S)}{Nc}.$$

This fixed point is also an equilibrium of the map $T$ provided that $E_1 \in \mathbb{D}^{(1)}$, in other words if $0 < z^*(E_1) < L$. Using the fact that $z^*(E_1) = \bar{x}_1$, this condition becomes

$$L > \frac{A\,(1 - \beta_S)\,(N - 1 + \beta_S)}{cN^2}.$$

Since

$$\frac{A\,(1 - \beta_S)\,(N - 1 + \beta_S)}{cN^2} < \frac{A}{4c}$$

always holds, being equivalent to $(N - 2\,(1 - \beta_S))^2 > 0$, we can summarize the existence results for an equilibrium as follows:

- If $L > \dfrac{A}{4c}$, then the region $\mathbb{D}^{(3)}$ is empty and the unique steady state is $E_1$.

- If $\dfrac{A(1-\beta_S)(N-1+\beta_S)}{cN^2} < L \leq \dfrac{A}{4c}$ then $\mathbb{D}^{(3)}$ is not empty, the unique steady state is $E_1$ (since $E_3 \in \mathbb{D}^{(1)}$) and all the trajectories starting in $\mathbb{D}^{(3)}$ enter $\mathbb{D}^{(1)}$.[2]
- If $L = \dfrac{A(1-\beta_S)(N-1+\beta_S)}{cN^2}$, then $E_1 = E_3$, and the equilibrium is located along the boundary that separates $\mathbb{D}^{(3)}$ from $\mathbb{D}^{(1)}$.
- If $L < \dfrac{A(1-\beta_S)(N-1+\beta_S)}{cN^2}$, then the unique steady state is $E_3$ (since $E_1 \in \mathbb{D}^{(3)}$).

With regard to the stability of the equilibria, it is easy to realize that whenever $E_3 \in \mathbb{D}^{(3)}$ it is a stable equilibrium, because the Jacobian matrix of the map $T|_{\mathbb{D}^{(3)}}$ is

$$\boldsymbol{J}^{(3)} = \begin{pmatrix} 1-a & 0 \\ N(1-a) & \beta_S \end{pmatrix}.$$

Hence its eigenvalues $1-a$ and $\beta_S$ are always less than one. In contrast, when $E_1 \in \mathbb{D}^{(1)}$, its stability is not as easily determined because this requires the study of the eigenvalues of the Jacobian matrix

$$\boldsymbol{J}^{(1)} = \begin{pmatrix} 1-aN+\frac{aA(N-1)}{2c\sqrt{\frac{A}{c}\left((N-1)\bar{x}+\beta_S\bar{S}\right)}} & a\beta_S\left[\frac{A}{2c\sqrt{\frac{A}{c}\left((N-1)\bar{x}+\beta_S\bar{S}\right)}}-1\right] \\ N\left[1-aN+\frac{aA(N-1)}{2c\sqrt{\frac{A}{c}\left((N-1)\bar{x}+\beta_S\bar{S}\right)}}\right] & \beta_S+Na\beta_S\left[\frac{A}{2c\sqrt{\frac{A}{c}\left((N-1)\bar{x}+\beta_S\bar{S}\right)}}-1\right] \end{pmatrix}$$

evaluated at $E_1$, which has the form

$$\boldsymbol{J}^{(1)}(E_1) = \begin{pmatrix} 1-\frac{aN(N-1+2\beta_S)}{2(N-1+\beta_S)} & \frac{a\beta_S(2(1-\beta_S)-N)}{2(N-1+\beta_S)} \\ N\left[1-\frac{aN(N-1+2\beta_S)}{2(N-1+\beta_S)}\right] & \beta_S+N\frac{a\beta_S(2(1-\beta_S)-N)}{2(N-1+\beta_S)} \end{pmatrix}.$$

Here the equilibrium condition

$$2c\sqrt{\frac{A}{c}(N-1)\bar{x}_1+\beta_S\bar{S}_1} = 2c\bar{S}_1 = 2A(N-1+\beta_S)/N$$

has been used. We can see that the stability of $E_1$ depends only on the parameters $N$, $a$ and $\beta_S$. Moreover, the matrix $\boldsymbol{J}^{(1)}(E_1)$ has the structure

$$\begin{pmatrix} A_{11} & A_{12} \\ NA_{11} & \beta_S+NA_{12} \end{pmatrix},$$

[2] Note that at $L = \dfrac{A}{4c}$ the region $\mathbb{D}^{(3)}$ reduce to the line $b = b_1 = b_2$, which is a set of measure zero in $\mathbb{R}^2$.

with

$$A_{11} = 1 - \frac{aN\,(N-1+2\beta_S)}{2\,(N-1+\beta_S)} \quad \text{and} \quad A_{12} = \frac{a\beta_S\,(2\,(1-\beta_S)-N)}{2\,(N-1+\beta_S)}.$$

The characteristic polynomial of this matrix is the quadratic

$$\lambda^2 - \lambda(A_{11} + \beta_S + NA_{12}) + \beta_S A_{11},$$

so the conditions for asymptotic stability are (see Appendix F)

$$(1-\beta_S)\,(1-A_{11}) - NA_{12} > 0, \tag{4.48}$$

$$(1+\beta_S)\,(1+A_{11}) + NA_{12} > 0, \tag{4.49}$$

$$\beta_S A_{11} < 1, \tag{4.50}$$

which reduce to the conditions,

$$(1-\beta_S)\,\frac{aN\,(N-1+2\beta_S)}{2\,(N-1+\beta_S)} - \frac{aN\beta_S\,(2\,(1-\beta_S)-N)}{2\,(N-1+\beta_S)} > 0, \tag{4.51}$$

$$(1+\beta_S)\left(2 - \frac{aN\,(N-1+2\beta_S)}{2\,(N-1+\beta_S)}\right) + \frac{aN\beta_S\,(2\,(1-\beta_S)-N)}{2\,(N-1+\beta_S)} > 0, \tag{4.52}$$

$$\frac{aN\beta_S\,(N-1+2\beta_S)}{2\,(N-1+\beta_S)} > \beta_S - 1. \tag{4.53}$$

It is obvious that due to the algebraic complexity of these stability conditions further analytical calculations will become quite involved. Therefore, instead we give a brief numerical study that will give us a flavor of the results one might expect to hold in general.

First of all we investigate the effect of the bifurcation that marks the exchange of the equilibrium $E_3$ with the equilibrium $E_1$ occurring at

$$L_{bif} = \frac{A\,(1-\beta_S)\,(N-1+\beta_S)}{cN^2}$$

along the boundary that separates the regions $\mathbb{D}^{(3)}$ and $\mathbb{D}^{(1)}$. This is not a usual transcritical bifurcation because $E_1$ and $E_3$ are fixed points of two different maps, and the merging occurs along a line of non-differentiability. Indeed, this is a typical border collision bifurcation, the effect of which is quite difficult to forecast. This is shown by three different bifurcation diagrams obtained for increasing values of the capacity limit $L$ across the bifurcation value. The first bifurcation diagram, shown in Fig. 4.13 is obtained with the set of parameters $N = 4$, $A = 2$, $c = 0.15$, $\beta_S = 0.6$, and $a = 0.5$, with $L$ in the range $[0.5, 2]$. For this set of parameters the bifurcation value is $L = 1.2$ and, as it can be seen in Fig. 4.13, when $L$ crosses the bifurcation

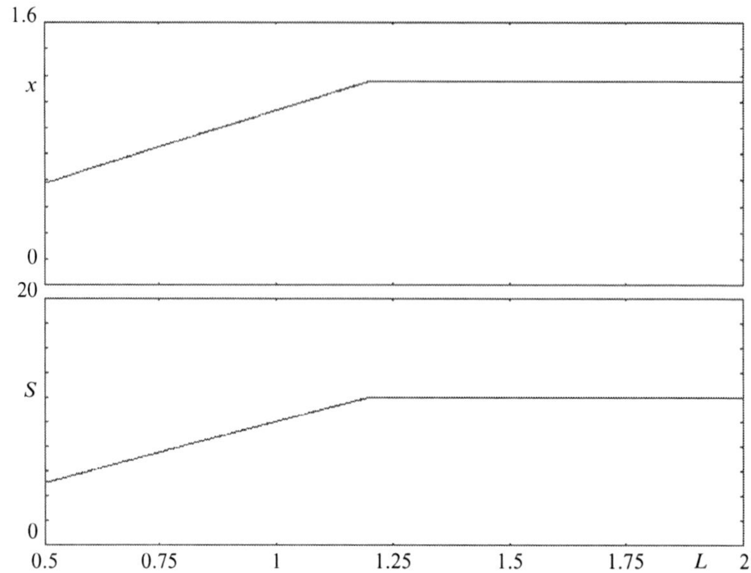


**Fig. 4.13** The oligopoly model with intertemporal demand interaction and adaptive adjustment in the discrete time case. The $N$-firm symmetric model with iso-elastic price function and linear cost functions. Bifurcation diagrams with respect to $L$ when the number of firms $N = 4$. Other parameter values are $A = 2, c = 0.15, \beta_S = 0.6, a = 0.5$. The bifurcation value is $L = 1.2$

value a simple change from the stable equilibrium $E_3$ to the stable equilibrium $E_1$ (that is independent of $L$) is observed.

In contrast to this, in the bifurcation diagram of Fig. 4.14, obtained with one more firm ($N = 5$) and $a = 1$ (the case of best reply dynamics) the bifurcation, now occurring at $L = 3.68/3.75 \simeq 0.98$, leads to the creation of a stable cycle of period 2. Indeed, by slight changes of the parameters, the creation of stable cycles of several different periods can be observed, as well as the sudden[3] creation of a chaotic attractor. This is shown in the bifurcation diagram of Fig. 4.15, obtained with parameters $N = 6$, $A = 2$, $c = 0.1$, $\beta_S = 0.6$, and $a = 0.7$ with bifurcation value $L = 4.48/3.6 \simeq 1.24$. However, we are mainly interested in the effect of the inertia parameter $\beta_S$ on the dynamic behavior of the model. The bifurcation diagram of Fig. 4.16 shows the role of increasing values of $\beta_S$, varying in the range $[0, 1]$, with the other parameters fixed at the values $N = 3$, $A = 1$, $c = 0.15$, $L = 2$ and $a = 1$. The equilibrium $E_1$ is stable for low values of $\beta_S$, then it loses stability and a stable cycle of period 2 appears. The amplitude of the oscillations increases for increasing values of $\beta_S$, until the lower periodic point reaches the constraint at $x = 0$. It is also interesting to study the impact of the number of firms in the market on the bifurcation with respect to $\beta_S$. This can be seen from the bifurcation diagram of Fig. 4.17 obtained with the same parameter values as in Fig. 4.16, but

[3] By "sudden" here we mean without the usual sequence of period doubling bifurcations.

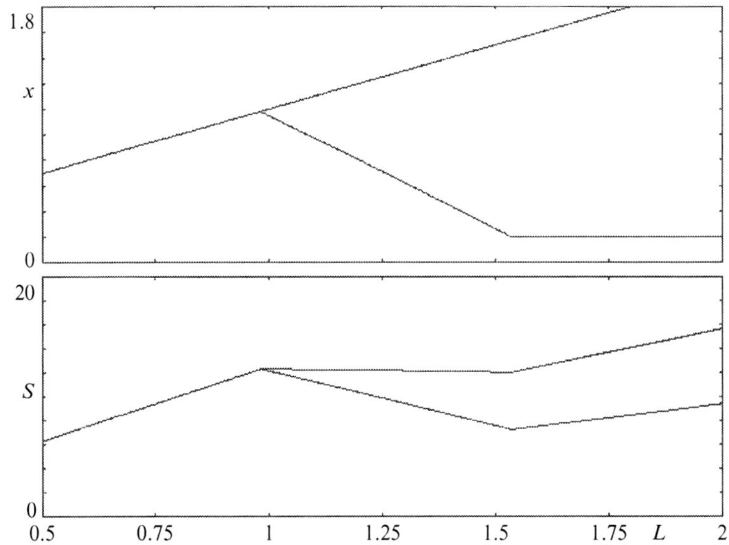


**Fig. 4.14** The oligopoly model with intertemporal demand interaction and adaptive adjustment in the discrete time case. The $N$-firm symmetric model with iso-elastic price function and linear cost functions. Bifurcation diagrams with respect to $L$ when the number of firms is increased to $N = 5$ and $a$ is increased to the value 1. Other parameters are as in Fig. 4.13. The bifurcation occurs at $L \simeq 0.98$, at which point a stable 2 cycle is born

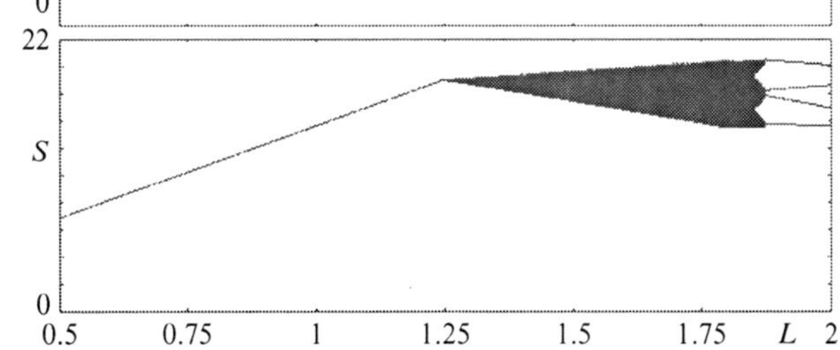


**Fig. 4.15** The oligopoly model with intertemporal demand interaction and adaptive adjustment in the discrete time case. The $N$-firm symmetric model with iso-elastic price function and linear cost functions. Bifurcation diagrams with respect to $L$ when the number of firms is increased further to $N = 6$. Other parametric changes with respect to Fig. 4.14 are $a = 0.7$ and $c = 0.1$, whilst $\beta$ remains at the value 0.6. The bifurcation now occurs at $L \simeq 1.24$, at which point a chaotic attractor appears

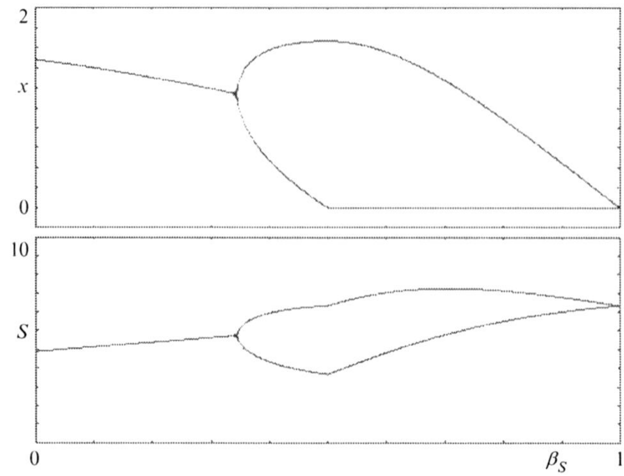


**Fig. 4.16** The oligopoly model with intertemporal demand interaction and adaptive adjustment in the discrete time case. The $N$-firm symmetric model with iso-elastic price function and linear cost functions. Bifurcation diagrams with respect to $\beta_S$ when the number of firms is $N = 3$. Other parameters are $A = 1, c = 0.15, L = 2, a = 1$. Stable 2-cycles are born at the bifurcation point

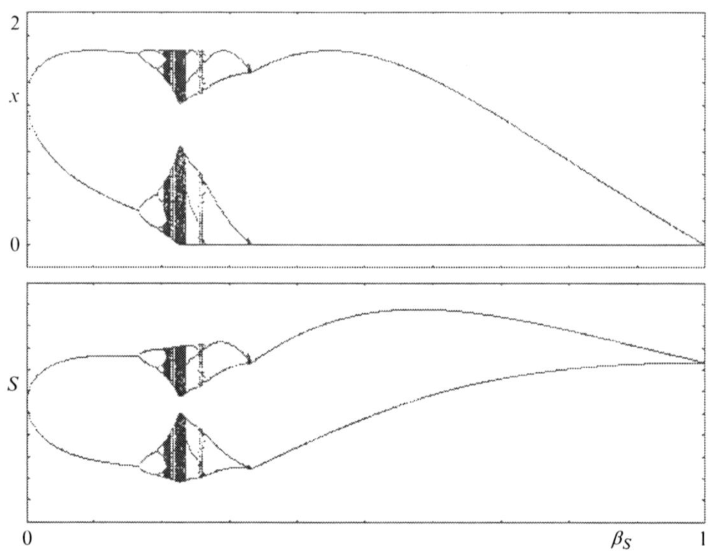


**Fig. 4.17** The oligopoly model with intertemporal demand interaction and adaptive adjustment in the discrete time case. The $N$-firm symmetric model with iso-elastic price function and linear cost functions. Bifurcation diagrams with respect to $\beta_S$ when the number of firms is increased to $N = 4$. Other parameter values are as in Fig. 4.16. Note that the stable 2-cycles are now born at $\beta_S = 0$, and also the sequence of period-doubling followed by period-halving at some intermediate values of $\beta_S$

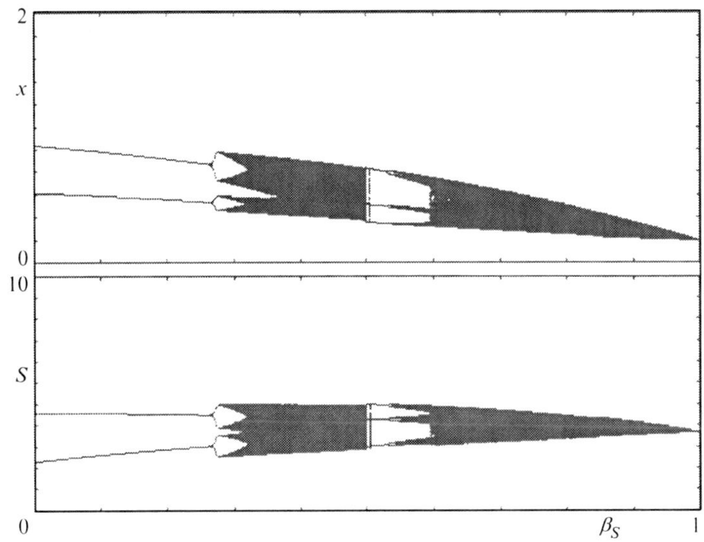


**Fig. 4.18** The oligopoly model with intertemporal demand interaction and adaptive adjustment in the discrete time case. The $N$-firm symmetric model with iso-elastic price function and linear cost functions. Bifurcation diagrams with respect to $\beta_S$ when the number of firms is increased to $N = 10$ and the speed of adjustment is decreased to $a = 0.5$. Other parameter values are as in Fig. 4.17. Note that the amplitude of the fluctuating attractors is much reduced compared to Fig. 4.16 and 4.17

with $N = 4$ instead of $N = 3$. Now we see that the stable 2-cycles are born at $\beta_S = 0$ and period-doubling followed by period halving occurs at intermediate values of $\beta_S$. As is commonly observed in adaptive models, the amplitude of the oscillations is reduced with decreasing values of the speed of adjustment $a$. This effect is shown in Fig. 4.18, obtained with the number of firms being increased to $N = 10$ and the speed of adjustment being decreased to $a = 0.5$. The other parameters remain as in Fig. 4.17.

When the asymptotic dynamics is chaotic, it is important to study the size and the shape of the chaotic attractors inside which the long-run dynamics are ultimately bounded. As already shown in the examples of the previous chapters (see also Appendix C), the boundaries of the chaotic sets can be obtained by taking the images of the folding curves, that may be critical curves (loci of vanishing Jacobian) or the lines of non-differentiability (the borders that separate the different regions $\mathbb{D}^{(i)}$). In our case, the candidates for the "folding curves" are:

- The curves of non-differentiability, which are the lines $(N-1)x + \beta_S S = A/c$ when $L > A/(4c)$ and the lines $(N-1)x + \beta_S S = z_1$ and $(N-1)x + \beta_S S = z_2$ when $L < A/(4c)$;
- The curves of vanishing Jacobian, given by $detJ^{(1)}(x, S) = 0$.

In Fig. 4.19 two chaotic attractors are shown. Figure 4.19a is obtained with the parameters $N = 8$, $A = 1$, $c = 0.15$, $L = 2$, $a = 0.8$, $\beta_S = 0.55$, for

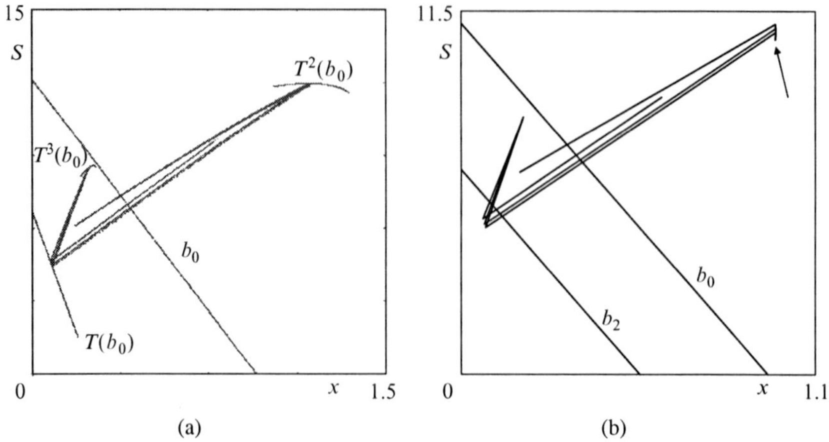


**Fig. 4.19** The oligopoly model with intertemporal demand interaction and adaptive adjustment in the discrete time case. The $N$-firm symmetric model with iso-elastic price function and linear cost functions. Parameter values are $N = 8, A = 1, c = 0.15, a = 0.8, \beta_S = 0.55$. Calculating the regions that delineate the chaotic attractors. (**a**) Here $L = 2$; the border $b_0$ and its images $T(b_0)$, $T^2(b_0)$ and $T^3(b_0)$ delineate the region within which the chaotic attractor lies. (**b**) Here $L = 1.2$; now new borders $b_1$ and $b_2$ appear. The crossing of the lower part of the chaotic attractor by $b_2$ leads to new foldings in the boundaries of the chaotic attractor, indicated by the *arrow*

which $L > A/4c$. In this case the chaotic attractor crosses the border $b_0$ between the regions $\mathbb{D}^1$ and $\mathbb{D}^2$ (see Fig. 4.12), and the images of the portion of $b_0$ that intersects the attractor, denoted by $T^k(b_0)$, $k = 1, 2, 3$ in the figure, give a delineation of the chaotic attractor (if the sequence of images is continued by representing $T^k(b_0)$ for increasing values of $k$, the whole boundary of the attractor will be obtained).

The chaotic attractor shown in Fig. 4.19b is obtained with $L = 1.2$, so that $L < A/4c$. In this case, borders $b_1$ and $b_2$ also exist, and $b_2$ crosses the lower portion of the chaotic attractor. This implies that its images determine new foldings in the boundaries of the chaotic attractor, as can be clearly seen in the upper part (indicated by the arrow) folded by $T(b_2)$. In the cases that we have examined here the second possible candidate for "folding curves," namely the locus of vanishing Jacobians plays no role. This is so since the curve of the vanishing Jacobian does not intersect the chaotic attractor, and so cannot be used to bound it.

### *4.3.3 Continuous Time Models*

In a similar fashion to the discussion of previous models the local asymptotic stability of the dynamical system (4.31)–(4.32) is examined by linearization. The Jacobian of the system at the equilibrium has the form

$$\begin{pmatrix} -a_1 & a_1 r_1 & \cdots & a_1 r_1 & a_1 r_1 \beta_S \\ a_2 r_2 & -a_2 & \cdots & a_2 r_2 & a_2 r_2 \beta_S \\ \vdots & \vdots & \ddots & \vdots & \vdots \\ a_N r_N & a_N r_N & \cdots & -a_N & a_N r_N \beta_S \\ 1 & 1 & \cdots & 1 & -\gamma_S \end{pmatrix},$$

where we use the notation of the previous section but for the sake of simplicity $\beta_S^*$ is now denoted by $\beta_S$, which is not necessarily the same as $\beta_S$ in the case of discrete time models. The eigenvalue equation of this matrix has the form

$$-a_k u_k + \sum_{l \neq k} a_k r_k u_l + a_k r_k \beta_S v = \lambda u_k, \qquad (1 \leq k \leq N), \tag{4.54}$$

$$\sum_{k=1}^{N} u_k - \gamma_S v = \lambda v, \tag{4.55}$$

where $\lambda$ is an eigenvalue and $(u_1, \ldots, u_N, v)$ is an associated eigenvector. By letting $U = \sum_{k=1}^{N} u_k$, these equations imply that

$$a_k r_k U = -a_k r_k \beta_S v + (\lambda + a_k + a_k r_k) u_k$$

and

$$U = (\lambda + \gamma_S) v. \tag{4.56}$$

Substituting this expression into the previous equation we obtain

$$u_k = \frac{a_k r_k (\lambda + \gamma_S + \beta_S)}{\lambda + a_k (1 + r_k)} v. \tag{4.57}$$

Here we assume that $\lambda \neq -a_k(1 + r_k)$, since in both the concave and isoelastic cases $r_k > -1$, so $-a_k(1 + r_k) < 0$ and negative eigenvalues cannot destroy local asymptotic stability. By summing (4.57) over all $k$ and using (4.56) we get for $v$ the single equation

$$\left( \sum_{k=1}^{N} \frac{a_k r_k (\lambda + \gamma_S + \beta_S)}{\lambda + a_k (1 + r_k)} - (\lambda + \gamma_S) \right) v = 0.$$

If $v = 0$, then from (4.57), $u_k = 0$ for all $k$, so the eigenvector becomes zero, which is impossible. Therefore $v \neq 0$, and the eigenvalue equation becomes

$$\sum_{k=1}^{N} \frac{a_k r_k}{\lambda + a_k (1 + r_k)} = \frac{\lambda + \gamma_S}{\lambda + \gamma_S + \beta_S}. \tag{4.58}$$

The following theorem provides results on the local stability of the equilibrium.

**Theorem 4.2.** *If* $-1 < r_k \leq 0$ *and* $a_k > 0$ *for all* $k$*, and furthermore* $\beta_S, \gamma_S > 0$*, then all roots of the eigenvalue equation (4.58) have negative real parts implying the local asymptotic stability of the equilibrium of the system (4.31)–(4.32).*

*Proof.* Let $\lambda = A + iB$ denote a root and suppose that $A \geq 0$. If $g(\lambda)$ and $h(\lambda)$ denote the left and right hand sides of (4.58), respectively, then

$$g(A+iB) = \sum_{k=1}^{N} \frac{a_k r_k}{A + iB + a_k(1+r_k)} = \sum_{k=1}^{N} \frac{a_k r_k (A + a_k(1+r_k) - iB)}{(A + a_k(1+r_k))^2 + B^2},$$

which has a non-positive real part under the stated assumptions. Similarly,

$$h(A+iB) = \frac{A + iB + \gamma_S}{A + iB + \gamma_S + \beta_S} = \frac{(A + \gamma_S + iB)(A + \gamma_S + \beta_S - iB)}{(A + \gamma_S + \beta_S)^2 + B^2},$$

the real part of which is given by

$$\frac{(A+\gamma_S)(A+\gamma_S+\beta_S) + B^2}{(A+\gamma_S+\beta_S)^2 + B^2} > 0.$$

The contradiction implies that $A < 0$ must hold. ∎

In the isoelastic case there is no guarantee that the $r_k$ values are non-positive. In this case the analysis can be performed in a similar fashion to the case shown earlier in Sect. 3.1.3. The details are not presented here.

In introducing time lags into the model (4.31)–(4.32) we assume that the firms react to delayed information about the value of $S$, and that the firms have identical delays. The more general non-symmetric case, when all information on the firms own outputs as well as on the outputs of the competitors are also delayed, can be discussed similarly, but the analysis becomes much more complicated.

In the simple case of delayed information about $S$ the system (4.31)–(4.32) becomes

$$\dot{x}_k(t) = \alpha_k \left( R_k \left( \sum_{l \neq k} x_l(t) + \beta_S \int_0^t w(t-s, T, m) S(s) \mathrm{d}s \right) - x_k(t) \right), \quad (4.59)$$

$$\dot{S}(t) = \sum_{l=1}^{N} x_l(t) - \gamma_S S(t),$$

where the weighting function is selected in the same way as in Sect. 2.6 for concave oligopolies. Linearizing the system we obtain

$$\dot{x}_{k\delta}(t) = a_k \left[ r_k \left( \sum_{l \neq k} x_{l\delta}(t) + \beta_S \int_0^t w(t-s, T, m) S_\delta(s) \mathrm{d}s \right) - x_{k\delta}(t) \right], \tag{4.60}$$

$$\dot{S}_\delta(t) = \sum_{l=1}^{N} x_{l\delta}(t) - \gamma_S S_\delta(t), \tag{4.61}$$

where $a_k = \alpha_k'(0), r_k = R_k'$ at the equilibrium, and $x_{k\delta}, S_\delta$ are respectively the deviations of $x_k$, $S$ from their equilibrium levels. Seeking solutions in the form of $x_{k\delta}(t) = u_k e^{\lambda t}$ and $S_\delta(t) = ve^{\lambda t}$, substituting these functions into (4.60)–(4.61) and letting $t \to \infty$ we have

$$(\lambda + a_k) u_k = a_k r_k \sum_{l \neq k} u_l + a_k r_k \beta_S \left( \int_0^\infty w(s, T, m) e^{-\lambda s} \mathrm{d}s \right) v, \tag{4.62}$$

$$(\lambda + \gamma_S) v = \sum_{l=1}^{N} u_l. \tag{4.63}$$

Nonzero solutions for $u_k$ $(1 \leq k \leq N)$ and $v$ exist if and only if the determinant of the matrix

$$\begin{pmatrix} -(\lambda + a_1) & a_1 r_1 & \cdots & a_1 r_1 & a_1 r_1 \beta_S I(\lambda) \\ a_2 r_2 & -(\lambda + a_2) & \cdots & a_2 r_2 & a_2 r_2 \beta_S I(\lambda) \\ \vdots & & & & \\ a_N r_N & a_N r_N & \cdots & -(\lambda + a_N) & a_N r_N \beta_S I(\lambda) \\ 1 & 1 & \cdots & 1 & -(\lambda + \gamma_S) \end{pmatrix},$$

is zero, where

$$I(\lambda) = \left( \frac{\lambda T}{p} + 1 \right)^{-(m+1)},$$

with

$$p = \begin{cases} 1 & \text{if} \quad m = 0, \\ m & \text{if} \quad m \geq 1. \end{cases}$$

Notice that in the special case of $T = 0$ (no time delay is present) this determinant reduces to the characteristic polynomial of the Jacobian of the model (4.31)–(4.32) since $I(\lambda) = 1$ in this case. In the general case of $T > 0$ we follow a similar path to that used in deriving equation (4.58) given earlier in this section. The (4.62) and (4.63) can be rewritten as

$$a_k r_k U = -a_k r_k \beta_S I(\lambda) v + (\lambda + a_k + a_k r_k) u_k \tag{4.64}$$

and

$$U = (\lambda + \gamma_S)v, \tag{4.65}$$

where $U = \sum_{k=1}^{N} u_k$ as before. Substituting (4.65) into (4.64) we obtain

$$u_k = \frac{a_k r_k(\lambda + \gamma_S + \beta_S I(\lambda))}{\lambda + a_k(1 + r_k)} v,$$

where we can assume again that $\lambda \neq -a_k(1 + r_k)$. By summing the last equation over all $k$ and using (4.65) we get for $v$ the single equation

$$\left(\sum_{k=1}^{N} \frac{a_k r_k(\lambda + \gamma_S + \beta_S I(\lambda)}{\lambda + a_k(1 + r_k)} - (\lambda + \gamma_S)\right) v = 0.$$

Noticing again that $v \neq 0$, since otherwise the eigenvector would become zero, we obtain the eigenvalue equation

$$\sum_{k=1}^{N} \frac{a_k r_k}{\lambda + a_k(1 + r_k)} = \frac{\lambda + \gamma_S}{\lambda + \gamma_S + \beta_S I(\lambda)}. \tag{4.66}$$

We have already seen in Theorem 4.2 that in the case $T = 0$ the equilibrium is locally asymptotically stable. The case $T > 0$ can be examined in a similar fashion to the cases discussed earlier in this book. In the general case computational methods are used to locate the eigenvalues and check stability. In the symmetric case however analytical results can be obtained.

Consider therefore the symmetric case of $a_1 = \cdots = a_N = a$ and $r_1 = \cdots = r_N = r$. Then (4.66) has the form

$$\frac{Nar}{\lambda + a(1 + r)} = \frac{\lambda + \gamma_S}{\lambda + \gamma_S + \beta_S \left(1 + \frac{T\lambda}{p}\right)^{-(m+1)}},$$

that is,

$$\frac{Nar}{\lambda + a(1 + r)} = \frac{(\lambda + \gamma_S)\left(1 + \frac{T\lambda}{p}\right)^{m+1}}{(\lambda + \gamma_S)\left(1 + \frac{T\lambda}{p}\right)^{m+1} + \beta_S},$$

or

$$\left(1 + \frac{T\lambda}{p}\right)^{m+1} (\lambda + \gamma_S)(\lambda + a(1 + r(1 - N))) - Nar\beta_S = 0. \tag{4.67}$$

The roots of this equation can be examined in a similar way to the case of (2.58), which was given in detail earlier in Sect. 2.6. The details are left as an exercise for the interested reader.

The result of this section are slight generalizations and extensions of those presented in Szidarovszky and Zhao (2004). Some simple results are discussed for multiproduct linear oligopolies in Okuguchi and Szidarovszky (1999).

## 4.4 Models with Production Adjustment Costs

This section will consider oligopolies in which the firms experience production adjustment costs. An early contribution along these lines is Howroyd and Rickard (1981); more recent work on oligopoly models with production adjustment costs has been carried out for example by Szidarovszky and Yen (1995) and Schoonbeek (1997). We will consider only discrete time scales as the continuous time case can be discussed in an analogous fashion. Just as was the case for the models introduced and discussed by Szidarovszky (1999), Chiarella and Szidarovszky (2008a) and Zhao and Szidarovszky (2008) consider an $N$-firm oligopoly without externalities, with price function $f$ and cost functions $C_k$, and assume that any increase or decrease in the outputs of the firms comes at some cost. Taking this additional cost component into account, at time period $t + 1$ the profit of firm $k$ can be written as

$$x_k f(x_k + Q_k^E(t + 1)) - C_k(x_k) - K_k(x_k - x_k(t)), \tag{4.68}$$

where $K_k$ is the additional cost component which depends on the amount $x_k - x_k(t)$ of output change from the previous time period, and $Q_k^E(t+1)$ is the expectation of the output of the rest of the industry by firm $k$. In addition to assumptions (A)–(C) stated at the beginning of Sect. 2.1 for concave oligopolies assume that $K_k$ is a twice continuously differentiable convex function.

Under the above conditions, the expression (4.68) is strictly concave in $x_k$ and if each firm has a finite capacity limit, then there is always a unique best response $R_k$, which depends on both $Q_k^E(t + 1)$ and $x_k(t)$ and can be 0, $L_k$, or an interior value.

Consider the case of an interior equilibrium. In a small neighborhood of it the best responses are also interior, and the first order condition implies that

$$f(x_k + Q_k^E) + x_k f'(x_k + Q_k^E) - C_k'(x_k) - K_k'(x_k - x_k(t)) = 0,$$

where we use the simplifying notation $Q_k^E$ for $Q_k^E(t + 1)$.

The left hand side of the last equation is strictly decreasing in $x_k$, so this equation has a unique solution, $x_k = R_k(Q_k^E, x_k(t))$, which depends on both $Q_k^E$ and $x_k(t)$. By implicit differentiation with respect to $Q_k^E$ one has

$$f'(R'_{kQ} + 1) + R'_{kQ} f' + x_k f''(R'_{kQ} + 1) - C_k'' R'_{kQ} - K_k'' R'_{kQ} = 0,$$

and with respect to $x_k(t)$,

$$f' R'_{kx} + R'_{kx} f' + x_k f'' R'_{kx} - C_k'' R'_{kx} - K_k''(R'_{kx} - 1) = 0,$$

where $R'_{kx} = \frac{\partial R_k}{\partial x_k}$, $R'_{kQ} = \frac{\partial R_k}{\partial Q_k^E}$. So

$$R'_{kQ} = -\frac{f' + x_k f''}{2f' + x_k f'' - C_k'' - K_k''}, \tag{4.69}$$

and

$$R'_{kx} = -\frac{K_k''}{2f' + x_k f'' - C_k'' - K_k''}. \tag{4.70}$$

It is easy to see that as in the concave case the derivatives of the reaction function of firm $k$ satisfy

$$-1 < R'_{kQ} \le 0 \le R'_{kx} < 1$$

and

$$-1 < R'_{kQ} - R'_{kx} \le 0.$$

Consider first the dynamic process (1.28)–(1.29) with adaptive expectations, which here assumes the form

$$x_k(t+1) = R_k\left(Q_k^E(t) + \alpha_k\left(\sum_{l\neq k} x_l(t) - Q_k^E(t)\right), x_k(t)\right), \tag{4.71}$$

$$Q_k^E(t+1) = Q_k^E(t) + \alpha_k\left(\sum_{l\neq k} x_l(t) - Q_k^E(t)\right). \tag{4.72}$$

The Jacobian of this system at the equilibrium has the special form

$$\begin{pmatrix} \bar{J}_{11} & \bar{J}_{12} \\ \bar{J}_{21} & \bar{J}_{22} \end{pmatrix}$$

where

$$\bar{J}_{11} = \begin{pmatrix} r_{1x} & r_{1Q}a_1 & \cdots & r_{1Q}a_1 \\ r_{2Q}a_2 & r_{2x} & \cdots & r_{2Q}a_2 \\ \vdots & \vdots & \ddots & \vdots \\ r_{NQ}a_N & r_{NQ}a_N & \cdots & r_{Nx} \end{pmatrix}, \bar{J}_{12} = \begin{pmatrix} r_{1Q}(1-a_1) & & & \\ & r_{2Q}(1-a_2) & & \\ & & \ddots & \\ & & & r_{NQ}(1-a_N) \end{pmatrix},$$

$$\bar{J}_{21} = \begin{pmatrix} 0 & a_1 & \cdots & a_1 \\ a_2 & 0 & \cdots & a_2 \\ \vdots & \vdots & & \vdots \\ a_N & a_N & \cdots & 0 \end{pmatrix}, \text{and} \quad \bar{J}_{22} = \begin{pmatrix} 1-a_1 & & & \\ & 1-a_2 & & \\ & & \ddots & \\ & & & 1-a_N \end{pmatrix},$$

and $r_{kQ} = \frac{\partial R_k}{\partial Q_k}$ and $r_{kx} = \frac{\partial R_k}{\partial x_k}$ evaluated at the equilibrium. If $\lambda$ is an eigenvalue and $(u_1, \ldots, u_N, v_1, \ldots, v_N)$ is an associated eigenvector, then the eigenvalue equation of the Jacobian has the form

$$r_{kx}u_k + r_{kQ}a_k \sum_{l \neq k} u_l + r_{kQ}(1 - a_k)v_k = \lambda u_k, \qquad (k = 1, 2, \ldots, N), \quad (4.73)$$

$$a_k \sum_{l \neq k} u_l + (1 - a_k)v_k = \lambda v_k. \qquad (4.74)$$

Subtract the $r_{kQ}$-multiple of the second equation from the first one to obtain

$$r_{kx}u_k = \lambda(u_k - r_{kQ}v_k),$$

implying that

$$r_{kQ}v_k = \frac{\lambda - r_{kx}}{\lambda}u_k,$$

where it is assumed that $\lambda \neq 0$. Note that a zero eigenvalue cannot destroy asymptotic stability. Substituting this relation into (4.73) it is found that

$$r_{kQ}a_k \sum_{l \neq k} u_l + \left( r_{kx} + \frac{(1 - a_k)(\lambda - r_{kx})}{\lambda} - \lambda \right) u_k = 0, \qquad (4.75)$$

for all $k$. A non-trivial solution exists if and only if the determinant of this system is zero. Notice that by introducing the notation

$$A_k(\lambda) = r_{kx} + \frac{(1 - a_k)(\lambda - r_{kx})}{\lambda} - \lambda, \quad B_k(\lambda) = a_k r_{kQ},$$

this determinant has the same structure as the one given by (E.2) in Appendix E. Therefore by using (E.3), the resulting determinantal equation becomes

$$\prod_{k=1}^{N} \left( r_{kx} + \frac{(1 - a_k)(\lambda - r_{kx})}{\lambda} - \lambda - a_k r_{kQ} \right)$$
$$\times \left[ 1 + \sum_{k=1}^{N} \frac{a_k r_{kQ}}{r_{kx} + \frac{(1 - a_k)(\lambda - r_{kx})}{\lambda} - \lambda - a_k r_{kQ}} \right] = 0. \qquad (4.76)$$

It is very complicated in general to find conditions that guarantee that the roots of (4.76) lie inside the unit circle, so instead of a general analysis we will here consider a particular example.

*Example 4.12.* Consider the case of symmetric firms, when $a_1 = \ldots = a_N = a, r_{1Q} = \ldots = r_{NQ} = r_Q, r_{1x} = \ldots = r_{Nx} = r_x$. The eigenvalues in this case are the roots of the equations

$$r_x + \frac{(1-a)(\lambda - r_x)}{\lambda} - \lambda - ar_Q = 0, \tag{4.77}$$

and

$$r_x + \frac{(1-a)(\lambda - r_x)}{\lambda} - \lambda - ar_Q + Nar_Q = 0. \tag{4.78}$$

Both of these equations can be written as the quadratic equations

$$\lambda^2 + \lambda(-1 + a(1 + r_Q) - r_x) + r_x(1-a) = 0,$$

and

$$\lambda^2 + \lambda(-1 + a(1 + (1-N)r_Q) - r_x) + r_x(1-a) = 0.$$

By using the result of Lemma F.1 (from Appendix F) and some simple algebra it can be seen that all roots are inside the unit circle if

$$r_x(1-a) < 1, \tag{4.79}$$

$$a(1 + r_Q - r_x) > 0, \tag{4.80}$$

$$2(1 + r_x) - a(1 + r_x + r_Q(1-N)) > 0, \tag{4.81}$$

where the first two inequalities always hold for $a > 0$, and the third is satisfied if $a$ is a sufficiently small positive value, that is, if the firms select a small common constant speed of adjustment, or have a small common derivative value $\alpha'(0)$.

To compare the results just obtained for the best reply dynamics with adaptive expectations, we now turn to partial adjustment towards the best response with naive expectations, namely

$$x_k(t+1) = x_k(t) + \alpha_k \left( R_k \left( \sum_{l \neq k} x_l(t), x_k(t) \right) - x_k(t) \right), \quad (1 \le k \le N), \tag{4.82}$$

which is a straightforward extension of the system (1.30) to take into account production adjustment costs. Conditions for the local asymptotic stability of the equilibrium are given in the following theorem.

**Theorem 4.3.** *Assume that $a_k > 0$ for all $k$, $C_k'' \ge 0$ for all $k$ and $x_k$, assumptions (A)–(C) of Sect. 2.1 of concave oligopolies hold, furthermore the conditions*

$$0 < a_j < \frac{2}{1 - r_{jx} + r_{jQ}} \quad (j = 1, 2, \ldots, N)$$

*are satisfied. Then the equilibrium of the system (4.82) is locally asymptotically stable, if*

$$1 + \sum_{k=1}^{N} \frac{a_k r_{kQ}}{2 - a_k(1 - r_{kx} + r_{kQ})} > 0.$$

*If this condition is violated with strict inequality, then the equilibrium is unstable.*

*Proof.* The Jacobian of the system (4.82) is given by

$$\begin{pmatrix} 1-a_1(1-r_{1x}) & a_1 r_{1Q} & \dots & a_1 r_{1Q} \\ a_2 r_{2Q} & 1-a_2(1-r_{2x}) & \dots & a_2 r_{2Q} \\ \vdots & \vdots & & \vdots \\ a_N r_{NQ} & a_N r_{NQ} & \dots & 1-a_N(1-r_{Nx}) \end{pmatrix},$$

where the previous notation of this section is used. The eigenvalue equation can be easily determined by using relation (E.5) and turns out to be

$$\prod_{k=1}^{N}(1-a_k(1-r_{kx}+r_{kQ})-\lambda)\left[1+\sum_{k=1}^{N}\frac{a_k r_{kQ}}{1-a_k(1-r_{kx}+r_{kQ})-\lambda}\right]=0. \tag{4.83}$$

As before, assume that $a_k > 0$ for all $k$, and let $1-a_j(1-r_{jx}+r_{jQ})$ ($j = 1,2,\dots,s$) denote the different $1-a_k(1-r_{kx}+r_{kQ})$ values and assume that they are repeated $m_1, m_2, \dots, m_s$ times among the $N$ firms. By adding the terms with identical denominators in the bracketed expression and denoting by $\theta_j$ the sum of the corresponding numerators $a_k r_{kQ}$, one obtains

$$\prod_{j=1}^{s}(1-a_j(1-r_{jx}+r_{jQ})-\lambda)^{m_j}\left[1+\sum_{j=1}^{s}\frac{\theta_j}{1-a_j(1-r_{jx}+r_{jQ})-\lambda}\right]=0, \tag{4.84}$$

where $\theta_j \le 0$ for all $j$. If $\theta_j = 0$ or $m_j \ge 2$, then $1-a_j(1-r_{jx}+r_{jQ})$ is an eigenvalue of the Jacobian, and this eigenvalue is between $-1$ and $+1$, if

$$0 < a_j < \frac{2}{1-r_{jx}+r_{jQ}}. \tag{4.85}$$

All other eigenvalues are the roots of the equation

$$1+\sum_{j=1}^{s}\frac{\theta_j}{1-a_j(1-r_{jx}+r_{jQ})-\lambda}=0,$$

when it is assumed that all $\theta_j$ values are nonzero. Under assumption (4.85), the graph of the left hand side is the same as the one shown in Fig. 2.1, so all roots are real, all poles are between $-1$ and $+1$, furthermore all roots are between $-1$ and $+1$, if

$$1+\sum_{k=1}^{N}\frac{a_k r_{kQ}}{2-a_k(1-r_{kx}+r_{kQ})} > 0. \tag{4.86}$$

■

*Example 4.13.* Consider the special case of symmetric firms, when $a_k \equiv a$, $r_{kQ} \equiv r_Q$ and $r_{kx} \equiv r_x$. Then condition (4.85) reduces to

$$0 < a < \frac{2}{1 - r_x + r_Q}$$

and condition (4.86) becomes

$$2 - a(1 - r_x + r_Q - N r_Q) > 0.$$

These relations hold if the value of $a$ is sufficiently small, in particular if

$$a < \frac{2}{1 + (r_Q - r_x) - N r_Q},$$

where the right hand side is always positive.

*Example 4.14.* Assume there are $N$ firms with identical capacity limit $L$, and assume that the price function is $f(Q) = LN - Q$, so the price is always nonnegative. If the firms have linear cost functions $C_k(x_k) = d_k + c_k x_k$ and quadratic output adjustment costs, $K_k(x_k - x_k(t)) = \gamma_k \cdot (x_k - x_k(t))^2$ where $x_k > x_k(t)$ and zero otherwise, then the profit of firm $k$ may be written as

$$x_k(LN - x_k - Q_k) - (d_k + c_k x_k) - \begin{cases} 0 & \text{if} \quad x_k \le x_k(t), \\ \gamma_k (x_k - x_k(t))^2 & \text{if} \quad x_k > x_k(t). \end{cases}$$

Assume an interior optimum for the profit maximization problem. In the first case ($x_k \le x_k(t)$) the first order condition can be written as

$$LN - 2x_k - Q_k - c_k = 0,$$

implying that

$$x_k = \frac{LN - Q_k - c_k}{2}.$$

This point is below $x_k(t)$ if and only if

$$LN - Q_k - c_k \le 2x_k(t).$$

In the second case ($x_k > x_k(t)$) the first order condition is

$$LN - 2x_k - Q_k - c_k - 2\gamma_k(x_k - x_k(t)) = 0,$$

the solution of which is

$$x_k = \frac{LN - Q_k - c_k + 2\gamma_k x_k(t)}{2 + 2\gamma_k}.$$

This value is larger than $x_k(t)$ if and only if

$$LN - Q_k - c_k > 2x_k(t).$$

Notice also that the profit function of firm $k$ has two parabolic segments, both are concave, and they have identical derivatives at $x_k(t)$. Consequently, define

$$z_k^* = \begin{cases} \dfrac{LN - Q_k - c_k}{2} & \text{if} \quad LN - Q_k - c_k \leq 2x_k(t), \\[2ex] \dfrac{LN - Q_k - c_k + 2\gamma_k x_k(t)}{2 + 2\gamma_k} & \text{otherwise,} \end{cases}$$

then the best response of firm $k$ is given by

$$R_k(Q_k, x_k(t)) = \begin{cases} 0 & \text{if} \quad z_k^* \leq 0, \\ L & \text{if} \quad z_k^* \geq L, \\ z_k^* & \text{otherwise.} \end{cases}$$

It is easy to see that in both cases the derivative of the best response with respect to $Q_k$ is between 0 and $-\frac{1}{2}$ so the local stability properties of the corresponding dynamical system are similar to the concave oligopoly case. Note that the best responses in this model are piece-wise linear. Therefore, the dynamical system based on partial adjustment towards the best response belongs to the same class as the models with linear and quadratic cost functions. Since we have analyzed the latter models in detail in Chap. 1, we abstain from presenting the details of a global analysis of the present model. Instead we leave such an analysis to the reader. Consider finally the symmetric case, when $a_k \equiv a$, $c_k \equiv c$, $d_k \equiv d$, $L_k = L$, $\gamma_k = \gamma$ and the initial outputs are identical. Then $Q_k = (N - 1)x$, and

$$z^* = \begin{cases} \dfrac{LN - (N-1)x - c}{2} & \text{if} \quad LN - (N-1)x - c \leq 2x, \\[2ex] \dfrac{LN - (N-1)x - c + 2\gamma x}{2 + 2\gamma} & \text{otherwise.} \end{cases}$$

Then the common best response of the firms is

$$R(x) = \begin{cases} 0 & \text{if} \quad z^* \leq 0, \\ L & \text{if} \quad z^* \geq L, \\ z^* & \text{otherwise.} \end{cases}$$

The dynamical system is therefore

$$x(t+1) = aR(x(t)) + (1-a)x(t).$$

This is again a piecewise linear model and can be analyzed in a similar way to the one-dimensional examples studied before. We leave the analysis as an exercise for the reader.

We mention here that Szidarovszky and Yen (1995) have introduced stability conditions in the linear case. These results and some extensions are presented in Okuguchi and Szidarovszky (1999). This latter book also discusses the continuous case where it is shown that the dynamical system is equivalent to a classical Cournot dynamics with modified speeds of adjustment.

## 4.5 Oligopolies with Partial Cooperation

It is well known that by selecting the Nash equilibrium quantities, firms in an oligopoly are trapped in a prisoner's dilemma situation, and a common way to increase their payoffs is for the firms to reach some sort of cooperation or collusion amongst each other. In this section we introduce partial cooperation into the framework of the oligopoly models studied earlier in the book.

The idea of partial cooperation was introduced and first explored by Cyert and DeGroot (1973). The survey paper of Szidarovszky et al. (2008) contains some special results.

As before, let $P = f(Q)$ denote the price of the common product that is produced by $N$ firms, and let $C_k$ denote the cost function of firm $k$. Then the profit of firm $k$ can be obtained as

$$\varphi_k(x_1, \ldots, x_N) = x_k f(Q) - C_k(x_k), \tag{4.87}$$

where $x_k$ is the output of firm $k$, $Q = \sum_{l=1}^{N} x_l$ is the total production of the industry and no cost externalities are considered. Cooperation among firms can be achieved if each firm takes the profits of its competitors into account. If, for example, a firm has equity positions in the other firms, it would certainly also care about the other firms' profits. In this case, cooperation among firms can be achieved, since each firm's objective function includes the profits of its competitors (see for example Clayton and Jorgensen (2005)). A similar effect occurs if firms are linked by partial equity interests and joint ventures (see for example Reynolds and Snapp (1986) and Bresnahan (1986)).

Let the parameters $\gamma_{kl} \in [0,1]$ denote the degree of cooperation of firm $k$ toward firm $l$ $(k,l \in \{1,2,\ldots,N\}, k \neq l)$, then we assume that firm $k$ maximizes $\varphi_k + \sum_{l \neq k} \gamma_{kl}\varphi_l$ instead of its own profit (see also Kopel and Szidarovszky (2006)). Thus the payoff function of firm $k$ becomes

$$\Psi_k(x_1,\ldots,x_N) = (x_k f(Q) - C_k(x_k)) + \sum_{l\neq k} \gamma_{kl}(x_l f(Q) - C_l(x_l)), \quad (4.88)$$

in the case of the classical Cournot model. Other model variants can be examined in a similar manner.

We can rewrite the payoff function of firm $k$ as

$$\Psi_k(x_1,\ldots,x_N) = (x_k + S_k) f(x_k + Q_k) - C_k(x_k) - \sum_{l\neq k} \gamma_{kl} C_l(x_l), \quad (4.89)$$

where

$$S_k = \sum_{l\neq k} \gamma_{kl} x_l.$$

Notice that

$$\frac{\partial \Psi_k}{\partial x_k} = f(x_k + Q_k) + (x_k + S_k) f'(x_k + Q_k) - C_k'(x_k)$$

and

$$\frac{\partial^2 \Psi_k}{\partial x_k^2} = 2f'(x_k + Q_k) + (x_k + S_k) f''(x_k + Q_k) - C_k''(x_k).$$

Assume that a slightly more restrictive set of conditions than that assumed for concave oligopolies (see Sect. 2.1) is satisfied, that is,

(A) $f'(Q) < 0$,

(B) $z f''(Q) + f'(Q) \leq 0$,

(C) $f'(Q) - C_k''(x_k) < 0$,

for all $k$, all feasible values of $x_k$ and $Q$, and $0 \leq z \leq \sum_{l=1}^N L_l$. Then $\Psi_k$ is strictly concave in $x_k$ with fixed values of $Q_k$ and $S_k$, since $x_k + S_k \leq Q$ and so $\partial^2 \Psi_k / \partial x_k^2$ is negative. As earlier, let $L_k$ denote the finite capacity limit of firm $k$, then it has a unique best response function, which depends on both $Q_k$ and $S_k$ and is given by

$$R_k(Q_k, S_k) = \begin{cases} 0 & \text{if} \quad f(Q_k) + S_k f'(Q_k) - C_k'(0) \leq 0, \\ L_k & \text{if} \quad f(L_k + Q_k) + (L_k + S_k) f'(L_k + Q_k) - C_k'(L_k) \geq 0, \\ z_k^* & \text{otherwise,} \end{cases}$$

where $z_k^*$ is the unique solution of the equation

$$f(z_k + Q_k) + (z_k + S_k) f'(z_k + Q_k) - C_k'(z_k) = 0 \quad (4.90)$$

inside the interval $(0, L_k)$.

Implicitly differentiating this equation with respect to $Q_k$ and $S_k$, and considering $z_k = R_k(Q_k, S_k)$ we have

$$f'(R'_{kQ} + 1) + R'_{kQ} f' + (z_k + S_k) f''(R'_{kQ} + 1) - C''_k R'_{kQ} = 0 \tag{4.91}$$

and

$$f' R'_{kS} + (R'_{kS} + 1) f' + (z_k + S_k) f'' R'_{kS} - C''_k R'_{kS} = 0, \tag{4.92}$$

where $R'_{kQ} = \partial R_k / \partial Q_k$ and $R'_{kS} = \partial R_k / \partial S_k$. Therefore the derivatives of the best response function are given by

$$R'_{kQ} = -\frac{f' + (z_k + S_k) f''}{2f' + (z_k + S_k) f'' - C''_k} \tag{4.93}$$

and

$$R'_{kS} = -\frac{f'}{2f' + (z_k + S_k) f'' - C''_k}, \tag{4.94}$$

implying that

$$-1 < R'_{kQ} \leq 0 \text{ and } R'_{kS} < 0. \tag{4.95}$$

If in addition, $f' + (z_k + S_k) f'' - C''_k \leq 0$, then $-1 \leq R'_{kS} < 0$. The payoff function $\Psi_k$ of each firm is concave in $x_k$, continuous, and if each firm has a finite capacity limit $L_k$, then the Nikaido–Isoda theorem (see for example Forgo et al. (1999)) implies the existence of at least one Nash equilibrium.

Before examining the dynamic extensions and investigating the asymptotic behavior of the resulting systems we will briefly discuss the effect of partial cooperation on the equilibrium quantities. Given that $R_k(Q_k, S_k)$ denotes the best response of firm $k$ with partial cooperation, then clearly $R_k(Q_k, 0)$ is its best response in the standard oligopoly with no cooperation. One can also consider the best responses as functions of the total production level $Q$, as is usual in oligopoly theory and which was also introduced in Chap. 2. Then clearly

$$\widetilde{R}_k(Q, S_k) = \begin{cases} 0 & \text{if } f(Q) + S_k f'(Q) - C'_k(0) \leq 0, \\ L_k & \text{if } f(Q) + (L_k + S_k) f'(Q) - C'_k(L_k) \geq 0, \\ z_k & \text{otherwise,} \end{cases} \tag{4.96}$$

where $z_k$ is the unique solution of the equation

$$f(Q) + (z_k + S_k) f'(Q) - C'_k(z_k) = 0 \tag{4.97}$$

inside the interval $(0, L_k)$.

**Lemma 4.1.** *The reaction function $\widetilde{R}_k(Q, S_k)$ defined above is a decreasing function of both variables $Q$ and $S_k$.*

*Proof.* Let $g(z_k, Q, S_k)$ denote the left hand side of (4.97), then

$$\frac{\partial g}{\partial z_k} = f'(Q) - C_k''(z_k) < 0,$$
$$\frac{\partial g}{\partial Q} = f'(Q) + (z_k + S_k)f''(Q) \leq 0,$$
$$\frac{\partial g}{\partial S_k} = f'(Q) < 0,$$

so $g$ is strictly decreasing in $z_k$ and $S_k$, and is non-increasing in $Q$. Assume first that $Q^{(1)} < Q^{(2)}$, and fix the value of $S_k$. If with $Q = Q^{(1)}$ the first case of (4.96) occurs then assumption (B) implies that the same hold also for $Q = Q^{(2)}$, so $\widetilde{R}_k$ remains 0, and if with $Q = Q^{(1)}$the second case occurs, then $\widetilde{R}_k$ cannot increase any further. If with $Q = Q^{(1)}$ the third case of (4.96) occurs, then the monotonicity of $g$ in $Q$ implies that with $Q = Q^{(2)}$ the second case is impossible so the best response with $Q = Q^{(2)}$ is either zero or the solution of (4.97). In the first case $\widetilde{R}_k$ clearly decreases. It will also be shown that if the third case of (4.96) occurs with both $Q = Q^{(1)}$ and $Q = Q^{(2)}$, then still $\widetilde{R}_k(Q^{(1)}, S_k) \geq \widetilde{R}_k(Q^{(2)}, S_k)$. Assume this is not the case, then

$$0 = g(\widetilde{R}_k(Q^{(1)}, S_k), Q^{(1)}, S_k) > g(\widetilde{R}_k(Q^{(2)}, S_k), Q^{(1)}, S_k)$$
$$\geq g(\widetilde{R}_k(Q^{(2)}, S_k), Q^{(2)}, S_k) = 0,$$

which is an obvious contradiction. Assume next that $S_k^{(1)} < S_k^{(2)}$, and fix the value of $Q$. If with $S_k = S_k^{(1)}$ the first case of (4.96) occurs, then $f' < 0$ implies that the same holds for $S_k = S_k^{(2)}$, so $\widetilde{R}_k$ remains 0, and if the second case occurs, then $\widetilde{R}_k$ cannot increase any further. If with $S_k = S_k^{(1)}$ the third case of (4.96) occurs, then $f' < 0$ implies that with $S_k = S_k^{(2)}$ the second case is impossible, so the best response is either zero or the solution of (4.97). In the first case $\widetilde{R}_k$ decreases. Assume finally that with both $S_k = S_k^{(1)}$ and $S_k = S_k^{(2)}$ the third case of (4.96) occurs. It can easily be shown that in this case $\widetilde{R}_k(Q, S_k^{(1)}) > \widetilde{R}_k(Q, S_k^{(2)})$. Assume not, then

$$0 = g(\widetilde{R}_k(Q, S_k^{(1)}), Q, S_k^{(1)}) > g(\widetilde{R}_k(Q, S_k^{(1)}), Q, S_k^{(2)})$$
$$\geq g(\widetilde{R}_k(Q, S_k^{(2)}), Q, S_k^{(2)}) = 0,$$

which is again a contradiction. ∎

Let now $(\bar{x}_1, \ldots, \bar{x}_N)$ be an equilibrium of the oligopoly without partial cooperation, and let $\bar{Q} = \sum_{k=1}^{N} \bar{x}_k$. Assume that $(\bar{x}'_1, \ldots, \bar{x}'_N)$ is an equilibrium with partial cooperation, and let $\bar{Q}' = \sum_{k=1}^{N} \bar{x}'_k$ and $\bar{S}'_k = \sum_{l \neq k} \gamma_{kl} \bar{x}'_l$. Then the following theorem holds:

**Theorem 4.4.** *Under conditions (A)–(C), $\bar{Q}' \leq \bar{Q}$, that is, partial cooperation decreases the total production level of the industry.*

*Proof.* Assume in contrary, that $\bar{Q}' > \bar{Q}$. Then

$$\begin{aligned}\bar{Q}' &= \sum_{k=1}^{N} R_k(\bar{Q}', \bar{S}'_k) \leq \sum_{k=1}^{N} R_k(\bar{Q}', 0) \\ &\leq \sum_{k=1}^{N} R_k(\bar{Q}, 0) = \bar{Q},\end{aligned}$$

which is a contradiction. ∎

In order to obtain more interesting results assume that each firm has identical cooperation levels towards its competitors, that is, $\gamma_{kl} \equiv \gamma_k$ for all $l \neq k$. We can then rewrite (4.96) as[4]

$$\widetilde{R}_k(Q, \gamma_k) = \begin{cases} 0, & \text{if} \quad f(Q) + \gamma_k Q f'(Q) - C'_k(0) \leq 0, \\ L_k, & \text{if} \quad f(Q) + ((1-\gamma_k)L_k + \gamma_k Q) f'(Q) - C'_k(L_k) \geq 0, \\ z_k, & \text{otherwise}, \end{cases} \tag{4.98}$$

where $z_k$ is the unique solution of the equation

$$f(Q) + ((1-\gamma_k)z_k + \gamma_k Q) f'(Q) - C'_k(z_k) = 0. \tag{4.99}$$

For the current situation we modify conditions (B) and (C) to read

(B′) $(1+\gamma_k) f' + v f'' \leq 0$,

(C′) $(1-\gamma_k) f' - C''_k < 0$

for all $Q, v \in \left[0, \sum_{k=1}^{N} L_k\right]$ and $z_k \in [0, L_k]$.

**Lemma 4.2.** *The reaction function $\widetilde{R}_k(Q, \gamma_k)$ defined above is a decreasing function of $Q$, and a decreasing function of $\gamma_k$ in the domain defined by*

$$\widetilde{R}_k(Q, \gamma_k) \leq Q.$$

---

[4] Here in the notation we emphasize the dependence of $\widetilde{R}_k$ on $Q$ and $\gamma_k$, since in this case $S_k = \gamma_k(Q - x_k)$, and use the same notation for this new form of the reaction function.

*Proof.* Now let $g(z_k, Q, \gamma_k)$ denote the left hand side of (4.99), then for all feasible $z_k$ and $Q$,

$$\frac{\partial g}{\partial z_k} = (1-\gamma_k)f'(Q) - C_k''(z_k) < 0,$$
$$\frac{\partial g}{\partial Q} = (1+\gamma_k)f'(Q) + ((1-\gamma_k)z_k + \gamma_k Q)f''(Q) \le 0,$$

and if $0 \le z_k \le Q$, then

$$\frac{\partial g}{\partial \gamma_k} = (Q - z_k)f'(Q) \le 0.$$

We will first prove that $\widetilde{R}_k(Q, \gamma_k)$ is decreasing in $Q$. Assume first that $Q^{(1)} < Q^{(2)}$, and fix the value of $\gamma_k$. If with $Q = Q^{(1)}$ the first case of (4.98) occurs, then condition $(B')$ implies that the same holds for $Q = Q^{(2)}$, so $\widetilde{R}_k$ remains the same. Assume next that the second case of (4.98) occurs with $Q = Q^{(1)}$, then $\widetilde{R}_k$ cannot increase further. If with $Q = Q^{(1)}$, the third case of (4.98) occurs, then with $Q = Q^{(2)}$ the second case of (4.98) is impossible, so either $\widetilde{R}_k(Q^{(2)}, \gamma_k)$ is zero or is the solution of (4.99). In the first case, $\widetilde{R}_k$ decreases, and it will easily be proven that this is the case even if the third case of (4.98) occurs with $Q = Q^{(2)}$. Assume not, that is, $\widetilde{R}_k(Q^{(1)}, \gamma_k) < \widetilde{R}_k(Q^{(2)}, \gamma_k)$. Then

$$\begin{aligned} 0 &= g(\widetilde{R}_k(Q^{(1)}, \gamma_k), Q^{(1)}, \gamma_k) > g(\widetilde{R}_k(Q^{(2)}, \gamma_k), Q^{(1)}, \gamma_k) \\ &\ge g(\widetilde{R}_k(Q^{(2)}, \gamma_k), Q^{(2)}, \gamma_k) = 0, \end{aligned}$$

which is a contradiction. Next it will be shown that $R_k(Q, \gamma_k)$ is also decreasing in $\gamma_k$ in the domain $\{R_k(Q, \gamma_k) \le Q\}$. Assume next that $\gamma_k^{(1)} < \gamma_k^{(2)}$ and fix the value of $Q$. If the first case of (4.98) occurs with $\gamma_k^{(1)}$, then the negativity of $f'$ implies that the same holds for $\gamma_k^{(2)}$, and if the second case occurs with $\gamma_k^{(1)}$, then $\widetilde{R}_k$ cannot increase further. Assume next that with $\gamma_k = \gamma_k^{(1)}$ the third case occurs. Then the second case is impossible for $\gamma_k = \gamma_k^{(2)}$, since if firm $k$ selects its maximal output level $L_k$, then $Q \ge L_k$. So either $\widetilde{R}_k(Q, \gamma_k^{(2)})$ is zero or the solution of (4.99). It will also be proven that even in this case, $\widetilde{R}_k(Q, \gamma_k^{(2)}) \le \widetilde{R}_k(Q, \gamma_k^{(1)})$. Assume not, then

$$\begin{aligned} 0 &= g(\widetilde{R}(Q, \gamma_k^{(1)}), Q, \gamma_k^{(1)}) > g(\widetilde{R}(Q, \gamma_k^{(2)}), Q, \gamma_k^{(1)}) \\ &\ge g(\widetilde{R}_k(Q_k, \gamma_k^{(2)}), Q, \gamma_k^{(2)}) = 0, \end{aligned}$$

which is again a contradiction. ■

Let $(\bar{x}_1, \ldots, \bar{x}_N)$ be an equilibrium of the oligopoly with partial cooperation levels $\bar{\gamma}_1, \ldots, \bar{\gamma}_N$. Assume that at least one player increases its cooperation level, so

$\bar{\gamma}'_k \geq \bar{\gamma}_k$ for all $k$ and let $(\bar{x}'_1, \ldots, \bar{x}'_N)$ be the new equilibrium. Let $\bar{Q} = \sum_{k=1}^{N} \bar{x}_k$ and $\bar{Q}' = \sum_{k=1}^{N} \bar{x}'_k$ be the total production levels of the industry in the two cases. Then the following theorem shows that any increase in cooperation levels results in a decrease of the industry output.

**Theorem 4.5.** *If conditions* $(A)$, $(B')$ *and* $(C')$ *hold, then* $\bar{Q} \geq \bar{Q}'$.

*Proof.* Assume on the contrary that $\bar{Q} < \widetilde{Q}$. Then

$$\begin{aligned} \bar{Q} = \sum_{k=1}^{N} R_k(\bar{Q}, \bar{\gamma}_k) &\geq \sum_{k=1}^{N} R_k(\bar{Q}', \bar{\gamma}_k) \\ &\geq \sum_{k=1}^{N} R_k(\bar{Q}', \bar{\gamma}'_k) = \bar{Q}', \end{aligned}$$

which is an obvious contradiction. ■

### 4.5.1 Local Stability Analysis

We will consider only continuous time scales since the discrete time case can be examined analogously to the model discussed in Sect. 4.4. In this subsection we consider the local stability of the equilibria and in the next subsection the global dynamics. In the current context the continuous time dynamical model (1.31) becomes

$$\dot{x}_k(t) = \alpha_k(R_k(\sum_{l \neq k} x_l, \sum_{l \neq k} \gamma_{kl} x_l) - x_k), \tag{4.100}$$

where $R_k(Q_k, S_k)$ is the best response function of firm $k$.

The Jacobian of this system at an equilibrium has the structure

$$\begin{pmatrix} -a_1 & a_1(R'_{1Q} + \gamma_{12} R'_{1S}) & \ldots & a_1(R'_{1Q} + \gamma_{1N} R'_{1S}) \\ a_2(R'_{2Q} + \gamma_{21} R'_{2S}) & -a_2 & \ldots & a_2(R'_{2Q} + \gamma_{2N} R'_{2S}) \\ \vdots & \vdots & \ddots & \vdots \\ a_N(R'_{NQ} + \gamma_{N1} R'_{NS}) & a_N(R'_{NQ} + \gamma_{N2} R'_{NS}) & \ldots & -a_N \end{pmatrix}. \tag{4.101}$$

In the following analysis we will consider this matrix at an interior equilibrium.

In the general case unfortunately, this Jacobian does not have any special structure that makes it possible to express its eigenvalue equation in a simple form. In some important special cases however it is possible to do so. In the general case computational methods are available to compute the eigenvalues and to check the stability conditions. Before turning to special cases, three sufficient conditions for stability will be presented for the general case.

**Theorem 4.6.** *Assume that for all* $k$, $a_k > 0$ *and*

$$(N-1)R'_{kQ} + \left(\sum_{l \neq k} \gamma_{kl}\right) R'_{kS} > -1, \tag{4.102}$$

*then the equilibrium is locally asymptotically stable.*

*Proof.* Notice that condition (4.102) implies that the Jacobian is strictly diagonally dominant in every row with negative diagonal elements. Then the Gerschgorin-cycle theorem (see for example, Szidarovszky and Yakowitz (1978)) implies that all eigenvalues have negative real parts. ■

Applying this result to the transpose of the Jacobian one easily obtains the following theorem.

**Theorem 4.7.** *Assume that for all* $k$, $a_k > 0$ *and*

$$a_k + \sum_{l \neq k} a_l (R'_{lQ} + \gamma_{lk} R'_{lS}) > 0, \tag{4.103}$$

*then the equilibrium is locally asymptotically stable.*

The application of Theorem B.7 given in Appendix B also offers a sufficient stability condition. Notice that the Jacobian matrix (4.101) can be factored as

$$\boldsymbol{A}(\boldsymbol{R}_Q + \boldsymbol{R}_S \boldsymbol{G}) \tag{4.104}$$

where $\boldsymbol{A} = diag(a_1, a_2, \ldots, a_N)$, $\boldsymbol{R}_S = diag(R'_{1S}, R'_{2S}, \ldots, R'_{NS})$,

$$\boldsymbol{R}_Q = \begin{pmatrix} -1 & R'_{1Q} & \cdots & R'_{1Q} \\ R'_{2Q} & -1 & \cdots & R'_{2Q} \\ \vdots & \vdots & \ddots & \vdots \\ R'_{NQ} & R'_{NQ} & \cdots & -1 \end{pmatrix} \text{ and } \boldsymbol{G} = \begin{pmatrix} 0 & \gamma_{12} & \cdots & \gamma_{1N} \\ \gamma_{21} & 0 & \cdots & \gamma_{2N} \\ \vdots & \vdots & & \vdots \\ \gamma_{N1} & \gamma_{N2} & \cdots & 0 \end{pmatrix},$$

and observe that $\boldsymbol{A}$ is positive definite, if $a_k > 0$ for all $k$. The above considerations make it possible to assert the following theorem:

**Theorem 4.8.** *Assume that* $a_k > 0$ *for all* $k$, $(\boldsymbol{R}_Q + \boldsymbol{R}_S \boldsymbol{G}) + (\boldsymbol{R}_Q + \boldsymbol{R}_S \boldsymbol{G})^T$ *is negative definite. Then the equilibrium is locally asymptotically stable.*

The condition of this theorem is satisfied, if $\boldsymbol{R}_Q + \boldsymbol{R}_Q^T$ is negative definite and $\boldsymbol{R}_S \boldsymbol{G} + \boldsymbol{G}^T \boldsymbol{R}_S$ is negative semi-definite.

Consider now the special case when $\gamma_{kl} \equiv \gamma_k$, that is, the cooperation levels of each firm are identical toward its competitors. In this case let

$$r_k = R'_{kQ} + \gamma_k R'_{kS} = -\frac{f'(1+\gamma_k) + (x_k + \gamma_k Q_k) f''}{2f' + (x_k + \gamma_k Q_k) f'' - C''_k},$$

then the Jacobian has exactly the same form as (2.46) for concave oligopolies.

Notice that under conditions (A), (B′) and (C′), there holds $-1 < r_k \leq 0$, similar to the case of concave oligopolies. Hence one can assert the following theorem:

**Theorem 4.9.** *Assume that $a_k > 0$ for all $k$. Under conditions (A), (B′), (C′) and with identical $\gamma_{kl}$ ($l \neq k$) values for all $k$, the equilibrium is locally asymptotically stable.*

The analysis of global asymptotic stability of the equilibrium with continuous time adjustment as well as the introduction of continuously distributed time lags can be carried out in a similar fashion to the cases considered in the previous chapters.

### 4.5.2 Global Dynamics

In this section we will illustrate the type of global dynamics that can arise in oligopolies with partial cooperation under a discrete time adjustment process by considering a specific example.

*Example 4.15.* We consider the hyperbolic price function $f(Q) = A/Q$ and linear cost functions $C_k(x_k) = d_k + c_k x_k \quad (k = 1, 2, \ldots, N)$. Assume that the firms have identical cooperation levels toward their competitors, so $\gamma_{kl} \equiv \gamma_k$ for all $l \neq k$. Then the payoff function (4.89) of firm $k$ can be written as

$$\Psi_k(x_1, \ldots, x_N) = \frac{(x_k + \gamma_k Q_k)A}{x_k + Q_k} - (d_k + c_k x_k) - \sum_{l \neq k} \gamma_k (d_l + c_l x_l). \quad (4.105)$$

Assuming an interior optimum, the first order conditions imply that

$$\frac{AQ_k(1-\gamma_k)}{(x_k + Q_k)^2} - c_k = 0,$$

from which we have the solution

$$z_k^* = \sqrt{\frac{AQ_k(1-\gamma_k)}{c_k}} - Q_k. \quad (4.106)$$

Let $L_k$ denote the capacity limit of firm $k$, then the strict concavity of the payoff function (4.105) implies that the best response of firm $k$ can be obtained as

$$R_k(Q_k) = \begin{cases} 0 & \text{if} \quad z_k^* \leq 0, \\ L_k & \text{if} \quad z_k^* \geq L_k, \\ z_k^* & \text{otherwise.} \end{cases} \tag{4.107}$$

Notice that the best response functions have the same shape as those considered, for example, in Example 1.5 and Example 3.4.

*Example 4.16.* In order to investigate the effect of a change in the cooperation levels $\gamma_k$ on the equilibrium values and on the global dynamics, following Example 3.4 we consider the semi-symmetric case. That is, we consider (4.106) with $Q_1 = (N-1)x_2$ and $Q_2 = x_1 + (N-2)x_2$, that is, the production decisions made by firm 1 and the identical firms $2, \ldots, N$ are captured by the two-dimensional dynamical system

$$T: \begin{cases} x_1(t+1) = (1-a_1)x_1(t) + a_1 R_1((N-1)x_2), \\ x_2(t+1) = (1-a_2)x_2(t) + a_2 R_2(x_1 + (N-2)x_2). \end{cases}$$

The unique positive equilibrium of this system may be written as

$$\begin{aligned} \bar{x}_1 &= \frac{(N-1)A(1-\gamma_1)(1-\gamma_2)\left[(N-1)(1-\gamma_1)c_2 - (N-2)(1-\gamma_2)c_1\right]}{(c_1(1-\gamma_2) + (N-1)(1-\gamma_1)c_2)^2}, \\ \bar{x}_2 &= \frac{Ac_1(N-1)(1-\gamma_1)(1-\gamma_2)^2}{(c_1(1-\gamma_2) + (N-1)(1-\gamma_1)c_2)^2}. \end{aligned} \tag{4.108}$$

It is interesting to note that if one of the cooperation levels equals 1 (full cooperation), then both equilibrium quantities vanish. This is due to the fact that a fully cooperative firm behaves like a profit maximizing monopolist (and therefore, due to the particular form of the isoelastic demand function, it selects a quantity close to zero, as mentioned in Example 1.5, whereas the other firm selects a small (close to zero) quantity as best reply. Notice also that if the degrees of cooperation of all firms are identical, so that $\gamma_k = \gamma$ for $k = 1, 2, \ldots, N$, then the firms' equilibrium quantities become

$$\begin{aligned} \bar{x}_1 &= \frac{(N-1)A(1-\gamma)\left[(N-1)c_2 - (N-2)c_1\right]}{(c_1 + (N-1)c_2)^2}, \\ \bar{x}_2 = \ldots = \bar{x}_N &= \frac{c_1(N-1)A(1-\gamma)}{(c_1 + (N-1)c_2)^2}. \end{aligned} \tag{4.109}$$

These expressions coincide with the expression we derived earlier in Example 4.3 if we let $\gamma = 0$ (no cooperation). Note that in the case of identical cooperation levels of all firms, from the expression of the individual equilibrium values (4.109) we can easily deduce that not only the total industry output decreases for increasing cooperation levels, but also the *individual* equilibrium quantities. However, because of the more complicated expressions of individual equilibrium output quantities (4.108)

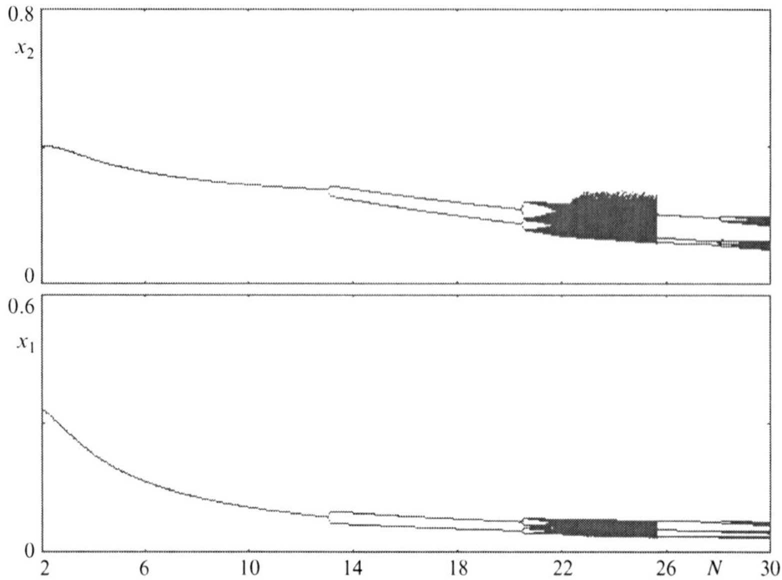


**Fig. 4.20** Example 4.15; the discrete time oligopoly under partial cooperation – the semi-symmetric case. Bifurcation diagrams of outputs with respect to the number of firms $N$. Here $A = 16, a_1 = 0.4, a_2 = 0.3, c_1 = 5, c_2 = 6, L_1 = L_2 = 2$ and $\gamma = 0.5$

in the case of non-identical cooperation levels, it is not clear if this property holds when cooperation levels $\gamma_1$ and $\gamma_2$ differ.

The dynamic behavior of the model with partial cooperation and identical cooperation levels is the same as the dynamical behavior of the model considered in Example 3.4, since the former can be obtained by the latter by just replacing the parameter $A$ with $A(1 - \gamma)$. From this property we can easily deduce that the stability condition (3.15), obtained for the semi-symmetric case with no cooperation, also holds for the model with partial cooperation with identical cooperation levels, since this stability condition is independent of the parameter $A$. Moreover, as numerical explorations suggest, the same statement holds even for the stability of periodic cycles. To illustrate this fact, in Fig. 4.20, obtained with parameters $A = 16, a_1 = 0.4, a_2 = 0.3, c_1 = 5, c_2 = 6, L_1 = L_2 = 2$ and $\gamma = 0.5$, we show bifurcations that occur as the number of firms increases. A simple comparison with Fig. 3.6, which has been obtained for the same set of parameters but without cooperation, shows that the losses in stability occur at the same values of the bifurcation parameter.

The bifurcation diagram in Fig. 4.21 shows the effect of changes in the levels of cooperation $\gamma_k$ on the stability of the positive equilibrium and the kind of asymptotic dynamics that can occur. Here the parameter $\gamma_2$ is taken as a bifurcation parameter and $N = 6$, $a_1 = 0.5$, $a_2 = 0.4, \gamma_1 = 0.5$. All the other parameters have the same values as in Fig. 4.20. In this case we can see that the equilibrium loses

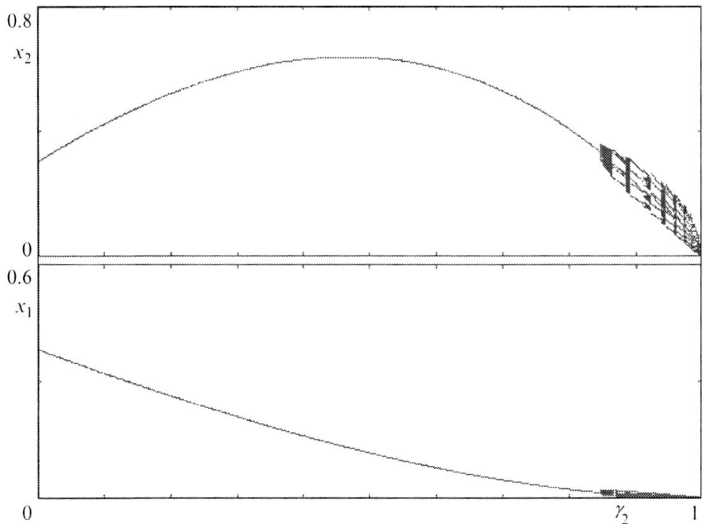


**Fig. 4.21** Example 4.15; the discrete time oligopoly under partial cooperation – the semi-symmetric case. Bifurcation diagrams of outputs with respect to $\gamma_2$ for $N = 6$. Here $a_1 = 0.5$, $a_2 = 0.4$, $\gamma_1 = 0.5$ and all other parameters are as in Fig. 4.20

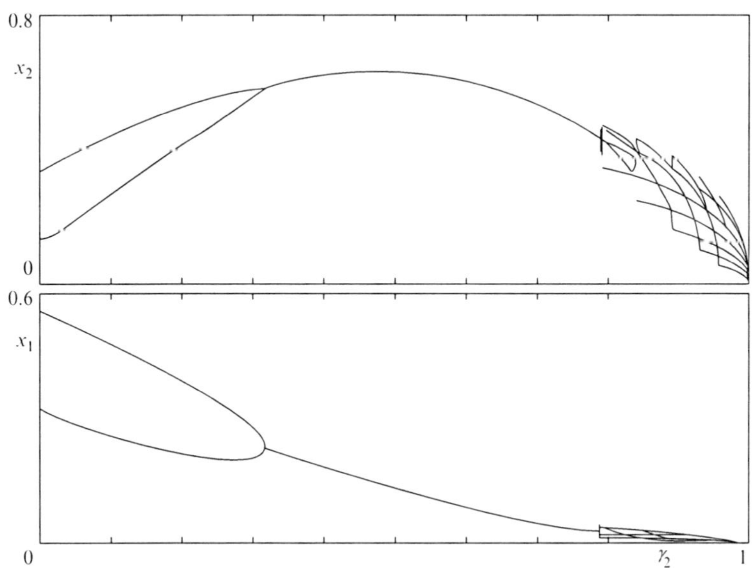


**Fig. 4.22** Example 4.15; the discrete time oligopoly under partial cooperation – the semi-symmetric case. Bifurcation diagrams of outputs with respect to $\gamma_2$ for $N = 8$. All other parameters are as in Fig. 4.21

stability as the bifurcation parameter $\gamma_2$ tends towards 1. This result is quite surprising as one would expect a fully cooperative firm to have a stabilizing effect. With respect to the monotonicity property of the individual quantities mentioned above, the bifurcation diagram reveals that for an increasing cooperation level of firm 2, the equilibrium quantity of firm 1 first increases, and then decreases, whereas firm 2's equilibrium quantity decreases throughout. The counterintuitive destabilizing effect of an increase in the cooperation level $\gamma_k$ of one of the firms is not a general property. This becomes obvious from the diagram shown in Fig. 4.22 (obtained with the same set of parameters as in Fig. 4.21; the only difference being that the number of firms has increased to $N = 8$). Here, an increase in the cooperation level of firm 2 leads first to a stabilization of the equilibrium through a period halving bifurcation. Then a further increase in the cooperation level finally causes destabilization and transition to periodic attractors of increasing period as well as to chaotic behavior.

We close this subsection by remarking that the bifurcation diagrams obtained with different values of $\gamma_1$ and $\gamma_2$ for increasing values of $N$ or increasing values of $a_k$ as bifurcation parameters are qualitatively very similar to those shown in Fig. 3.6 and 3.7 respectively. The only difference is that with higher cooperation levels both individual quantities and their fluctuations are generally reduced.

# Chapter 5
# Oligopolies with Misspecified and Uncertain Price Functions, and Learning

The previous chapters have already dealt with the behavior of boundedly rational firms in an oligopoly. Although the firms know the true demand relationship, we have assumed that they do not know their competitors' quantity choices. Instead they form expectations about these quantities and they base their own decisions on these beliefs. In particular, we have focused on several adjustment processes that firms might use to determine their quantity selections and we have investigated the circumstances under which such adjustment processes might lead to convergence to the Nash equilibrium of the static oligopoly game. However, the information that firms have about the environment may be incomplete on several accounts. For example, players may misspecify the true demand function or just misestimate the slope of the demand relationship, the reservation price, or the market saturation point. However, if firms base their decisions on such wrong estimates, they will realize that their beliefs are incorrect, since the market data they observe (for example, market prices or quantities) will be different from their predictions. Obviously, firms will try to update their beliefs on the demand relationship and this will give rise to an adjustment process. In other words, firms will try to learn the game they are playing. Following this line of thought, in this chapter we study oligopoly models under the assumption that firms either use misspecified price functions (Sect. 5.1) or do not know certain parameters of the market demand (Sect. 5.2). The main questions we want to answer are the following. If we understand an equilibrium in a game as a steady state of some non-equilibrium process of adjustment and "learning," what happens if the players use an incorrect model of their environment? Does a reasonable adaptive process (for example, based on the best response) converge to anything? If so, to what does it converge? Is the limit that can be observed when the players play their perceived games (close to) an equilibrium of the underlying true model? Is the observed situation consistent with the (limit) beliefs of the players?

In Sect. 5.1 we consider a framework based on the idea of Léonard and Nishimura (1999), who derive similar insights for a simple Cournot duopoly model with decreasing reaction functions. We demonstrate that in situations where players choose their actions based on a misspecified model of the environment, additional self-confirming steady states may emerge, despite the fact that the Nash equilibrium of the game under perfect knowledge is unique. We will derive (sufficient) conditions for the local and global stability of these steady states. For discrete time

G.I. Bischi et al., *Nonlinear Oligopolies*, DOI 10.1007/978-3-642-02106-0_5, 

models the various steady states of the game may have quite complicated basins of attraction and, as a consequence, the long run outcome of the game may be highly dependent upon initial conditions. We will also study the continuous time version introduced by Chiarella and Szidarovszky (2001*b*) and investigate how the asymptotic properties are altered by time delays in obtaining and implementing information on the output of the rivals.

A weakness of the framework presented in Sect. 5.1 is that it does not take into account the fact that players might want to change their subjective (misspecified) demand functions. Such an approach can be justified by pointing out that the firms do not know the cost functions of their competitors and therefore are not able to derive the output decisions of their competitors. This implies that they are not able to estimate the whole quantity sold in the market. So, the price they observe does not convey sufficient information for them to realize that they are using a misspecified demand function. An alternative to such a setup is presented in Sect. 5.2, where the firms use a local linear approximation of the price function based on only their own outputs. Two types of dynamics are examined. First believed best response dynamics are examined, and then the case of adaptive adjustment processes is discussed.

Three special adaptive learning models are introduced and examined in Sect. 5.3. Based on their beliefs of the price function each firm computes its believed equilibrium output and price. There the observed discrepancy between the price estimate and the realized market price not only allows the players to conclude that they are using an incorrect estimate of the demand function, but they are also allowed to adaptively adjust the believed demand (or price) function. To be more precise, an $N$-firm single-product Cournot oligopoly where the demand and cost functions are linear is considered. Cost functions are completely known by all firms and, although they know that the (inverse) demand relationship is linear, they either do not know the slope, or the reservation price. Each firm has its own estimate of the unknown market parameter and, by solving a static game, determines its own production quantity as well as an expectation on the production quantity of the rest of the industry (and hence, an expected industry output and an expected price). While firms will never observe the realized industry output, they can see if the realized market price differs from their expected price. This will make players aware that their estimate of the market parameter is wrong, leading them to update their estimate. Our main goal in this section is to investigate the conditions under which such a simple learning process has a unique steady state determined by the true market parameters and if the adjustment process converges to this steady state. In other words, we are interested in situations in which the firms *learn* the true demand. We will also see that adjustment processes of these kinds are not always convergent, and we examine their global dynamics including their basins of attraction.

Section 5.4 introduces the case of uncertain price functions, when each firm believes in a randomized function. It is assumed that they want to maximize their expected profits and minimize their variances. By introducing a linear utility function it is shown that the game can be reduced to deterministic oligopolies with misspecified price functions.

## 5.1 Misspecified Price Functions

As before, let $x_k$ denote the output of firm $k$ $(1 \le k \le N)$, $C_k(x_k)$ its cost, so no externalities are assumed, and $p = f(Q)$ the true price function, where $Q = \sum_{k=1}^{N} x_k$. Assume that the firms only have estimates of the price function $f$ and let $\widetilde{f_k}$, denote firm $k$'s belief about $f$. In order to be realistic, $\widetilde{f_k}$ is assumed to be strictly decreasing. In the case of full information each firm $k$ knows the true price function. Hence, given the observed market price $p$, from the equation

$$p = f(Q_k + x_k) \tag{5.1}$$

each firm would be able to obtain the output of the rest of the industry, namely

$$Q_k = f^{-1}(p) - x_k.$$

In the case of misspecified price functions, the true price is again given by $p = f(Q_k + x_k)$. However firm $k$ believes that it is $p = \widetilde{f_k}(Q_k + x_k)$, so its estimate $\widetilde{Q}_k$ about the output of the rest of the industry satisfies the equation

$$\widetilde{f_k}(\widetilde{Q}_k + x_k) = p = f(Q_k + x_k),$$

implying that

$$\widetilde{Q}_k = (\widetilde{f_k}^{-1} o f)(Q_k + x_k) - x_k. \tag{5.2}$$

Firm $k$ also believes that its profit at any time period $t + 1$ is

$$\widetilde{\varphi}_k = x_k \widetilde{f_k}(x_k + Q_k^E(t+1)) - C_k(x_k), \tag{5.3}$$

where $Q_k^E(t+1)$ is its expectation of the output of the rest of the industry in period $t + 1$. The best response $\widetilde{R}_k$ of this firm given its belief can be obtained in the same way as it was shown by relation (1.3) in Chap. 1. However in this case, the true price function $f$ has to be replaced by the believed price function $\widetilde{f_k}$.

Assuming discrete time scales, using (5.2) and the fact that the firms form adaptive expectations on the output of the rest of the industry we obtain the dynamic model

$$x_k(t+1) = \widetilde{R}_k\left(Q_k^E(t) + \alpha_k\left((\widetilde{f_k}^{-1} o f)\left(\sum_{l=1}^{N} x_l(t)\right) - x_k(t) - Q_k^E(t)\right)\right), \tag{5.4}$$

$$Q_k^E(t+1) = Q_k^E(t) + \alpha_k\left((\widetilde{f_k}^{-1} o f)\left(\sum_{l=1}^{N} x_l(t)\right) - x_k(t) - Q_k^E(t)\right), \tag{5.5}$$

for $k = 1, 2, \ldots, N$, where the firm's adaptive expectations are based on its belief $\widetilde{Q}_k$ on the output of the rest of the industry. Here $\alpha_k$ is a sign-preserving function for all $k$.

The corresponding discrete time model with partial adjustment towards the best response has the form

$$x_k(t+1) = x_k(t) + \alpha_k \left( \widetilde{R}_k \left( (\widetilde{f}_k^{-1} o f)(\sum_{l=1}^{N} x_l(t)) - x_k(t) \right) - x_k(t) \right) \quad (5.6)$$

for $k = 1, 2, \ldots, N$.

Under the assumption of continuous time scales and that each firm adjusts its output in the direction toward its believed best response we have the model

$$\dot{x}_k(t) = \alpha_k \left( \widetilde{R}_k \left( (\widetilde{f}_k^{-1} o f)(\sum_{l=1}^{N} x_l(t)) - x_k(t) \right) - x_k(t) \right). \quad (5.7)$$

Notice that in the case of full knowledge of the price function we have $\widetilde{R}_k = R_k$ and $\widetilde{f}_k = f$ for all $k$, so that $(\widetilde{f}_k^{-1} o f)$ is the identity function and the models (5.4)–(5.7) formally reduce to the models (1.28)–(1.31) introduced earlier in Chap. 1. For the sake of simplicity we introduce the notation $H_k = (\widetilde{f}_k^{-1} o f)$. If $\widetilde{f}_k$ is a good approximation of $f$, then $H_k$ is a good approximation of the identity function. In the subsequent parts of this section, the asymptotic behavior of systems (5.4)–(5.7) will be examined. Since usually $\widetilde{f}_k^{-1} o f$ differs from the identity function, and the best response functions $\widetilde{R}_k$ are different from the full information best responses $R_k$, the steady states of these systems are usually different from the Nash equilibria of the full information case. Even if any of these systems is asymptotically stable, the outputs of the firms will not converge to the Nash equilibria. The trajectories will instead converge to the steady state of the system, which can be called a *believed* or *subjective equilibrium.*

*Example 5.1.* Assume that the true price function is isoelastic, $f(Q) = A/Q$, but firm $k$ believes that it is $\widetilde{f}_k(Q) = A_k/Q$, where $A_k \neq A$. Since firm $k$ does not know the true price function $f$, it is not able to derive the true value of $Q_k$. After having observed a particular price $p$, it is only able to estimate this quantity by using the relationship given in (5.2), which in the present case becomes

$$\frac{A_k}{\widetilde{Q}_k + x_k} = p = \frac{A}{Q_k + x_k}.$$

So firm $k$ believes that the output of the rest of the industry is

$$\widetilde{Q}_k = \frac{A_k - p x_k}{p}.$$

Using the right hand equality in the first equation of this example the above quantity can be rewritten in terms of the unknown value $A$ as

$$\widetilde{Q}_k = \frac{(A_k - A)x_k + A_k Q_k}{A}.$$

Firm $k$ can now use this value $\widetilde{Q}_k$ to determine its best response. The expected profit of firm $k$ is

$$\widetilde{\varphi}_k = \frac{A_k x_k}{x_k + \widetilde{Q}_k} - (d_k + c_k x_k),$$

where we assume a linear cost function, which is known by the firm. The best response is therefore given by

$$\widetilde{R}_k(\widetilde{Q}_k) = \begin{cases} 0 & \text{if } z_k^* \leq 0, \\ L_k & \text{if } z_k^* \geq 0, \\ z_k^* & \text{otherwise,} \end{cases}$$

where

$$z_k^* = \sqrt{\frac{A_k \widetilde{Q}_k}{c_k}} - \widetilde{Q}_k.$$

We will next determine the subjective equilibrium in the case when it is interior and we will realize that it does not coincide with the full information Nash equilibrium. The quantities in the subjective equilibrium satisfy the equation

$$x_k = \sqrt{\frac{A_k}{c_k A}((A_k - A)x_k + A_k Q_k)} - \frac{1}{A}((A_k - A)x_k + A_k Q_k),$$

where we have simply used the expressions provided above. Using the relation $Q = x_k + Q_k$, this implies that

$$\frac{A_k Q}{A} = \sqrt{\frac{A_k}{c_k A}(A_k Q - A x_k)},$$

so that the market quantity of firm $k$ can be expressed in terms of the realized industry output $Q$ as

$$x_k = \frac{A_k(AQ - c_k Q^2)}{A^2}.$$

By adding up the expressions for the individual quantities offered by the firms $k = 1, 2, \ldots, N$, we obtain a simple equation for the realized industry output $Q$,

$$Q = \frac{Q}{A}\sum_{k=1}^{N} A_k - \frac{Q^2}{A^2}\sum_{k=1}^{N} A_k c_k.$$

This further implies that the realized industry output at the subjective equilibrium can be expressed as

$$Q = \frac{A\left(\sum_{k=1}^{N} A_k - A\right)}{\sum_{k=1}^{N} A_k c_k}.$$

In Example 1.5 we determined the output of the industry in the full information case as

$$Q = \frac{(N-1)A}{\sum_{k=1}^{N} c_k}.$$

It is important to realize that the quantities in the subjective equilibrium differ from the quantities in the full information case. Of course, if $A_k = A$ for each $k$, that is all firms know the true price function, then the expressions above coincide. ▼

Oligopolies without full information have been examined by several authors. Okuguchi (1976) investigated discrete time dynamic models without full information, and his stability analysis was based on the contraction mapping theorem. Szidarovszky and Okuguchi (1990) discussed the asymptotic properties of dynamic oligopolies with perceived marginal costs. Kirman (1975, 1983), and Gates et al. (1982) should also be mentioned as early contributions. Léonard and Nishimura (1999) assumed that the firms know the shape of the demand function but they misspecify its scale. They show that if players (slightly) over- or underestimate the true demand, then an adaptive process based on the best replies converges towards a unique steady state that differs from the full-information (Nash) equilibrium. They also demonstrate that this steady state may lose stability as the misspecification error (of one firm) becomes larger. The general case has been briefly analyzed in Szidarovszky et al. (2008) for the concave case.

In his early paper Kirman (1975) considers a simple duopoly model, where he assumes that the duopolists are not aware that their demand depends on each other's action. The players choose their quantities such that the expected profit of the next period is maximized and the duopolists update their estimates of the parameters of the (misspecified) perceived model. Within this simple framework, he shows that instead of converging to the "true" situation, the beliefs of the agents may drive the model towards some other outcome. In addition to the result that agents are not able to learn the true equilibrium, it is also shown in Kirman (1975, 1983) that if convergence to the full information equilibrium fails, the process may become path dependent, that is the particular equilibrium that can be observed depends on the starting conditions. Furthermore, Brousseau and Kirman (1993) find regions of stability as well as complicated dynamics in their simulations, whilst Kirman (1995) makes some remarks on basins of attraction. Schinkel et al. (2002) consider an oligopolistic price setting model where firms do not know the market demand but have demand conjectures instead. They analyze the global dynamics and show that the particular equilibrium that is reached in the long run depends on the initial beliefs. These observations are interesting, since they once again stress the fact that we need to study the global dynamics of the market game, in particular the characteristics of its possible long run outcomes and their respective basins of attraction.

For further results on continuous time models we refer the reader to Chiarella and Szidarovszky (2004), where firms may also misspecify the shape of the demand function and not only its scale.

### 5.1.1 Discrete Time Models and Local Stability

We consider first the model (5.4)–(5.5), and note that it is the mathematically equivalent to the dynamical system (1.28)–(1.29) introduced in Chap. 1. In Chap. 2 we have determined the Jacobian of this system as the matrix given in (2.15), which in the current situation assumes the particular form

$$\begin{pmatrix} \bar{J}_{11} & \bar{J}_{12} \\ \bar{J}_{21} & \bar{J}_{22} \end{pmatrix},$$

where

$$\bar{J}_{11} = \begin{pmatrix} r_1 a_1 (h_1 - 1) & r_1 a_1 h_1 & \dots & r_1 a_1 h_1 \\ r_2 a_2 h_2 & r_2 a_2 (h_2 - 1) & \dots & r_2 a_2 h_2 \\ \vdots & \vdots & & \vdots \\ r_N a_N h_N & r_N a_N h_N & \dots & r_N a_N (h_N - 1) \end{pmatrix},$$

$$\bar{J}_{12} = diag\left(r_1(1-a_1), r_2(1-a_2), \dots, r_N(1-a_N)\right),$$

$$\bar{J}_{21} = \begin{pmatrix} a_1 (h_1 - 1) & a_1 h_1 & \dots & a_1 h_1 \\ a_2 h_2 & a_2 (h_2 - 1) & \dots & a_2 h_2 \\ \vdots & \vdots & \ddots & \vdots \\ a_N h_N & a_N h_N & \dots & a_N (h_N - 1) \end{pmatrix},$$

$$\bar{J}_{22} = diag\left(1-a_1, 1-a_2, \dots, 1-a_N\right),$$

with $r_k = \widetilde{R}'_k$, $h_k = H'_k = (\widetilde{f_k} o f)'$ at the steady state of the system and $a_k = \alpha'_k(0)$. Since both $f$ and $\widetilde{f_k}$ are strictly decreasing, $H_k$ is strictly increasing, so it is reasonable to assume that $h_k > 0$. The eigenvalue equation of the Jacobian can be written similarly to (2.17)–(2.18) as

$$r_k a_k (h_k - 1) u_k + r_k a_k h_k \sum_{l \neq k} u_l + r_k (1 - a_k) v_k = \lambda u_k, \tag{5.8}$$

$$a_k (h_k - 1) u_k + a_k h_k \sum_{l \neq k} u_l + (1 - a_k) v_k = \lambda v_k \qquad (k = 1, 2, \ldots, N). \tag{5.9}$$

Subtract the $r_k$-multiple of the second equation from the first one to obtain

$$\lambda (u_k - r_k v_k) = 0.$$

We may assume that $\lambda \neq 0$, so $u_k = r_k v_k$. If we substitute this relation into (5.8) we see that

$$(r_k a_k (h_k - 1) + (1 - a_k))\, u_k + r_k a_k h_k \sum_{l \neq k} u_l = \lambda u_k,$$

which is the eigenvalue equation of the $N \times N$ matrix

$$H = \begin{pmatrix} r_1 a_1 (h_1 - 1) + 1 - a_1 & r_1 a_1 h_1 & \ldots & r_1 a_1 h_1 \\ r_2 a_2 h_2 & r_2 a_2 (h_2 - 1) + 1 - a_2 & & r_2 a_2 h_2 \\ \vdots & \vdots & \ddots & \vdots \\ r_N a_N h_N & r_N a_N h_N & \ldots & r_N a_N (h_N - 1) + 1 - a_N \end{pmatrix}. \tag{5.10}$$

Observe that this matrix is the Jacobian of system (5.6), the model with partial adjustment towards the best response. Therefore if local asymptotic stability is our concern, then the stability conditions of the systems (5.4)–(5.5) and (5.6) are again equivalent, and the eigenvalues of the matrix (5.10) determine whether or not an equilibrium is stable.

Assume that functions $f$, $\widetilde{f}_k$ and $C_k$ for all $k$ are twice continuously differentiable, furthermore,

(A) $f'(Q) < 0$, $\widetilde{f}_k'(Q) < 0$,

(B) $x_k \widetilde{f}_k''(Q) + \widetilde{f}_k'(Q) \leq 0$,

(C) $\widetilde{f}_k'(Q) - C_k''(x_k) < 0$,

for all feasible values of $x_k$ and $Q$ (see Chap. 2, Sect. 2.1, for an economic interpretation of these conditions).

The main stability result of this section is given in the following theorem.

**Theorem 5.1.** *Assume that $a_k > 0$ for all $k$, and conditions (A)–(C) are satisfied.*

*(i) The equilibrium is locally asymptotically stable if for all $k$,*

$$a_k (1 + r_k) < 2 \tag{5.11}$$

*and*

$$\sum_{k=1}^{N} \frac{r_k a_k h_k}{2 - a_k(1 + r_k)} > -1. \tag{5.12}$$

*(ii) The equilibrium is unstable if for at least one k,*

$$a_k(1 + r_k) \geq 2$$

*or*

$$\sum_{k=1}^{N} \frac{r_k a_k h_k}{2 - a_k(1 + r_k)} < -1.$$

*Proof.* The structure of matrix $\boldsymbol{H}$ is the same as (E.4) shown in Appendix E. Therefore the eigenvalue equation of $\boldsymbol{H}$ can be written as (E.5), which here has the special form

$$\prod_{k=1}^{N} (1 - a_k(1 + r_k) - \lambda) \cdot \left[ 1 + \sum_{k=1}^{N} \frac{r_k a_k h_k}{1 - a_k(1 + r_k) - \lambda} \right] = 0. \tag{5.13}$$

If we assume that both $f'$ and $\widetilde{f}_k'$ are negative, then $H_k'$ is positive, so $h_k > 0$.

Conditions (B) and (C) imply that the believed price functions satisfy the conditions of concave oligopolies stated at the beginning of Sect. 2.1. Under these conditions

$$-1 < r_k \leq 0$$

for all $k$, which can be proved similarly to (2.7) when $f'$ and $f''$ are replaced by $\widetilde{f}_k'$ and $\widetilde{f}_k''$. Assume that the firms are numbered in such a way that the different $a_k(1 + r_k)$ values are

$$a_1(1 + r_1) > a_2(1 + r_2) > \ldots > a_s(1 + r_s)$$

and they are repeated $m_1, m_2, \ldots, m_s$ times. By adding the terms with identical denominators in the bracketed expression and denoting by $\theta_j$ the sum of the corresponding numerators $r_k a_k h_k$, we can rewrite (5.13) as (2.24). Therefore the proof of Theorem 2.1 can be applied to show the assertion. ■

In the full information case $\widetilde{f}_k \equiv f$ for all $k$, so $H_k$ is the identity function, and $h_k = 1$ for all $k$. In this special case, the matrix (5.10) reduces to (2.20), and Theorem 5.1 specializes to Theorem 2.1.

*Example 5.2.* Assume as in Example 5.1 that the price function is isoelastic. Let $f(Q) = A/Q$ be the true price function and assume that firm $k$ believes that the price function is $\widetilde{f}_k(Q) = A_k/Q$, where $A_k > 0$ is a constant. The believed best response function of firm $k$ has the same derivative as given by (3.3) with the only difference being that $A$ is replaced by $A_k$, so it has the same properties as in the full information case. The local asymptotic stability of the equilibrium then can be examined by following the argument of Sect. 3.1.1, with identical conclusions. ▼

### 5.1.2 Discrete Time Models and Global Dynamics

In order to illustrate the asymptotic behavior of oligopolies with misspecified price functions and to show the dependence of the believed equilibrium on the way in which firms misspecify the price function we study a particular example.

*Example 5.3.* Assume again that the true price function is $f(Q) = A/Q$, but firm $k$ believes that the price function is $\widetilde{f}_k(Q) = A/(Q + \varepsilon_k)$ with some $\varepsilon_k \geq 0$. Assume that the firms have linear cost functions denoted by $C_k(x_k) = d_k + c_k x_k$. Let $L_k$ denote the capacity limit of firm $k$. Firm $k$'s expected profit is (assuming naive expectations)

$$\widetilde{\varphi}_k = \frac{Ax_k}{x_k + \widetilde{Q}_k + \varepsilon_k} - (d_k + c_k x_k).$$

Therefore, the best reply is given by

$$\widetilde{R}_k(\widetilde{Q}_k) = \begin{cases} 0 & \text{if} \quad z_k^* \leq 0, \\ L_k & \text{if} \quad z_k^* \geq 0, \\ z_k^* & \quad \text{otherwise}, \end{cases}$$

where

$$z_k^* = \sqrt{\frac{A(\widetilde{Q}_k + \varepsilon_k)}{c_k}} - (\widetilde{Q}_k + \varepsilon_k). \tag{5.14}$$

Firm $k$ is able to observe the market price $p$, so it can determine the believed output of the rest of the industry based on its price function estimate. So from the equation

$$\frac{A}{x_k + \widetilde{Q}_k + \varepsilon_k} = p,$$

the firm believes that

$$\widetilde{Q}_k = \frac{A}{p} - x_k - \varepsilon_k.$$

Then the best response of this firm the $z_k^*$ is given by (5.14). However the price is $p = A/(x_k + Q_k)$, where $Q_k$ is the true output of the rest of the industry, so

$$\widetilde{Q}_k = (x_k + Q_k) - x_k - \varepsilon_k = Q_k - \varepsilon_k$$

and so

$$z_k^* = \sqrt{\frac{AQ_k}{c_k}} - Q_k,$$

which is the best response of the firm with full information. This observation means that even if firms misspecify the price function in this way, they still take the right decision with their best responses. ▼

The global asymptotic stability of the models (5.4)–(5.5) for best reply dynamics with adaptive expectations and (5.6) for the partial adjustment dynamics can be discussed similarly to the other cases. The Jacobian of system (5.6) has the special form (5.10). By using Lemma B.2 of Appendix B, we see that the equilibrium is globally asymptotically stable if for all $k$ and all feasible values of $x_1, \ldots, x_N$,

$$|r_k a_k(h_k - 1) + 1 - a_k| + (N - 1)\,|r_k a_k h_k| < 1, \tag{5.15}$$

the feasible output sets are compact and all functions $R_k$, $\alpha_k$ and $H_k$ are continuously differentiable on these sets. Under conditions (A)–(C) and by assuming that $h_k > 0$, this inequality can be written as the pair of inequalities

$$r_k a_k(h_k - 1) + 1 - a_k - (N - 1)r_k a_k h_k < 1$$

and

$$-r_k a_k(h_k - 1) - 1 + a_k - (N - 1)r_k a_k h_k < 1.$$

These relations can be rewritten as

$$a_k\Big(r_k h_k(2 - N) - r_k - 1\Big) < 0 \tag{5.16}$$

and

$$a_k\Big(1 + r_k(1 - N h_k)\Big) < 2. \tag{5.17}$$

Consider the first relation (5.16). Since $-1 < r_k \le 0$, for $N = 2$ it always holds, and for $N = 3$ it holds if

$$-r_k(h_k + 1) - 1 < 0$$

that is, when

$$r_k > \frac{-1}{h_k + 1}.$$

If $N$ becomes larger, then $r_k h_k(2 - N)$ becomes a large positive number, so (5.16) no longer holds, and stability is lost. Consider next relation (5.17). Since

$$1 + r_k(1 - Nh_k) > 0 \qquad \text{for} \quad -1 < r_k \leq 0,$$

it holds if

$$a_k < \frac{2}{1 + r_k(1 - Nh_k)}.$$

Notice that if $r_k < 0$, then the right hand side converges to zero as $N \to \infty$. So this condition becomes very restrictive for large values of $N$.

The previous derivation holds if all trajectories are interior. In the general case, when boundary points occur we can use Theorem B.3. For each best response the feasible output set is divided into three regions depending on the zero, $L_k$, or interior value of the response function. In the first two cases $r_k = 0$, so in each subregion the Jacobian has the same form as (5.10) with the difference that at least one $r_k$ value equals zero. Consider now a given value of $k$. If $r_k \neq 0$, then (5.15) remains the same, so the conclusions are also the same as given above. If $r_k = 0$, then the left hand side of (5.15) is $|1 - a_k|$, which is below unity if $a_k < 2$. After these rather general consideration about global stability we now present an example which demonstrates that multiple "believed" equilibria may occur. Therefore, global stability is lost and the tools of global analysis have to be employed to gain further insights.

*Example 5.4.* In this example we restrict our attention to a duopoly game and study its global dynamics. Again, the true inverse demand relationship is given by

$$p = f(Q), \tag{5.18}$$

but firms do not know this relationship. Instead they subjectively believe, as before, that the price function is $\widetilde{f}_k(Q), k = 1, 2$, that is firm $k$ believes that

$$p = \widetilde{f}_k(Q), \qquad (k = 1, 2). \tag{5.19}$$

Considering the cost side, we allow for externalities. That is, the costs $C_k$ of firm $k$ may not only depend on the firm's own quantity $x_k$ but also on the quantity of the other firm (see Chap. 1, Sect. 1.1, for a motivation as to why positive externalities of this kind might occur). We assume that players are able to take this effect on their own costs into account. More precisely, at the beginning of period $t$ firm $k$ chooses $x_k(t)$ such that the expected profit

$$\widetilde{\varphi}_k = x_k \widetilde{f}_k(x_k + Q_k^{E,prior}) - C_k(x_k, Q_k^{E,prior}), \tag{5.20}$$

is maximized. Here the quantity $Q_k^{E,prior}$ denotes firm $k$'s belief in period $t$ about the quantity chosen by its rival and the additional superscript *prior* is used to indicate that at the time when this expectation is formed, firm 1 knows the previous price $p(t-1)$, but *does not yet know* the new price $p(t)$. Using its subjective demand relationship $\widetilde{f}_k$ and its belief on the competitor's quantity $Q_k^{E,prior}$, the duopolists determine their best responses $\widetilde{R}_1(x_2^{E,prior})$ and $\widetilde{R}_2(x_1^{E,prior})$. In general, these best responses are different from the reaction functions in the usual sense, that is if firms were to posses full information of the demand relationship. At the end of period $t$, after both firms have sold their selected quantities at the market, they observe the realized market price $p(t)$, but do not observe the industry output or the quantity supplied by its rival. They use the realized price $p(t)$ to update their belief on the rival's choice. The updated belief $Q_k^{E,post}(t)$ for firm $k$ is derived from relationship

$$p(t) = \widetilde{f}_k(x_k(t) + Q_k^{E,post}(t)), \tag{5.21}$$

or equivalently by

$$Q_k^{E,post}(t) = \widetilde{f}_k^{-1}(p(t)) - x_k(t). \tag{5.22}$$

If we assume naive expectations following Léonard and Nishimura (1999) and Bischi et al. (2004b), we have

$$Q_k^{E,prior}(t+1) = Q_k^{E,post}(t), \tag{5.23}$$

which by using (5.22) yields

$$Q_k^{E,prior}(t+1) = \widetilde{f}_k^{-1}(p(t)) - x_k(t). \tag{5.24}$$

Summarizing, the dynamics of the duopoly with misspecified demand can be described as follows. The firms start with initial expectations about their rival's output and given their subjective demand relationships, the duopolists derive their best replies $\widetilde{R}_1(x_2^{E,prior})$ and $\widetilde{R}_2(x_1^{E,prior})$, with $Q_1^{E,prior} = x_2^{E,prior}$ and $Q_2^{E,prior} = x_1^{E,prior}$. The firms use the best replies to determine their quantity choices, for example, by using a partial adjustment towards the best response process with constant speeds of adjustment $a_1$ and $a_2$, so that

$$\begin{aligned} x_1(t+1) &= x_1(t) + a_1\left(\widetilde{R}_1(x_2^{E,prior}(t+1)) - x_1(t)\right), \\ x_2(t+1) &= x_2(t) + a_2\left(\widetilde{R}_2(x_1^{E,prior}(t+1)) - x_2(t)\right). \end{aligned} \tag{5.25}$$

The price that clears the market is determined by the true (but unknown) price function according to (5.18). After observing the current price, the firms use the relation (5.24) to update their beliefs on the rival's quantity. The expectation-feedback cycle then repeats itself. Using (5.18) and (5.24), the partial adjustment process can be written as

$$x_1(t+1) = x_1(t) + a_1\left(\widetilde{R}_1[\widetilde{f}_1^{-1}(f(x_1(t) + x_2(t))) - x_1(t)] - x_1(t)\right), \tag{5.26}$$

$$x_2(t+1) = x_2(t) + a_2\left(\widetilde{R}_2[\widetilde{f}_2^{-1}(f(x_1(t) + x_2(t))) - x_2(t)] - x_2(t)\right). \tag{5.27}$$

In general, this dynamical system involves the best replies which are based on misspecified beliefs. However, in the special case where firm $k$ mistakenly over- or underestimates the actual demand by a factor of $\varepsilon_k$, so that

$$\widetilde{f_k}^{-1}(p) = \varepsilon_k f^{-1}(p), \tag{5.28}$$

the dynamical system can be expressed in terms of the true reaction functions $R_1$ and $R_2$ (see again Léonard and Nishimura (1999), Bischi et al. (2004b)). This can be seen as follows. First note that with this misspecification we obtain $\widetilde{f_k}(Q) = f(\varepsilon_k^{-1}Q)$, and hence $\widetilde{f_k}'(Q) = \varepsilon_k^{-1}f'(\varepsilon_k^{-1}Q)$. Therefore, assuming an interior solution, the first order condition for firm $k$ can be written as

$$f(\varepsilon_k^{-1}x_k + \varepsilon_k^{-1}Q_k^{E,prior}) + \varepsilon_k^{-1}x_k f'(\varepsilon_k^{-1}x_k + \varepsilon_k^{-1}Q_k^{E,prior}) - C_k' = 0.$$

If we contrast this equation with the first order condition in the full information case ($\varepsilon_k = 1$) which implicitly defines the relation $x_k = R_k(Q_k^{E,prior})$, we can conclude that the first order condition together with assumption (5.28) implicitly defines the relation $\varepsilon_k^{-1}x_k = R_k(\varepsilon_k^{-1}Q_k^{E,prior})$, where $R_k$ denotes the reaction function in the full information case. Obviously, it follows further that the relations between the reaction functions $\widetilde{R}_k$ and the full information reaction functions are given by

$$\widetilde{R}_k(Q_k^{E,prior}) = \varepsilon_k R_k(\varepsilon_k^{-1}Q_k^{E,prior}). \tag{5.29}$$

Consequently, if we take (5.28) and (5.29) into account, the dynamical system based on partial adjustment towards the best response can be rewritten as

$$\begin{aligned} x_1(t+1) &= x_1(t) + a_1\left(\varepsilon_1 R_1[\frac{\varepsilon_1 - 1}{\varepsilon_1}x_1(t) + x_2(t)] - x_1(t)\right), \\ x_2(t+1) &= x_2(t) + a_2\left(\varepsilon_2 R_2[x_1(t) + \frac{\varepsilon_2 - 1}{\varepsilon_2}x_2(t)] - x_2(t)\right). \end{aligned} \tag{5.30}$$

Notice that if both players know the true demand ($\varepsilon_1 = \varepsilon_2 = 1$), then (5.30) reduces to the partial adjustment process introduced in Chap. 1. On the other hand, our derivations show that even if players over- or underestimate the demand by a certain factor $\varepsilon_k \neq 1$, the dynamics of the repeated duopoly game with misspecified demand is still governed by equations only involving the reaction functions of the full information case. It is clear that this property makes this particular type of misspecification quite appealing for further analysis. We will now briefly describe the effects of mistaken beliefs on the long-run properties of the dynamical system. More precisely, we will focus on the existence and stability of steady states if firms misspecify the demand relationship and we will study the extent and topological structure of the basins of attractions of these steady states. To consider a particular example, we follow the model of Sect. 3.2 and specify the full information reaction functions as

$$x_1 = R_1(x_2) = \mu_1 x_2 (1 - x_2), \quad x_2 = R_2(x_1) = \mu_2 x_1 (1 - x_1). \tag{5.31}$$

As before, the quantities $(x_1, x_2)$ are selected in the strategy space $[0, 1]^2$ and $\mu_k \in (1, 4]$. For simplicity we restrict our analysis to the best reply dynamics, that is we set $a_1 = a_2 = 1$, and obtain from (5.30) the dynamical system

$$\widetilde{T} : \begin{cases} x_1 (t+1) = & \mu_1 \left[ \varepsilon_1 x_2 (t) (1 - x_2 (t)) \right. \\ & \left. + (\varepsilon_1 - 1) x_1 (t) \left( 1 - \frac{\varepsilon_1 - 1}{\varepsilon_1} x_1 (t) - 2 x_2 (t) \right) \right], \\ x_2 (t+1) = & \mu_2 \left[ \varepsilon_2 x_1 (t) (1 - x_1 (t)) \right. \\ & \left. + (\varepsilon_2 - 1) x_2 (t) \left( 1 - \frac{\varepsilon_2 - 1}{\varepsilon_2} x_2 (t) - 2 x_1 (t) \right) \right], \end{cases} \tag{5.32}$$

where the map $\widetilde{T}$ defined above generates the dynamics of the game. Observe that if both firms know the true demand function so that, $\varepsilon_1 = \varepsilon_2 = 1$, then (5.32) reduces to the Cournot best reply dynamics given by

$$T : \begin{cases} x_1(t+1) = \mu_1 x_2(t) (1 - x_2(t)), \\ x_2(t+1) = \mu_2 x_1(t)(1 - x_1(t)), \end{cases} \tag{5.33}$$

which is a special case of the adjustment dynamics already investigated in Sect. 3.2 (see also Kopel (1996) and Bischi et al. (2000a)). In a similar fashion to the analysis presented in Sect. 3.2, for (5.32) a complete description of the stability regions of the emerging equilibria can be obtained for the case of homogenous firms $\mu_1 = \mu_2$ and $\varepsilon_1 = \varepsilon_2$. In fact, analytic expressions of the curves that constitute the boundaries of such regions can be obtained from a standard analysis of the eigenvalues of the Jacobian matrix (for details, see Bischi et al. (2004b)). Here we will instead focus on the model with heterogeneous firms where, although a rigorous analytical characterization cannot be given, the global dynamical properties of (5.32) can still be studied by a mixture of analytical and numerical methods. Figure 5.1a depicts the full information reaction function in a situation (for a choice of $\mu_1 \neq \mu_2$) where a unique Nash equilibrium exists. The question is, what happens if players misspecify the demand? Will the adjustment dynamics still converge to a steady state close to the Nash equilibrium of the true game? As will become clear, this depends on the global dynamics of the system. First observe that the steady states of the dynamical system with misspecified beliefs (5.32) are the real solutions of the algebraic system $\widetilde{T}(x_1, x_2) = (x_1, x_2)$, which yields the equations

$$\begin{aligned} \mu_1(\varepsilon_1 - 1)^2 x_1^2 + 2\mu_1\varepsilon_1 (\varepsilon_1 - 1) x_1 x_2 + \mu_1 \varepsilon_1^2 x_2^2 \\ + \varepsilon_1 (1 - \mu_1 (\varepsilon_1 - 1)) x_1 - \mu_1 \varepsilon_1^2 x_2 = 0, \\ \mu_2(\varepsilon_2 - 1)^2 x_2^2 + 2\mu_2\varepsilon_2 (\varepsilon_2 - 1) x_1 x_2 + \mu_2 \varepsilon_2^2 x_1^2 \\ + \varepsilon_2 (1 - \mu_2 (\varepsilon_2 - 1)) x_2 - \mu_2 \varepsilon_2^2 x_1 = 0. \end{aligned}$$

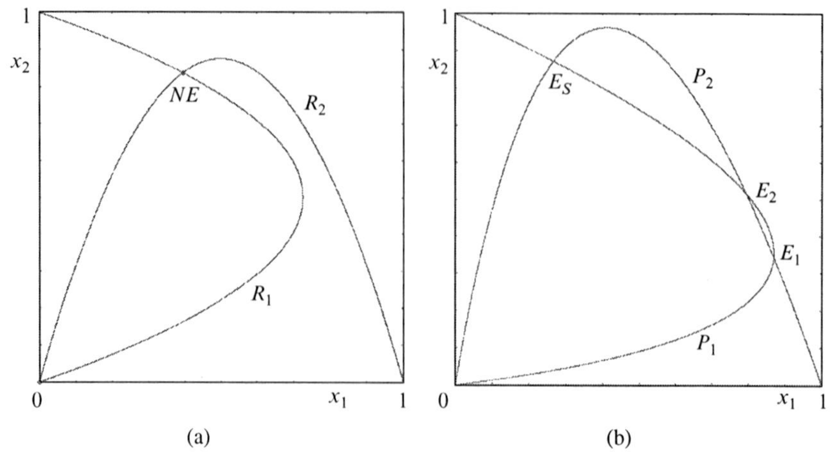


**Fig. 5.1** Occurrence of multiple steady states due to misspecified demand in the duopoly game with logistic reaction functions. (**a**) Reaction functions and the unique Nash equilibrium NE in the full information case, $\varepsilon_1 = 1, \varepsilon_2 = 1$. (**b**) Now $\varepsilon_1 \neq 1, \varepsilon_2 \neq 1$ and multiple steady states are obtained by overestimating the true demand

These equations represent two parabolas $\mathcal{P}_1$ and $\mathcal{P}_2$ in the $(x_1, x_2)$-plane. Both of them pass through the origin $(0,0)$ and the parabola $\mathcal{P}_1$ $(\mathcal{P}_2)$ intersects the vertical axis in the point $(0,1)$ (intersects the horizontal axis in the point $(1,0)$). So, besides the trivial fixed point $E_0 = (0,0)$, we may obtain one or three positive steady states which are located at the intersections of the two parabolas and are obtained as the real solutions of a cubic equation. Obviously, for $\varepsilon_k = 1$ the parabola $\mathcal{P}_k$ coincides with the reaction curve $R_k$. If $\varepsilon_k \neq 1$ $(k = 1,2)$, then the parabolas $\mathcal{P}_1$ and $\mathcal{P}_2$ no longer coincide with the reaction curves $R_1$ and $R_2$ and intersection points do not correspond to Nash equilibria of the "true" game, but to "subjective" equilibria of the "perceived" game. The qualitative representation in Fig. 5.1b illustrates the basic mechanism for the emergence of several subjective equilibria when one or both error parameters $\varepsilon_k$ are varied. In this situation there are three intersections of the two parabolas $\mathcal{P}_1$ and $\mathcal{P}_2$ in the interior of the unit square and, consequently, three potential long run outcomes of the game emerge (depending on their stability properties). If, for example, both $E_S$ and $E_1$ are locally asymptotically stable, then the adjustment process might guide the players to the (subjective) equilibrium $E_S$ and, hence, close to the Nash equilibrium of the true game. However, players might in the long run also end up in a situation represented by $E_1$ and, hence, far away from the Nash equilibrium of the true game.

To illustrate these arguments, we consider the following numerical example. Let $\mu_1 = 2.8$ and $\mu_2 = 2.9$. For these parameter values the Nash equilibrium of the true game is unique and globally stable for the adjustment process (5.33), which is obtained in the full information case, with $\varepsilon_1 = \varepsilon_2 = 1$. Starting from such a selection, if the misspecification parameter $\varepsilon_1$ is increased, first a stable cycle of period 2 appears (due to a saddle-node bifurcation). This is illustrated in Fig. 5.2a

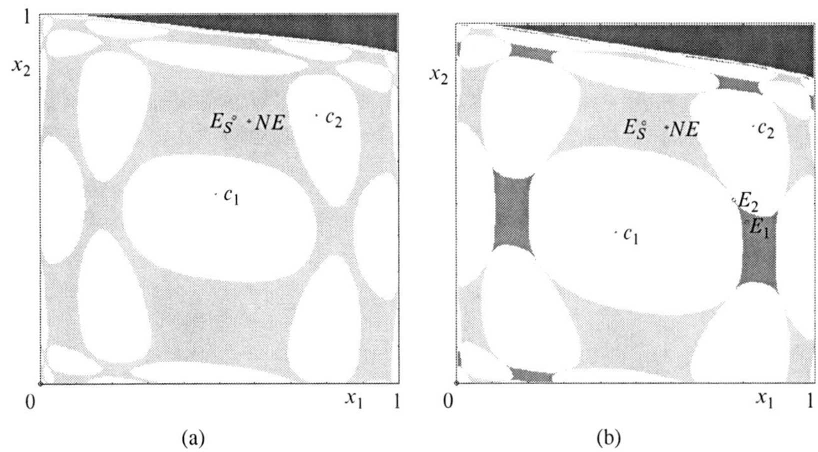


**Fig. 5.2** The duopoly game with logistic reaction functions. Firm 2 has full information but firm 1 has misspecified demand. (**a**) The misspecification parameter of firm 1 is $\varepsilon_1 = 1.12$, $E_S$ is the unique stable steady state (*light grey* basin of attraction) coexisting with a stable cycle of period 2 (*white basin*). Other parameters are $\mu_1 = 2.8, \mu_2 = 2.9, \varepsilon_1 = 1.12, \varepsilon_2 = 1$. (**b**) The misspecification parameter of firm 1 increases to $\varepsilon_1 = 1.18$. Two new subjective equilibria $E_1$, $E_2$ emerge, of which $E_1$ is stable (*dark grey* basin of attraction) and $E_2$ is unstable. Other parameters are $\mu_1 = 2.8, \mu_2 = 2.9, \varepsilon_1 = 1.18, \varepsilon_2 = 1$

with $\varepsilon_1 = 1.12$ and $\varepsilon_2 = 1$. In this case the strategy space consists of the basins of two coexisting attractors, namely the subjective equilibrium $E_S$ and a 2-cycle $C_2 = (c_1, c_2)$ (as well as a small portion of the basin of infinity). If the misspecification parameter $\varepsilon_1$ is further increased, two new steady states are created, denoted by $E_1$ and $E_2$ in Fig. 5.2b (obtained for $\varepsilon_1 = 1.18$). These new subjective equilibria are created via a saddle-node bifurcation (through a mechanism similar to the one shown in Fig. 5.1) and, as a result, they appear far away from $E_S$. The subjective equilibrium $E_1$ is stable (a stable node) and $E_2$ is unstable. Furthermore, a stable cycle $C_2$ coexists.

Observe that the Nash equilibrium $NE$ of the true game is located in the basin of $E_S$ and is quite near to $E_S$. If the initial quantities of the firms are located in the basin of $E_S$, then the adjustment process leads to a situation where the long run outcome is close to the Nash equilibrium of the true game. On the other hand, if the trajectories converge to $E_1$, then the adjustment process based on misspecified demand relationships leads the firms to an equilibrium which is quite different from the true Nash equilibrium. It is interesting to notice that Fig. 5.1b shows that in $E_1$ firm 1 has a higher market share, and it turns out that it also has a higher profit than firm 2 ($\varphi_1 = 0.465, \varphi_2 = 0.2$). Despite the fact that firm 2 knows the true demand, firm 1 (although unwittingly) achieves not only market dominance, but – with regard to the full information case – gains a higher profit, whereas firm 2's profit is reduced by more than 50%. ▼

### 5.1.3 Continuous Time Models

Consider the continuous model (5.7) and assume that conditions (A)–(C) hold, furthermore $h_k > 0$. We will first prove the following result.

**Theorem 5.2.** *Under assumption (A)–(C) and by assuming that $h_k > 0$ and $a_k > 0$ for all $k$, the equilibrium is always locally asymptotically stable under the continuous adjustment process (5.7).*

*Proof.* The Jacobian of system (5.7) can be written as

$$\begin{pmatrix} a_1[r_1(h_1-1)-1] & a_1 r_1 h_1 & \cdots & a_1 r_1 h_1 \\ a_2 r_2 h_2 & a_2[r_2(h_2-1)-1] & \cdots & a_2 r_2 h_2 \\ \vdots & \vdots & \ddots & \vdots \\ a_N r_N h_N & a_N r_N h_N & \cdots & a_N[r_N(h_N-1)-1] \end{pmatrix} \tag{5.34}$$

which is a straightforward extension of the Jacobian (2.46) of the full information case, since if $\widetilde{f}_k \equiv f$ for all $k$, then $H_k$ is the identity map with $h_k = H_k' = 1$. The eigenvalue equation has now the form

$$\prod_{k=1}^{N}[-a_k(1+r_k)-\lambda]\cdot\left[1+\sum_{k=1}^{N}\frac{a_k r_k h_k}{-a_k(1+r_k)-\lambda}\right]=0. \tag{5.35}$$

This equation is equivalent to (2.48) with the only difference that the $\theta_j$ values are now the sums of numerators $a_k r_k h_k$ with identical denominators.

Under conditions (A)–(C) of Sect. 5.1.1, Theorem 2.2 remains true, that is, the equilibrium is locally asymptotically stable. ∎

The case of isoelastic price function also can be examined in the same way as was demonstrated in Chap. 3 for the full information case, and the conclusions are also identical.

The global asymptotic stability of the equilibrium based on Lyapunov functions can be similarly discussed to the full information case. The details are omitted. We will only examine the effect of delayed information on the stability of the equilibrium.

Assume next that there is a time delay in obtaining and implementing information on the market price that is used in (5.2) by the firms to form their expectations on the output of the rest of the industry. By assuming the same type of weighting function as in Sect. 2.6 for the full information case, the dynamic model (5.7) becomes

$$\dot{x}_k(t)=\alpha_k\left(\widetilde{R}_k\left(\widetilde{f}_k^{-1}\left(\int_0^t w(t-s,T_k,m_k)f\left(\sum_{l=1}^{N}x_l(s)\right)\mathrm{d}s\right)-x_k(t)\right)-x_k(t)\right), \tag{5.36}$$

for $k = 1, 2, \ldots, N$, since firm $k$ observes the market price $f(Q)$ with delay. Similarly to the full information case we linearize this equation around the equilibrium to have

$$\dot{x}_{k\delta}(t) = a_k \left( r_k \left( h_k \int_0^t w(t-s, T_k, m_k) \sum_{l=1}^{N} x_{l\delta}(s)\mathrm{d}s - x_{k\delta}(t) \right) - x_{k\delta}(t) \right),$$

where $x_{k\delta}$ denotes the deviation of $x_k$ from its equilibrium level, and $a_k$, $r_k$ and $h_k$ are the same as in Sect. 5.1.1. By seeking the solution as $x_{k\delta} = v_k e^{\lambda t}$, substituting it into the linearized equation and letting $t \to \infty$, we obtain the equation

$$(\lambda + a_k(r_k + 1))v_k - \left( a_k r_k h_k \int_0^\infty w(s, T_k, m_k) e^{-\lambda s} \mathrm{d}s \right) \sum_{l=1}^{N} v_l = 0.$$

We can further simplify this equation by using the limiting values of the integrals (D.3) derived in Appendix D to obtain

$$A_k(\lambda)v_k + B_k(\lambda) \sum_{l \neq k} v_l = 0 \qquad (k = 1, 2, \ldots, N), \tag{5.37}$$

where

$$A_k(\lambda) = \lambda + a_k(r_k + 1) - a_k r_k h_k \left( 1 + \frac{\lambda T_k}{p_k} \right)^{-(m_k+1)}$$

and

$$B_k(\lambda) = a_k r_k h_k \left( 1 + \frac{\lambda T_k}{p_k} \right)^{-(m_k+1)},$$

with

$$p_k = \begin{cases} 1 & \text{if} \quad m_k = 0, \\ m_k & \text{if} \quad m_k > 0, \end{cases}$$

as before. System (5.37) has a non-trivial solution if its determinant is zero. Notice that the determinant has the same structure as (2.54) in the full information case, in addition its characteristic polynomial can also be expressed similarly to (2.55), which here has the form

$$\prod_{k=1}^{N} (\lambda + a_k(r_k + 1)) \left[ 1 - \sum_{k=1}^{N} \frac{a_k r_k h_k}{(\lambda + a_k(r_k + 1)) \left( 1 + \frac{\lambda T_k}{p_k} \right)^{m_k+1}} \right] = 0.$$

In the concave and isoelastic cases $r_k > -1$ for all $k$, so if $a_k > 0$ for all $k$, then the values $-a_k(r_k + 1)$ are negative, so in order to examine stability we have only to analyze the locations of the roots of the equation

$$1 - \sum_{k=1}^{N} \frac{a_k r_k h_k}{(\lambda + a_k(r_k + 1))\left(1 + \frac{\lambda T_k}{p_k}\right)^{m_k+1}} = 0. \tag{5.38}$$

In the general case computational methods can be used to locate the roots. In order to obtain analytic results we will consider the case of symmetric firms, when $a_k \equiv a, r_k \equiv r, h_k \equiv h, T_k \equiv T, m_k \equiv m$ and so $p_k \equiv p$. In both concave and isoelastic cases $-1 < r \leq 0, a > 0$ and $h > 0$. Notice that in the isoelastic case at most one firm can have a positive $r_k$ value, but in the symmetric case all firms would have positive derivatives $r_k$, which is impossible. Then (5.38) becomes the polynomial equation

$$(\lambda + a(r + 1))\left(1 + \frac{\lambda T}{p}\right)^{m+1} - Narh = 0. \tag{5.39}$$

Our results can be summarized in the following theorem.

**Theorem 5.3.** *Assume symmetric firms and that $h \geq 1/N$. The equilibrium with information lag is locally asymptotically stable if $T = 0$, or $T > 0$ and $m = 0$. If $T > 0$ and $m = 1$, then*

*(i) The equilibrium is locally asymptotically stable, if*

$$Nhr + 8(r + 1) > 0.$$

*(ii) The equilibrium is locally asymptotically stable for all $aT \neq 1 + \frac{8}{Nh}$, when*

$$Nhr + 8(r + 1) = 0.$$

*(iii) Otherwise the equilibrium is locally asymptotically stable if*

$$aT < (aT)_1^* \text{ or } aT > (aT)_2^*,$$

*where $(aT)_1^*$ and $(aT)_2^*$ ($(aT)_1^* < (aT)_2^*$) are given in (5.43), and the equilibrium is unstable, if*

$$(aT)_1^* < aT < (aT)_2^*.$$

*At the critical values $aT = (aT)_1^*$, and $aT = (aT)_2^*$ Hopf bifurcations occur giving the possibility of the birth of limit cycles around the equilibrium.*

*Proof.* Assume first that $T = 0$, that is, there is no time lag. Then (5.39) becomes

$$\lambda + a(1 + r(1 - Nh)) = 0,$$

with the only root $\lambda = -a(1 + r(1 - Nh))$. If $\widetilde{f_k}$ is a reasonable approximation of $f$, then $h \simeq 1$, so $Nh \geq 1$ implying that $\lambda < 0$ and the equilibrium is locally asymptotically stable.

Assume next that $T > 0$ and $m = 0$. Then (5.39) is reduces to the quadratic equation

$$\lambda^2 T + \lambda(1 + a(r + 1)T) + a(1 + r(1 - Nh)) = 0.$$

Since all coefficients are positive by assuming again that $Nh \geq 1$, the equilibrium is locally asymptotically stable.

Consider next the case of $T > 0$ and $m = 1$. Then (5.39) becomes the cubic equation

$$\lambda^3 T^2 + \lambda^2 (a(r+1)T^2 + 2T) + \lambda(1 + 2a(r+1)T) + a(1 + r(1 - Nh)) = 0. \quad (5.40)$$

All coefficients are positive if $Nh \geq 1$, and the Routh–Hurwitz stability criterion shows that the roots have negative real parts if and only if

$$(a(r+1)T^2 + 2T)(1 + 2aT(r+1)) > T^2 a(1 + r(1 - Nh)), \quad (5.41)$$

which is equivalent to the quadratic inequality

$$2(r+1)^2(aT)^2 + (aT)(4(r+1) + Nhr) + 2 > 0. \quad (5.42)$$

The discriminant of the left hand side of (5.42) is

$$(4(r+1) + Nhr)^2 - 16(r+1)^2 = Nhr(Nhr + 8(r+1)).$$

The first factor is negative, so we have the following cases:

Case 1. If $Nhr + 8(r+1) > 0$, then the discriminant is negative, so (5.42) always holds and the equilibrium is locally asymptotically stable.

Case 2. If $Nhr + 8(r+1) = 0$, then (5.42) holds for all values of $aT$ except the single root of the quadratic polynomial. So the equilibrium is locally asymptotically stable unless

$$aT = \frac{-4(r+1) - Nhr}{4(r+1)^2} = \frac{4(r+1)}{4(r+1)^2} = \frac{1}{r+1} = 1 + \frac{8}{Nh}.$$

Case 3. If $Nhr + 8(r+1) < 0$, then the quadratic polynomial (5.42) has two real roots,

$$(aT)^*_{1,2} = \frac{-4(r+1) - Nhr \pm \sqrt{Nhr(Nhr + 8(r+1))}}{4(r+1)^2}. \quad (5.43)$$

Since

$$-4(r+1) - Nhr = -(Nhr + 8(r+1)) + 4(r+1) > 0,$$

both roots are positive. Hence the equilibrium is locally asymptotically stable if

$$aT < (aT)^*_1 \text{ or } aT > (aT)^*_2$$

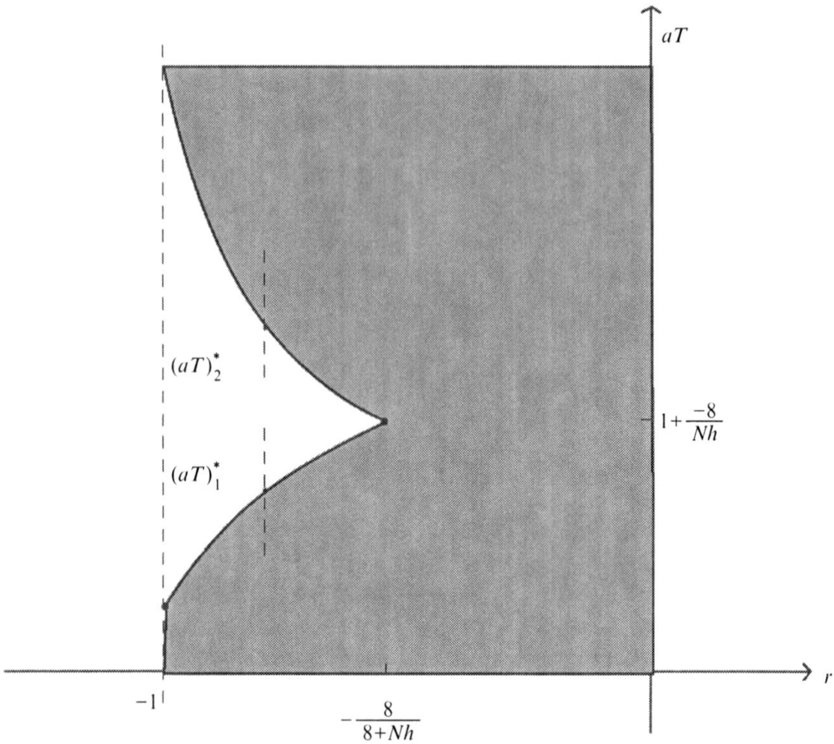


**Fig. 5.3** The continuous time model with symmetric firms and misspecified demand. The stability region (*shaded*) in the $(r, aT)$ plane. The parameter $r$ is the slope of the reaction function at the steady state, $a = \alpha'(0)$ and $T$ is the time lag in the weighting function. Notice how stability is lost then regained as $aT$ increases along the *dashed* vertical line. Hopf bifurcations may occur at the points $(aT)_1^*$ and $(aT)_2^*$

where $(aT)_1^* < (aT)_2^*$. The equilibrium is unstable if

$$(aT)_1^* < aT < (aT)_2^*.$$

The stability region is shown in Figure 5.3. Assume now that $-1 < r < -\frac{8}{8+Nh}$, then the equilibrium is asymptotically stable with small and large values of $aT$. With fixed such values of $r$, if $aT$ is gradually increasing from a very small value and crosses $(aT)_1^*$, then stability is lost. The instability holds until the value of $aT$ reaches $(aT)_2^*$, and after crossing this value stability is regained. We will next show, that at these critical points Hopf bifurcations occur. Select $T$ as the bifurcation parameter similarly to the full information case. Then the eigenvalues are functions of $T$, so $\lambda = \lambda(T)$. At the critical values of $T$, the inequality (5.41) becomes an inequality, so the cubic equation (5.40) can be rewritten as

$$\lambda^3T^2 + \lambda^2(a(r+1)T^2 + 2T) + \lambda\frac{T^2a(1+r(1-Nh))}{a(1+r)T^2+2T} + a(1+r(1-Nh))$$
$$= (\lambda T^2 + (a(r+1)T^2 + 2T))\left(\lambda^2 + \frac{a(1+r(1-Nh))}{a(1+r)T^2+2T}\right) = 0.$$

Therefore the eigenvalues are

$$\lambda_{1,2} = \pm i\sqrt{\frac{a(1+r(1-Nh))}{a(r+1)T^2+2T}}$$

and

$$\lambda_3 = -\frac{a(r+1)T^2+2T}{T^2} < 0.$$

So we have a pair of pure complex eigenvalues, and the third eigenvalue is negative. Differentiating equation (5.40) implicitly with respect to $T$ and using the notation $\dot{\lambda} = \frac{d\lambda}{dT}$ we have

$$\begin{aligned}3\lambda^2\dot{\lambda}T^2 &+ 2\lambda^3T + 2\lambda\dot{\lambda}(a(r+1)T^2+2T)\\ &+\lambda^2(2a(r+1)T+2) + \dot{\lambda}(1+2a(r+1)T) + \lambda 2a(r+1) = 0\end{aligned}$$

implying that

$$\dot{\lambda} = \frac{-2\lambda^3T - \lambda^2(2a(r+1)T+2) - 2\lambda a(r+1)}{3\lambda^2T^2 + 2\lambda(a(r+1)T^2+2T) + (1+2a(r+1)T)}. \tag{5.44}$$

For the sake of simplicity introduce the notation

$$\alpha^2 = \frac{a(1+r(1-Nh))}{a(r+1)T^2+2T}\left(= \frac{1+2aT(r+1)}{T^2}\right)$$

then $\lambda_{1,2} = \pm\alpha i$ and at these values

$$\begin{aligned}\dot{\lambda} &= \frac{\pm 2\alpha^3 iT + \alpha^2(2a(r+1)T+2) \mp 2a\alpha(r+1)i}{-3\alpha^2T^2 \pm 2\alpha i(a(r+1)T^2+2T) + (1+2a(r+1)T)}\\ &= \frac{\alpha^2(2a(r+1)T+2) + (\pm 2\alpha^3T \mp 2a\alpha(r+1))i}{-2\alpha^2T^2 \pm 2\alpha i(a(r+1)T^2+2T)}\end{aligned}$$

with real part

$$Re\dot{\lambda} = \frac{4\alpha^4T^2 - 4a^2\alpha^2(r+1)^2T^2 - 8a\alpha^2T(r+1)}{(-2\alpha^2T^2)^2 + 4\alpha^2(a(r+1)T^2+2T)^2}. \tag{5.45}$$

The numerator can be simplified to

$$4\alpha^2 T[T(\alpha^2 - a^2(r+1)^2) - 2a(r+1)].$$

Here the first factor is positive, and the second factor can be rewritten as

$$\frac{1-(aT)^2(r+1)^2}{T} \neq 0,$$

since it is easy to see that in Case 3

$$(aT)_1^* < \frac{1}{r+1} < (aT)_2^*. \tag{5.46}$$

Hence all conditions of the Hopf bifurcation theorem are satisfied, and therefore there is the possibility of the birth of limit cycles around the equilibrium. ■

Consider again the general equation (5.39) and assume very shallow best response functions so that

$$r > \frac{-1}{1+Nh}.$$

We can easily show that in this case all roots of this equation have negative real parts, so the equilibrium is asymptotically stable. On the contrary assume that $Re\lambda \geq 0$, then

$$|\lambda + a(r+1)| \geq a(r+1)$$

and

$$|1 + \frac{\lambda T}{p}| \geq 1,$$

so

$$\left|(\lambda + a(r+1))\left(1+\frac{\lambda T}{p}\right)^{m+1}\right| \geq a(r+1) > -Narh = |-Narh|,$$

and hence $\lambda$ cannot be a solution of (5.39).

In the general case, higher values of $m$ in (5.39) require the use of computational methods to locate the eigenvalues. We note again that the case when for all $k$, $\widetilde{f_k} \equiv f$ (that is, the full information case) is the special case of model (5.36) by selecting $H_k$ as the identity mapping with $h_k = 1$ for all $k$. However there are slight differences between the full information model presented in Sect. 2.6 and the model shown in this section. In the full information case for all firms we assumed time delays in the information on the output of the rest of the industry, and also in the firms' own output levels. In this section we assumed that the firms receive information only on the price, and they compute the output of the rest of the industry by using the observed market price. Here they use delayed information on the function values $f(Q_k + x_k)$ (which is the actual market price), but they use their most current output levels for $x_k$.

## 5.2 Cournot Oligopolies with Local Monopolistic Approximation

In this section we consider again a classical Cournot oligopoly model, where quantity setting firms have incomplete information about the price function. In particular, the firms do not know the shape of the true price function, although at each time step they are able to get a correct estimate of the local slope of the price function. Using this information, they solve the corresponding profit maximization problem by assuming that the true demand function is a linear function with that slope, and in addition by ignoring any effects of the competitors' outputs. As we shall see, despite such a rough approximation, which has been called "Local Monopolistic Approximation" (LMA) in Bischi et al. (2007), the adjustment process may converge to a Nash equilibrium of the game under the assumption of full information. For further work along these lines, see Negishi (1961), Silvestre (1977), and Tuinstra (2004).

### 5.2.1 *Adjustments with Local Monopolistic Approximation*

Let the price function $f$ and the cost functions $C_k$, $k = 1, \ldots, N$, be twice continuously differentiable. Assume that through market experiments at any time period each firm is able to get a correct estimate of the partial derivative

$$\frac{\partial f(x_k(t) + Q_k(t))}{\partial x_k} = f'(Q(t)), \tag{5.47}$$

which is used to obtain a simple "rule of thumb" for the computation of the expected price

$$p^e(t+1) = p(t) + f'(Q(t))(x_k(t+1) - x_k(t)) \tag{5.48}$$

where $p(t) = f\,(Q(t))$.

Of course, the approximation (5.48) is obtained more easily than complete information about the demand function (that involves values of the price or quantity that may be quite different from the current observations). Indeed, the estimate of $f'(Q)$ at time $t$ may be obtained by computing the effects of small price or quantity variations. For example, introducing a small output variation $\Delta x_k$ at time $t$, firm $k$ can compute

$$\frac{f(x_k(t) + Q_k(t) + \Delta x_k) - f\,(x_k(t) + Q_k(t))}{\Delta x_k}, \tag{5.49}$$

and we assume that this allows firm $i$ to get a correct estimate of $f'(Q)$. It is worth noting that such an estimate can also be obtained through small price variations since

$$\frac{df(Q)}{dQ} = \left[\frac{dQ(p)}{dp}\right]^{-1}, \tag{5.50}$$

which shows that information about the current price elasticity of demand is sufficient to obtain such an estimate. In practice, price experiments are commonly carried out by firms through price discounts, hence this information is usually readily available.

Notice that (5.48) is not a linear approximation of $f$, as firm $k$ neglects the influence of the competitor's production in the computation of the expected price. Needless to say that this is a very rough approximation. However, this might not be far from reality, as many authors point out (see for example Kirman (1975)). Moreover, as we shall see below, even if in the computation of the expected price the firms neglect the influence of competitors' outputs, the dynamic process generated by such an adjustment procedure may lead to convergence to the same equilibria as the best reply dynamics.

If firm $k$ uses (5.48) to compute the expected price, the expected profit for the next time period is approximated by

$$x_k(t+1)\left(f(Q(t))+f'(Q(t))(x_k(t+1)-x_k(t))\right)-C_k(x_k(t+1)) \quad (k=1,\ldots,N),$$

and the optimal response of firm $k$, under this information set, is computed as

$$\begin{aligned}&\widetilde{R}_k\left(Q_k(t),x_k(t)\right)\\&=\arg\max_{x_k\geq 0}\left\{x_k\left(f(x_k(t)+Q_k(t))+f'(x_k(t)+Q_k(t))(x_k-x_k(t))\right)-C_k(x_k)\right\}\end{aligned} \tag{5.51}$$

for $k=1,\ldots,N$. By assuming a positive optimum, the first order condition implies that

$$f(Q(t))+2f'(Q(t))x_k-f'(Q(t))x_k(t)-C_k'(x_k)=0 \quad (k=1,\ldots,N). \tag{5.52}$$

These first order conditions, computed at the equilibrium, are the same as the first order conditions obtained for the Cournot game with perfect knowledge of the price function $f$. Consequently, the steady states of the optimization problem with local monopolistic approximation are also Cournot–Nash equilibria of the Cournot game with complete knowledge of the price function. It is important to point out that this distinguishes the oligopoly models based on LMA from the oligopoly models with misspecified demand functions, which we have considered in the previous section of this chapter. Whereas with misspecified demand functions the steady states are no longer Nash equilibria of the true game, in the case of LMA the repeated decisions of boundedly rational players who do not know the global shape of the demand function may lead to convergence to a Nash equilibrium. Of course, the more refined the decision-making process and the corresponding decision rule, the more expensive it is likely to be to obtain data for such a rule. Therefore, especially when a (single) decision is not of crucial importance, no more than an approximate solution may be justified. Some authors denote such decisions which are based on simple and inexpensive computations as "optimally imperfect decisions" (see for example Baumol and Quandt (1964)). Notice that, in order to solve the optimization problem (5.51) at any time period $t$ firm $k$ needs only the following information: (1) Its current output $x_k(t)$; (2) The current price $p(t)$; (3) The current derivative $f'(Q(t))$; (4) Its own cost function $C_k(x_k)$.

A global study of the dynamic properties of the adjustment process based on the local monopolistic approximation of the demand function, is possible if the implicit equation (5.52) can be written in the form of an explicit discrete time dynamical system (that is if one can uniquely compute $x_k$ from (5.52) based on the knowledge of the state variables at time $t$). This outcome can be obtained if we consider suitable cost functions, such as:

1. Linear cost functions $C_k(x_k) = d_k + c_k x_k$, so that $C_k'(x_k) = c_k$ and (5.52) gives

$$x_k(t+1) = \frac{1}{2}x_k(t) - \frac{f(Q(t)) - c_k}{2f'(Q(t))} \qquad (k = 1, \ldots, N); \tag{5.53}$$

2. Quadratic cost functions: $C_k(x_k) = d_k + e_k x_k^2$, so that $C_k'(x_k) = 2e_k x_k$, and so (5.52) gives

$$x_k(t+1) = \frac{x_k(t)f'(Q(t)) - f(Q(t))}{2\left[f'(Q(t)) - e_k\right]} \qquad (k = 1, \ldots, N). \tag{5.54}$$

In the following examples we assume that the demand function is isoelastic and we study the dynamic properties of the corresponding model.

*Example 5.5.* Let us consider a duopoly model with the isoelastic price function,

$$p = f(Q) = \frac{1}{Q^\alpha}, \qquad \alpha > 0\,, \tag{5.55}$$

and linear cost functions $C_k = d_k + c_k x_k$. Notice that for $\alpha = 1$ we obtain again the hyperbolic price function already considered in several examples in this book. The model (5.53) with $N = 2$ and inverse demand function (5.55) becomes a two dimensional dynamical system, defined by the iterated map

$$\begin{aligned} x_1(t+1) &= \frac{1}{2}x_1(t) - \frac{1}{2\alpha}\left(x_1(t) + x_2(t)\right)\left(c_1\left(x_1(t) + x_2(t)\right)^\alpha - 1\right), \\ x_2(t+1) &= \frac{1}{2}x_2(t) - \frac{1}{2\alpha}\left(x_1(t) + x_2(t)\right)\left(c_2\left(x_1(t) + x_2(t)\right)^\alpha - 1\right). \end{aligned} \tag{5.56}$$

The equations for the determination of the fixed points, obtained by setting $x_k = x_k(t+1) = x_k(t)$ in (5.56), become

$$\begin{aligned} x_1 + \tfrac{1}{\alpha}\left(x_1 + x_2\right)\left(c_1\left(x_1 + x_2\right)^\alpha - 1\right) = 0, \\ x_2 + \tfrac{1}{\alpha}\left(x_1 + x_2\right)\left(c_2\left(x_1 + x_2\right)^\alpha - 1\right) = 0. \end{aligned} \tag{5.57}$$

After adding these two equations, we obtain

$$(x_1 + x_2)^\alpha = \frac{2-\alpha}{c_1 + c_2}.$$

This equilibrium condition shows us that a realistic non-vanishing steady state exists only if $\alpha < 2$. We can use this equation to substitute, for example, $x_2 = -x_1 + \left(\frac{2-\alpha}{c_1+c_2}\right)^{1/\alpha}$ in one of equations (5.57), from which we get the unique non-vanishing equilibrium $\bar{\mathbf{x}} = (\bar{x}_1, \bar{x}_2)$, with

$$\begin{aligned} \bar{x}_1 &= \frac{1}{\alpha}\left(\frac{2-\alpha}{c_1+c_2}\right)^{\frac{1}{\alpha}}\left(\frac{c_2 - c_1(1-\alpha)}{c_1+c_2}\right), \\ \bar{x}_2 &= \frac{1}{\alpha}\left(\frac{2-\alpha}{c_1+c_2}\right)^{\frac{1}{\alpha}}\left(\frac{c_1 - c_2(1-\alpha)}{c_1+c_2}\right). \end{aligned} \tag{5.58}$$

This equilibrium is positive if $\alpha > 1 - \min\{c_1/c_2, c_2/c_1\}$. The study of the stability of this equilibrium is particularly easy, because the Jacobian matrix for the map (5.56) given by

$$J(x_1, x_2) = \begin{pmatrix} \frac{1}{2} - \frac{1}{2\alpha}\left[(\alpha+1)c_1(x_1+x_2)^\alpha - 1\right] & -\frac{1}{2\alpha}\left[c_1(\alpha+1)(x_1+x_2)^\alpha - 1\right] \\ -\frac{1}{2\alpha}\left[c_2(\alpha+1)(x_1+x_2)^\alpha - 1\right] & \frac{1}{2} - \frac{1}{2\alpha}\left[(\alpha+1)c_2(x_1+x_2)^\alpha - 1\right] \end{pmatrix},$$

computed at the equilibrium becomes

$$J(\bar{x}_1, \bar{x}_2) = \begin{pmatrix} \frac{1}{2} - \frac{1}{2\alpha}\left[(\alpha+1)c_1\frac{2-\alpha}{c_1+c_2} - 1\right] & -\frac{1}{2\alpha}\left[c_1(\alpha+1)\frac{2-\alpha}{c_1+c_2} - 1\right] \\ -\frac{1}{2\alpha}\left[c_2(\alpha+1)\frac{2-\alpha}{c_1+c_2} - 1\right] & \frac{1}{2} - \frac{1}{2\alpha}\left[(\alpha+1)c_2\frac{2-\alpha}{c_1+c_2} - 1\right] \end{pmatrix}$$

and has the simple characteristic equation

$$\lambda^2 - \frac{1+\alpha}{2}\lambda + \frac{\alpha}{4} = 0.$$

Hence the eigenvalues are $\lambda_1 = 1/2$ and $\lambda_2 = \alpha/2$. This implies that the equilibrium $(\bar{x}_1, \bar{x}_2)$ is locally asymptotically stable for each $\alpha$ in the range $0 < \alpha < 2$.

This contrasts with the results obtained for the best reply dynamics with complete knowledge of the demand function as discussed by Puu (1991), in Example 3.4 with $\alpha = 1$. It has been shown there that the unique Nash equilibrium is given by (5.58) with $\alpha = 1$, that is

$$\bar{\mathbf{x}} = (\bar{x}_1, \bar{x}_2) = \left(\frac{c_2}{(c_1+c_2)^2}, \frac{c_1}{(c_1+c_2)^2}\right). \tag{5.59}$$

Furthermore, as we have demonstrated in this example the local stability of this equilibrium under the best reply dynamics depends on the ratio between the marginal costs $c_1/c_2$. Feasible, that is bounded and non-negative, trajectories of the best reply dynamics are obtained provided that $c_1/c_2 \in [4/25, 25/4] = [0.16, 6.25]$. Moreover, the Nash equilibrium (5.59) is stable if and only if $c_1/c_2 \in (3 - 2\sqrt{2}, 3 + 2\sqrt{2}) \simeq (0.17, 5.83)$. If $c_1/c_2$ exits this interval then the Nash equilibrium loses stability via a period doubling bifurcation. If $c_1/c_2$ falls outside the interval $(3 - 2\sqrt{2}, 3 + 2\sqrt{2})$ then the asymptotic dynamics may converge to periodic cycles or even exhibit chaotic motion around the Nash equilibrium. Consequently, in terms of the cost parameters convergence to the Nash equilibrium is obtained for a wider range of parameters in the model with LMA than in the case where firms know the true nonlinear demand and at each time step play the best reply. This insight could be summarized in the statement that *less information implies more stability*. However, it should be noticed that this result is obtained through a comparison of the stability region in the space of unit cost parameters $(c_1, c_2)$ in the following sense: the Nash equilibrium $\bar{\mathbf{x}}$ is stable for each selection of the parameters $(c_1, c_2)$ for the model with LMA, whereas stability only holds in the subset $c_1/c_2 \in (3 - 2\sqrt{2}, 3 + 2\sqrt{2})$ in the case of best reply adjustment. Quite different conclusions may be reached if we compare the basins of attraction. In fact, with cost parameters such that the Nash equilibrium is stable under both adjustment mechanisms, larger basins of attraction can be observed for the model with best reply. This is illustrated in Fig. 5.4. The white regions represent the basins of attraction of the corresponding stable Nash equilibrium in the best reply model (case (a), where we also depict the best replies) and the LMA model (case (b)). The grey regions represent the set of initial conditions that generate infeasible trajectories. Obviously, the basin is larger in the former case.

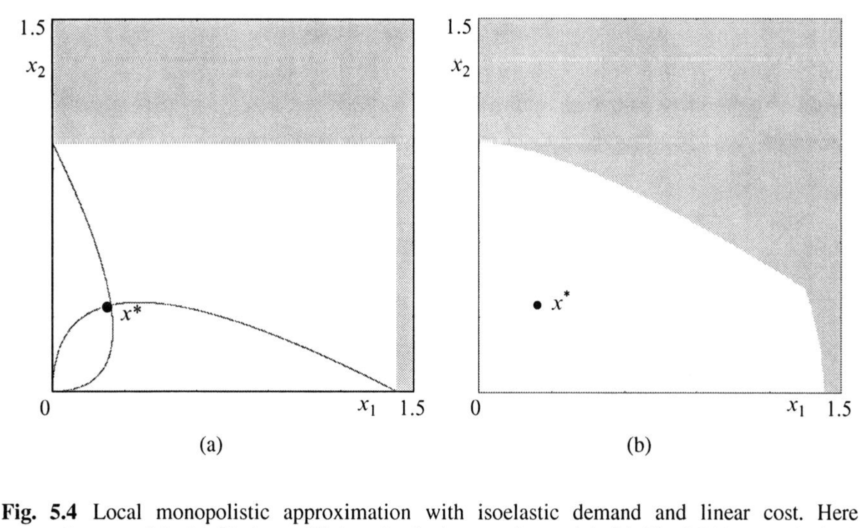


**Fig. 5.4** Local monopolistic approximation with isoelastic demand and linear cost. Here $c_1 = 1, c_2 = 0.7$. (**a**) Nash equilibrium in the best reply model. (**b**) The LMA case. In both cases, the *white region* represents the basin of attraction of the stable equilibrium, initial values in the *grey region* generate infeasible trajectories

▼

*Example 5.6.* We consider now the duopoly model with the same isoelastic price function $p = 1/Q$ as before and with quadratic cost functions $C_k = d_k + e_k x_k^2$. The best reply dynamics with complete knowledge of the demand function cannot be expressed by a simple dynamical system. In fact, the profit of player $k$ is $\varphi_k = x_k/(x_1 + x_2) - d_k - e_k x_k^2$, and the first order conditions for profit maximization give rise to third degree algebraic equations. For example, the condition for the reaction function of player 1 becomes

$$2e_1x_1^3 + 4e_1x_2x_1^2 + 2e_1x_2^2x_1 - x_2 = 0.$$

Since the left hand side strictly increases in $x_1$, it is easy to see that a unique positive solution $x_1 = R_1(x_2)$ exists, however its precise form is not easily obtained. On the other hand, if we consider the dynamics with LMA, a simple two-dimensional dynamical system is obtained based on the two-dimensional iterated map

$$\begin{aligned} x_1(t+1) &= \frac{2x_1(t) + x_2(t)}{2(1 + e_1(x_1(t) + x_2(t))^2)}, \\ x_2(t+1) &= \frac{x_1(t) + 2x_2(t)}{2(1 + e_2(x_1(t) + x_2(t))^2)}, \end{aligned} \tag{5.60}$$

which can be derived from (5.54). The equations for the determination of the fixed points, obtained by setting $x_i(t+1) = x_i(t)$ in (5.60), become

$$\begin{aligned} 2e_1x_1(x_1 + x_2)^2 &= x_2, \\ 2e_2x_2(x_1 + x_2)^2 &= x_1. \end{aligned} \tag{5.61}$$

By dividing the first equation by the second we have $x_1 = \sqrt{\frac{e_2}{e_1}}x_2$. Substituting this into (5.61) we calculate the unique non-vanishing equilibrium

$$\begin{aligned} \bar{x}_1 &= \frac{\sqrt{e_2}}{\sqrt{e_1} + \sqrt{e_2}} \frac{1}{\sqrt{2\sqrt{e_1e_2}}}, \\ \bar{x}_2 &= \frac{\sqrt{e_1}}{\sqrt{e_1} + \sqrt{e_2}} \frac{1}{\sqrt{2\sqrt{e_1e_2}}}. \end{aligned} \tag{5.62}$$

This equilibrium is always locally asymptotically stable. In fact, sufficient conditions for its stability are easily obtained from the computation of the Jacobian matrix $\mathbf{J} = \left(J_{ij}\right)$ of (5.60) at the equilibrium. The diagonal entries are

$$J_{11} = -\frac{\bar{x}_1 f''(\bar{Q})}{2\left(f'(\bar{Q}) - e_1\right)} = \frac{3e_1\sqrt{e_2}}{3e_1\sqrt{e_2} + e_1\sqrt{e_1} + 2e_2\sqrt{e_1}},$$

$$J_{22} = -\frac{\bar{x}_2 f''\left(\bar{Q}\right)}{2\left(f'(\bar{Q}) - e_2\right)} = \frac{3e_2\sqrt{e_1}}{3e_2\sqrt{e_1} + e_2\sqrt{e_2} + 2e_1\sqrt{e_2}},$$

and the off-diagonal entries are

$$J_{12} = -\frac{\bar{x}_1 f''\left(\bar{Q}\right) + f'(\bar{Q})}{2\left(f'(\bar{Q}) - e_1\right)} = \frac{e_1\sqrt{e_2} - e_2\sqrt{e_1}}{3e_1\sqrt{e_2} + e_1\sqrt{e_1} + 2e_2\sqrt{e_1}},$$

$$J_{21} = -\frac{\bar{x}_2 f''\left(\bar{Q}\right) + f'(\bar{Q})}{2\left(f'(\bar{Q}) - e_2\right)} = \frac{e_2\sqrt{e_1} - e_1\sqrt{e_2}}{3e_2\sqrt{e_1} + e_2\sqrt{e_2} + 2e_1\sqrt{e_2}}.$$

Hence, the trace of the Jacobian matrix at the equilibrium is

$$Tr = \frac{3e_1\sqrt{e_2}}{3e_1\sqrt{e_2} + e_1\sqrt{e_1} + 2e_2\sqrt{e_1}} + \frac{3e_2\sqrt{e_1}}{3e_2\sqrt{e_1} + e_2\sqrt{e_2} + 2e_1\sqrt{e_2}}$$

and the determinant is

$$Det = \frac{e_1e_2(7\sqrt{e_1e_2} + e_1 + e_2)}{\left(3e_1\sqrt{e_2} + e_1\sqrt{e_1} + 2e_2\sqrt{e_1}\right)\left(3e_2\sqrt{e_1} + e_2\sqrt{e_2} + 2e_1\sqrt{e_2}\right)}.$$

A set of sufficient conditions for the stability of $\bar{\mathbf{x}}$ (that is for the eigenvalues to be located inside the unit circle of the complex plane) is given by

$$1 + Tr + Det > 0; \quad 1 - Tr + Det > 0; \quad Det < 1 \tag{5.63}$$

(see Appendix F). These conditions become trivial in our case. In fact, given that $Tr$ and $Det$ are both positive, the first condition is always satisfied. Moreover

$$1 - Tr + Det = \frac{2\sqrt{e_1e_2}\left(e_1^2 + e_2^2 + 6e_1e_2\right)(e_1 + e_2)}{\left(3e_1\sqrt{e_2} + e_1\sqrt{e_1} + 2e_2\sqrt{e_1}\right)\left(3e_2\sqrt{e_1}+e_2\sqrt{e_2}+2e_1\sqrt{e_2}\right)} > 0$$

and $Det < 1$ since

$$\begin{aligned} e_1e_2(7\sqrt{e_1e_2} + e_1 + e_2) < {} & \left(3e_1 \times \sqrt{e_2} + e_1\sqrt{e_1} + 2e_2\sqrt{e_1}\right) \times \\ & \times \left(3e_2\sqrt{e_1} + e_2\sqrt{e_2} + 2e_1\sqrt{e_2}\right). \end{aligned}$$

▼

*Example 5.7.* We now turn back to the case of isoelastic price and linear cost functions and consider the case of $N$ firms. As mentioned in previous chapters, one question that is often discussed in the literature on oligopoly games deals with the effect of the number of players on the stability properties of the equilibrium. In general it is not a straightforward matter to find an answer to this question, since increasing the number of players implies increasing the dimension of the dynamical system. To obtain some insight into this problem, let us consider the model (5.53)

$$x_k(t+1) = \frac{1}{2}x_k(t) + \frac{c_k - f(Q(t))}{2f'(Q(t))} \quad (k = 1, \ldots, N). \tag{5.64}$$

This $N$-dimensional dynamical system in the state variables $x_k$ can be reduced to a one-dimensional dynamical system in the total quantity $Q(t)$ by summing up the equations (5.64) to yield

$$Q(t+1) = \frac{1}{2}Q(t) + \frac{\gamma - Nf(Q(t)}{2f'(Q(t))} \tag{5.65}$$

where $\gamma = \sum_{k=1}^{N} c_k$. The dynamic equation (5.65) for the aggregate production includes the number of players $N$ as a parameter. Therefore, we can investigate the effects of this parameter on the dynamics of the global production. It is trivial to see that if $(\bar{x}_1, \ldots, \bar{x}_N)$ is a steady state of the disaggregated dynamical system (5.64), then $\bar{Q} = \sum_{k=1}^{N} \bar{x}_k$ is a steady state of the aggregated dynamical system (5.65). In particular, if $(\bar{x}_1, \ldots, \bar{x}_N)$ is a Nash equilibrium, then it is a fixed point of (5.64) and consequently it corresponds to a fixed point of (5.65). However, the converse is not true in general because a fixed point $\bar{Q}$ of (5.65) can correspond to several different arrangements of $(x_1, \ldots, x_N)$, that do not correspond to fixed points of (5.64). If we consider the model (5.53) with $N$ firms and isoelastic inverse demand function

$$p = f(Q) = \frac{1}{Q}, \tag{5.66}$$

then the dynamical system (5.64) becomes

$$x_k(t+1) = \frac{1}{2}\left[x_k(t) + Q(t) - c_k Q^2(t)\right] \quad (k = 1, \ldots, N), \tag{5.67}$$

and the one-dimensional map (5.65) that describes the time evolution of the aggregated output $Q(t)$ becomes

$$Q(t+1) = \frac{1}{2}\left[1 + N - \gamma Q(t)\right] Q(t), \tag{5.68}$$

where $\gamma = \sum_{k=1}^{N} c_k$. This is a quadratic one-dimensional map which is topologically conjugate to the standard logistic map $x(t+1) = \mu x(t)(1 - x(t))$ through the linear homeomorphism $Q = x(1 + N)/\gamma$ and with the parameters related by $\mu = (1 + N)/2$. The time evolution of the aggregate production can be deduced from well-known properties of the logistic map (see e.g., Devaney (1989)). In particular, here we are interested in the role of the integer parameter $N$. First of all, we notice that the dynamics of (5.68) converge to the positive steady state $\bar{Q} = (1 + N - 2)/\gamma$ provided that $N \leqslant 5$, corresponding to the well known condition $\mu \leqslant 3$. The convergence is monotone if $N \leqslant 3$, whereas it exhibits damped oscillations if $4 \leqslant N \leqslant 5$. With 6 firms we have $\mu = 3.5$; hence we have stable oscillations of period 4 since $\mu > 1 + \sqrt{6}$. The case of $N = 7$ competitors gives rise to fully developed chaos, as it corresponds with $\mu = 4$. Hence, stability is obtained for a limited number of oligopolists, namely $N \leqslant 5$, and as usual instability occurs as the number of players increases. ▼

### 5.2.2 Dynamics Under Adaptive Adjustment

If we assume that in a neighborhood of the equilibrium the left hand side of (5.52) is strictly decreasing in $x_k$, then a unique best response (under the assumptions of LMA) is obtained, and it is a continuously differentiable function $\widetilde{R}_k(Q_k(t), x_k(t))$. By implicit differentiation of (5.52) and noticing that $Q(t) = Q_k(t) + x_k(t)$ we have

$$f' + 2f''x_k + 2f'\widetilde{R}'_{kQ} - f''x_k(t) - C''_k \cdot \widetilde{R}'_{kQ} = 0$$

and

$$f' + 2f''x_k + 2f'\widetilde{R}'_{kx} - f''x_k(t) - f' - C''_k \widetilde{R}'_{kx} = 0,$$

where we use the notation $\widetilde{R}'_{kQ} = \frac{\partial \widetilde{R}_k}{\partial Q_k(t)}$ and $\widetilde{R}'_{kx} = \frac{\partial \widetilde{R}_k}{\partial x_k(t)}$. So we have

$$\widetilde{R}'_{kQ} = -\frac{f' + 2f''x_k - f''x_k(t)}{2f' - C''_k} \tag{5.69}$$

and

$$\widetilde{R}'_{kx} = -\frac{2f''x_k - f''\, x_k(t)}{2f' - C''_k}. \tag{5.70}$$

The discrete dynamical process with adaptive adjustments based on the above subjective best responses has the usual form

$$x_k(t+1) = x_k(t) + \alpha_k \left( \widetilde{R}_k(Q_k(t), x_k(t)) - x_k(t) \right), \tag{5.71}$$

where $\alpha_k$ is a sign preserving function. It is easy to see that the interior equilibria under full information are steady states of this system.

The local asymptotic stability of the equilibrium depends on the location of the eigenvalues of the Jacobian of the system, which has the form

$$\boldsymbol{J} = \begin{pmatrix} 1 + a_1(r_{1x} - 1) & a_1 r_{1Q} & \dots & a_1 r_{1Q} \\ a_2 r_{2Q} & 1 + a_2(r_{2x} - 1) & \dots & a_2 r_{2Q} \\ \vdots & \vdots & & \vdots \\ a_N r_{NQ} & a_N r_{NQ} & \dots & 1 + a_N(r_{Nx} - 1) \end{pmatrix},$$

where $a_k = \alpha'_k(0)$, and $r_{kQ}$ and $r_{kx}$ are the partial derivatives of the subjective best response function at the equilibrium, given by

$$r_{kQ} = -(f' + f''\bar{x}_k)/(2f' - C''_k)$$

and

$$r_{kx} = -(f''\bar{x}_k)/(2f' - C_k'').$$

All eigenvalues are inside the unit circle, if any norm of the Jacobian is less than unity. By selecting the $\| \,.\, \|_\infty$, $\| \,.\, \|_1$ and Frobenius norms, we get three sufficient conditions, namely

$$\max_k \left\{ |1 + a_k(r_{kx} - 1)| + (N-1)a_k|r_{kQ}| \right\} < 1 \tag{5.72}$$

or

$$\max_k \left\{ |1 + a_k(r_{kx} - 1)| + \sum_{l \neq k} a_l |r_{lQ}| \right\} < 1 \tag{5.73}$$

or

$$\sum_{k=1}^{N} \left\{ (1 + a_k(r_{kx} - 1))^2 + (N-1)a_k^2 r_{kQ}^2 \right\} < 1. \tag{5.74}$$

*Example 5.8.* Assume linear cost functions and (subjective) best response dynamics, then $C_k'' = 0$, so $r_{kx} - r_{kQ} = \frac{1}{2}$ and $a_k = 1$ for all $k$. In this special case

$$1 + a_k(r_{kx} - 1) = r_{kx} = -\frac{f''\bar{x}_k}{2f'} \tag{5.75}$$

at the equilibrium, and

$$a_k r_{kQ} = -\frac{f' + f''\bar{x}_k}{2f'},$$

so (5.72) reduces to

$$\bar{x}_k |f''(\bar{Q})| + (N-1)|f'(\bar{Q}) + \bar{x}_k f''(\bar{Q})| < 2|f'(\bar{Q})| \tag{5.76}$$

at the equilibrium for all $k$. Assume next quadratic cost functions $C_k(x_k) = d_k + e_k x_k^2$ and (subjective) best response dynamics. Then $C_k'(x_k) = 2e_k x_k$ and $C_k''(x_k) = 2e_k$, so

$$1 + a_k(r_{kx} - 1) = r_{kx} = -\frac{f''\bar{x}_k}{2(f' - e_k)}$$

and

$$a_k r_{kQ} = -\frac{f' + f''\bar{x}_k}{2(f' - e_k)},$$

therefore condition (5.72) can be rewritten as

$$\bar{x}_k |f''(\bar{Q})| + (N-1)|f'(\bar{Q}) + \bar{x}_k f''(\bar{Q})| < 2|f'(\bar{Q}) - e_k|, \tag{5.77}$$

for all $k$. The other two matrix norms can be used to obtain similar conditions, the details are left as easy exercises for the reader. ▼

In the following part of this section we will show that the characteristic polynomial of the Jacobian can be derived in a simple form, and therefore the eigenvalues can be easily located. Therefore a more accurate stability condition can be derived which is "almost" sufficient and necessary.

Notice that this Jacobian has the special structure (E.4), so its characteristic polynomial can be written in the special form given by (E.5), namely

$$\varphi(\lambda) = \prod_{k=1}^{N} \left(1+a_k \cdot (r_{kx} - r_{kQ} - 1) - \lambda\right) \left[1+ \sum_{k=1}^{N} \frac{a_k r_{kQ}}{1+a_k \cdot (r_{kx} - r_{kQ} - 1)-\lambda}\right]. \tag{5.78}$$

Consider first the concave oligopolies of Chap. 2, where we have assumed that

(A) $f' < 0$,

(B) $x_k f'' + f' \leq 0$,

(C) $f' - C_k'' < 0$,

for all $k$ and feasible output levels. Then

$$r_{kQ} \leq 0 \quad \text{and} \quad 0 < r_{kx} - r_{kQ} < 1$$

for all $k$. Note that in the special case of linear cost functions, $r_{kx} - r_{kQ} = \frac{1}{2}$. The eigenvalues are $1 + a_k(r_{kx} - r_{kQ} - 1)$ together with the roots of the equation

$$g(\lambda) = 1 + \sum_{k=1}^{N} \frac{a_k r_{kQ}}{1 + a_k(r_{kx} - r_{kQ} - 1) - \lambda} = 0. \tag{5.79}$$

The values $1 + a_k(r_{kx} - r_{kQ} - 1)$ are inside the unit circle if and only if

$$a_k < \frac{2}{1 - (r_{kx} - r_{kQ})}. \tag{5.80}$$

Since $0 < a_k \leq 1$ is assumed, this inequality always holds.

The poles of the left hand side of (5.79) are between $-1$ and $+1$ under assumption (5.80), its derivative is negative (unless all $r_{kQ} = 0$). The graph of the left hand side is the same as the one shown in Fig. 2.1. Under condition (5.80) all roots are real, and they are between $-1$ and $+1$ if and only if

$$g(-1) = 1 + \sum_{k=1}^{N} \frac{a_k r_{kQ}}{2 + a_k(r_{kx} - r_{kQ} - 1)} > 0. \tag{5.81}$$

If this inequality is violated with strict opposite inequality, then the equilibrium is unstable.

*Example 5.9.* Assume again linear cost functions and (subjective) best response dynamics. Then $a_k = 1$, $r_{kx} - r_{kQ} = \frac{1}{2}$ and $C_k'' = 0$ for all $k$. So (5.81) can be simplified to

$$\sum_k r_{kQ} = \frac{-Nf' - f'' \cdot \bar{Q}}{2f'} > -\frac{3}{2},$$

or

$$(N-3)f' + f'' \cdot \bar{Q} > 0. \tag{5.82}$$

Notice that under conditions (A) and (B) the norm-based sufficient condition (5.76) assumes the form

$$(N-3)f' + (N-1)\bar{x}_k f'' - \bar{x}_k |f''| > 0 \tag{5.83}$$

for all $k$. If $f''(\bar{Q}) < 0$, then these conditions may hold only for $N = 2$, and have the following forms

$$-f' + f'' \cdot \bar{Q} > 0$$

and

$$-f' + 2\bar{x}_k f'' > 0 \quad (k = 1, 2).$$

The second inequality is slightly stronger than the first one unless the $\bar{x}_k$ values are identical. Assume next that $f''(Q) \geq 0$. In the case of $N = 2$ both conditions (5.82) and (5.83) are satisfied. If $N = 3$, then the two conditions have the special forms

$$f'' \cdot \bar{Q} > 0 \quad \text{and} \quad f'' \cdot \bar{x}_k > 0,$$

where the second inequality is stronger again. If $N > 3$, then we have the two conditions

$$(N-3)f' + f'' \cdot \bar{Q} > 0$$

and

$$(N-3)f' + (N-2)\bar{x}_k f'' > 0,$$

where the second condition is again stronger then the first one (by taking $\bar{x}_k = \min_l \{\bar{x}_l\}$, the left hand side of the second inequality is smaller than that of the first one). Very similar conditions can be obtained by assuming quadratic cost functions and in comparing conditions (5.77) and (5.81). ▼

*Example 5.10.* Consider next the case of an isoelastic price function. Then $f(Q) = A/Q$ and therefore

$$f' = -\frac{A}{Q^2}, \quad f'' = \frac{2A}{Q^3}, \quad r_{kQ} = \frac{A(2\bar{x}_k - \bar{Q})}{\bar{Q}(2A + \bar{Q}^2 C_k'')} \quad \text{and} \quad r_{kx} = \frac{2A\bar{x}_k}{\bar{Q}(2A + \bar{Q}^2 C_k'')}.$$

If there is no dominant firm which produces more than the rest of the industry, then $2\bar{x}_k - \bar{Q} \leq 0$ for all $k$, so at the equilibrium

$$r_{kQ} \leq 0 \quad \text{and} \quad 0 < r_{kx} - r_{kQ} \leq \frac{1}{2} < 1$$

by assuming that $C_k'' \geq 0$ for all $k$. Therefore all results derived previously for the concave case remain valid. If there is a dominant firm, then the method shown above for the concave case cannot be applied, since the monotonicity of the function in (5.79) cannot be guaranteed. However the sufficient stability conditions (5.72)–(5.74) remain applicable without any changes. ▼

*Example 5.11.* Consider finally the case of a general duopoly, when $N = 2$. From the special form of the Jacobian we see that the characteristic polynomial is

$$(1 + a_1(r_{1x} - 1) - \lambda)(1 + a_2(r_{2x} - 1) - \lambda) - a_1 a_2 r_{1Q} r_{2Q}$$

which becomes the quadratic equation

$$\begin{aligned} &\lambda^2 + \lambda\,(-2 + a_1(1 - r_{1x}) + a_2(1 - r_{2x})) \\ &\quad + (1 + a_1(r_{1x} - 1))(1 + a_2(r_{2x} - 1)) - a_1 a_2 r_{1Q} r_{2Q} = 0. \end{aligned}$$

By using the stability condition introduced in Appendix F we see that the roots are inside the unit circle if and only if

$$(1 + a_1(r_{1x} - 1))(1 + a_2(r_{2x} - 1)) - a_1 a_2 r_{1Q} r_{2Q} < 1,$$

$$\begin{aligned} &-2 + a_1(1 - r_{1x}) + a_2(1 - r_{2x}) \\ &\quad + (1 + a_1(r_{1x} - 1))(1 + a_2(r_{2x} - 1)) - a_1 a_2 r_{1Q} r_{2Q} + 1 > 0, \end{aligned}$$

$$\begin{aligned} &2 - a_1(1 - r_{1x}) - a_2(1 - r_{2x}) \\ &\quad + (1 + a_1(r_{1x} - 1))(1 + a_2(r_{2x} - 1)) - a_1 a_2 r_{1Q} r_{2Q} + 1 > 0. \end{aligned}$$

Instead of examining this in the general case, assume that $a_1 = a_2 = 1$. Then we have the conditions

$$r_{1x} r_{2x} - r_{1Q} r_{2Q} < 1,$$

$$\begin{aligned} 1 - r_{1x} - r_{2x} + r_{1x} r_{2x} - r_{1Q} r_{2Q} &> 0, \\ 1 + r_{1x} + r_{2x} + r_{1x} r_{2x} - r_{1Q} r_{2Q} &> 0, \end{aligned}$$

showing that

$$r_{1x} r_{2x} - 1 < r_{1Q} r_{2Q} < \min\{1 - r_{1x} - r_{2x} + r_{1x} r_{2x},\, 1 + r_{1x} + r_{2x} + r_{1x} r_{2x}\}.$$

For example, in the case of linear cost functions, $C_1'' = C_2'' = 0$, we have

$$r_{kQ} = -\frac{f' + f''\bar{x}_k}{2f'} \quad \text{and} \quad r_{kx} = -\frac{f''\bar{x}_k}{2f'}$$

and the stability condition becomes

$$\frac{\bar{x}_1\bar{x}_2 f''^2}{4f'^2} - 1 < \frac{(f' + \bar{x}_1 f'')(f' + \bar{x}_2 f'')}{4f'^2}$$
$$< \min\left\{1 + \frac{f'' \cdot (\bar{x}_1 + \bar{x}_2)}{2f'} + \frac{\bar{x}_1\bar{x}_2 f''^2}{4f'^2},\quad 1 - \frac{f'' \cdot (\bar{x}_1 + \bar{x}_2)}{2f'} + \frac{\bar{x}_1\bar{x}_2 f''^2}{4f'^2}\right\}.$$

It is very easy to check if this condition holds or not. ▼

## 5.3 Other Learning Processes

In this section we consider other adaptive learning processes. We study situations where firms adaptively adjust their beliefs about the price function based on the discrepancies between their predicted and actually observed market prices. For the sake of mathematical simplicity, we will assume in this section that no cost externalities are present, and that the (inverse) demand and cost functions are linear.

Hence, in this section we consider the demand relationship

$$p = f(Q) = A - BQ \quad \text{or} \quad Q(p) = f^{-1}(p) = \frac{A}{B} - \frac{1}{B}p$$

and the cost function

$$C_k(x_k) = d_k + c_k x_k$$

for $k = 1, 2, \ldots, N$, where $x_k$ is the output of firm $k$ and $Q = \sum_{l=1}^{N} x_l$ is the industry output. We assume that the firms have only limited knowledge of the price function $p = f(Q)$ and over time they repeatedly update their estimates.

We consider three scenarios. First, regarding the demand function of the general form $Q(p)$ we assume that the particular values of the reservation price $A$ and the slope of the price function $B$ are unknown, but firms know the value of the market saturation point, that is, they know the value of $A/B$. In the second scenario we consider the inverse demand function and assume that the firms know the slope $B$ but do not know the reservation price $A$. Finally, we study a situation where firms know the reservation price $A$, but do not know the slope $B$. As we will demonstrate, the possibility of learning as well as the asymptotic behavior of the learning process strongly depends on the firm's knowledge about the demand parameters and also on the updating procedure the firms use.

The learning schemes discussed in this chapter have been introduced by Szidarovszky (2003) and extended by Szidarovszky and Krawczyk (2005) They are also briefly discussed in Szidarovszky et al. (2008). The model studied in Sect. 5.3.2 has been analyzed in Szidarovszky (2004) where the effect of information delay has also been investigated.

### 5.3.1 Unknown Slope with Known Market Saturation Point

We assume first that the firms know that the demand function is linear and decreasing. Firms know the value of the market saturations point $A/B$, that is the total output level that renders the price zero, but they have only a misspecified estimate of the slope $1/B$ of the demand function. Suppose that in period $t$ firm $k$ has an estimate of this slope, which we write as $1/\varepsilon_k(t)$. Then, in this case this is equivalent to saying that firm $k$'s estimate of the price function is $\widetilde{f_k}(Q) = \varepsilon_k(A/B - Q)$, where the factor $\varepsilon_k(t)$ is adjusted over time on the basis of observed price data.

Let us consider the situation from the point of view of an arbitrary firm, say firm $k$. Given $\varepsilon_k$, each firm $k$ solves the static game. It believes that the profit of each firm $l$ (including itself) is given as

$$\widetilde{x_l}\varepsilon_k\left(\frac{A}{B} - \widetilde{Q_l} - \widetilde{x_l}\right) - (c_l\widetilde{x_l} + d_l). \tag{5.84}$$

Based on this belief firm $k$ it is able to calculate the believed equilibrium outputs and the equilibrium price. Then this believed price will be compared to the actual market price the firm receives, and based on the discrepancy between the believed and actual prices firm $k$ can adjust the shape estimate $\varepsilon_k$.

Assuming an interior optimum, firm $k$ believes that the best response of firm $l$ is

$$\widetilde{x_l} = \frac{A}{2B} - \frac{c_l}{2\varepsilon_k} - \frac{\widetilde{Q_l}}{2},$$

implying that

$$\widetilde{x_l} = \frac{A}{B} - \frac{c_l}{\varepsilon_k} - \widetilde{Q}. \tag{5.85}$$

By summing these equations for all firms we have

$$\widetilde{Q} = \frac{NA}{B} - \frac{1}{\varepsilon_k}\sum_{l=1}^{N} c_l - N\widetilde{Q},$$

so firm $k$ believes that the total output of the industry is

$$\widetilde{Q}^k = \frac{1}{N+1}\left(\frac{NA}{B} - \frac{1}{\varepsilon_k}\sum_{l=1}^{N} c_l\right). \tag{5.86}$$

Therefore firm $k$ will produce the output

$$\begin{aligned} x_k &= \frac{A}{B} - \frac{c_k}{\varepsilon_k} - \frac{1}{N+1}\left(\frac{NA}{B} - \frac{1}{\varepsilon_k}\sum_{l=1}^{N} c_l\right) \\ &= \frac{A}{(N+1)B} - \frac{c_k}{\varepsilon_k} + \frac{1}{(N+1)\varepsilon_k}\sum_{l=1}^{N} c_l, \end{aligned} \tag{5.87}$$

and the equilibrium price is believed to be

$$\widetilde{p}_k = \widetilde{f}_k(\widetilde{Q}^k) = \frac{1}{N+1}\left(\frac{A\varepsilon_k}{B} + \sum_{l=1}^{N} c_l\right), \tag{5.88}$$

which are the consequences of (5.85) with $l = k$, and (5.86). Note that the "tilde" indicates that we are dealing with expected quantities based on firm $k$'s estimated price function. In reality, however, each firm thinks in the same way independently of each other, and each firm's expected (or believed) price and actually produced amount depend on its own price function estimate. Therefore the actual total output of the industry becomes

$$Q = \sum_{k=1}^{N} x_k = \frac{NA}{(N+1)B} - \sum_{k=1}^{N} \frac{c_k}{\varepsilon_k} + \left(\frac{\sum_{l=1}^{N} c_l}{N+1}\right)\sum_{k=1}^{N}\frac{1}{\varepsilon_k},$$

with the corresponding actual market price

$$p = A - BQ = \frac{A}{N+1} + B\sum_{k=1}^{N}\frac{c_k}{\varepsilon_k} - \frac{B}{N+1}\left(\sum_{l=1}^{N} c_l\right)\left(\sum_{k=1}^{N}\frac{1}{\varepsilon_k}\right) \tag{5.89}$$

being what the firms receive. The actual prices are usually different than the expected prices of the firms. For firm $k$, the discrepancy between the actual and believed prices is

$$\Delta p_k = p - \widetilde{p}_k = \frac{A}{N+1}\left(1 - \frac{\varepsilon_k}{B}\right) + B\sum_{k=1}^{N}\frac{c_k}{\varepsilon_k} - \frac{1}{N+1}\left(\sum_{l=1}^{N} c_l\right)\left(B\sum_{k=1}^{N}\frac{1}{\varepsilon_k} + 1\right). \tag{5.90}$$

Based on this discrepancy, firm $k$ develops the following adjustment process. If $\Delta p_k = 0$, then there is no discrepancy, so firm $k$ believes that its price estimate is correct. If $\Delta p_k > 0$, then the believed price is too low, so firm $k$ wants to increase its price estimate by increasing the value of $\varepsilon_k$. If $\Delta p_k < 0$, then the believed price is too high, so firm $k$ wants to decrease its price estimate by decreasing the value of $\varepsilon_k$. If the time scale is discrete, then this adjustment concept can be modeled as

$$\varepsilon_k(t+1) = \varepsilon_k(t) + a_k \Delta p_k(t) \quad (k = 1, 2, \ldots, N) \tag{5.91}$$

where $a_k > 0$ is the speed of adjustment of firm $k$ (see Bischi et al. (2008)). Here we assume linear adjustments for the sake of mathematical simplicity. If the time scales are continuous, then the dynamic process becomes

$$\dot{\varepsilon}_k = a_k \Delta p_k \quad (k = 1, 2, \ldots, N). \tag{5.92}$$

First we prove that both systems (5.91) and (5.92) have the unique steady state $\bar{\varepsilon}_k = B$ for all $k$, which corresponds to the full knowledge case. Notice first that if $\Delta p_k = 0$ for all $k$, then the $\bar{\varepsilon}_k$ values are identical. Let $\bar{\varepsilon}$ denote their common value, then

$$\begin{aligned}
0 &= \frac{A}{N+1}\left(1 - \frac{\bar{\varepsilon}}{B}\right) + \frac{B}{\bar{\varepsilon}} \sum_{k=1}^{N} c_k - \frac{1}{N+1}\left(\sum_{l=1}^{N} c_l\right)\left(\frac{NB}{\bar{\varepsilon}} + 1\right) \\
&= \frac{A}{N+1}\left(1 - \frac{\bar{\varepsilon}}{B}\right) + \left(\sum_{k=1}^{N} c_k\right)\left(\frac{B}{\bar{\varepsilon}} - 1\right)\frac{1}{N+1}.
\end{aligned}$$

If $\bar{\varepsilon} > B$, then both terms are negative, and if $\bar{\varepsilon} < B$, then both terms are positive. If $\bar{\varepsilon} = B$, then both terms are equal to zero. Hence $\bar{\varepsilon} = B$ is the only steady state.

It is important to notice that this unique steady state, $\bar{\varepsilon}_k = B$ for each $k$, corresponds to the situation where all the believed demand functions coincide with the true market demand. If the adjustment process converges to such a unique steady state, then we can say that all the firms learn the true demand, although they start from misspecified (and different) initial guesses about the slope of the demand function. In what follows we provide conditions for the stability of the steady state, that is we identify the sets of parameters which ensure the convergence of the adjustment process. Furthermore, we also examine some bifurcations that lead to instability of the steady state.

The local asymptotic stability of the dynamical systems (5.91) and (5.92) can be examined by linearization. Notice first that

$$\frac{\partial \Delta p_k}{\partial \varepsilon_k} = -\frac{A}{(N+1)B} + \frac{B}{\varepsilon_k^2}\left(-c_k + \frac{1}{N+1}\sum_{l=1}^{N} c_l\right) \tag{5.93}$$

and for $l \neq k$

$$\frac{\partial \Delta p_k}{\partial \varepsilon_l} = \frac{B}{\varepsilon_l^2}\left(-c_l + \frac{1}{N+1}\sum_{k=1}^{N} c_k\right). \tag{5.94}$$

In order to make the notation as simple as possible, let

$$\gamma_k = \frac{1}{N+1} \sum_{l=1}^{N} c_l - c_k.$$

The Jacobian of the discrete time system (5.91) has the special structure

$$\boldsymbol{I} + \boldsymbol{J}$$

where

$$\boldsymbol{J} = \begin{pmatrix} a_1 \left( -\frac{A}{(N+1)B} + \frac{B\gamma_1}{\varepsilon_1^2} \right) & a_1 \frac{B\gamma_2}{\varepsilon_2^2} & \cdots & a_1 \frac{B\gamma_N}{\varepsilon_N^2} \\ \frac{a_2 B\gamma_1}{\varepsilon_1^2} & a_2 \left( -\frac{A}{(N+1)B} + \frac{B\gamma_2}{\varepsilon_2^2} \right) & \cdots & a_2 \frac{B\gamma_N}{\varepsilon_N^2} \\ \vdots & \vdots & \ddots & \vdots \\ \frac{a_N B\gamma_1}{\varepsilon_1^2} & \frac{a_N B\gamma_2}{\varepsilon_2^2} & \cdots & a_N \left( -\frac{A}{(N+1)B} + \frac{B\gamma_N}{\varepsilon_N^2} \right) \end{pmatrix}, \tag{5.95}$$

so it can be written as $\boldsymbol{D} + \boldsymbol{a}\boldsymbol{b}^T$ with

$$\boldsymbol{D} = diag\left(1 - \frac{a_1 A}{(N+1)B}, \ldots, 1 - \frac{a_N A}{(N+1)B}\right), \quad \boldsymbol{a} = (a_1 B, \ldots, a_N B)^T,$$

and

$$\boldsymbol{b}^T = \left(\frac{\gamma_1}{\varepsilon_1^2}, \ldots, \frac{\gamma_N}{\varepsilon_N^2}\right).$$

Therefore the characteristic equation of the Jacobian can be rewritten as

$$\begin{aligned} &det(\boldsymbol{D} + \boldsymbol{a}\boldsymbol{b}^T - \lambda \boldsymbol{I}) = det(\boldsymbol{D} - \lambda \boldsymbol{I}) det(\boldsymbol{I} + (\boldsymbol{D} - \lambda \boldsymbol{I})^{-1} \boldsymbol{a}\boldsymbol{b}^T) \\ &= \prod_{k=1}^{N} \left(1 - \frac{a_k A}{(N+1)B} - \lambda\right) \left[1 + \sum_{k=1}^{N} \frac{\frac{a_k B\gamma_k}{\varepsilon_k^2}}{1 - \frac{a_k A}{(N+1)B} - \lambda}\right] = 0, \end{aligned} \tag{5.96}$$

where we have used the results of Appendix E. Assume that

$$\gamma_k = \frac{1}{N+1} \sum_{l=1}^{N} c_l - c_k \le 0 \tag{5.97}$$

for all $k$, which is satisfied if the marginal costs $c_l$ are close to each other. By repeating the proof of Theorem 2.1 and noticing that at the steady state $\varepsilon_k = B$ for all $k$, we have the following result.

**Theorem 5.4.** *Under assumption (5.97) the steady state of system (5.91) is locally asymptotically stable if for all k*

$$\frac{a_k A}{(N+1)B} < 2 \tag{5.98}$$

*and*

$$\sum_{k=1}^{N} \frac{a_k \gamma_k (N+1)}{2(N+1)B - a_k A} > -1. \tag{5.99}$$

*If for at least one k,*

$$\frac{a_k A}{(N+1)B} \geq 2$$

*or*

$$\sum_{k=1}^{N} \frac{a_k \gamma_k (N+1)}{2(N+1)B - a_k A} < -1,$$

*then the steady state is unstable.*

Notice that both stability conditions (5.98) and (5.99) are satisfied if the speeds of adjustment $a_k$ are sufficiently small for all firms.

Consider now the special case of symmetric firms, when $a_k \equiv a$ and $c_k \equiv c$. Notice that in this case

$$\gamma_k \equiv \gamma = \frac{Nc}{N+1} - c = -\frac{c}{N+1} < 0,$$

so condition (5.97) is satisfied. Relation (5.98) can be rewritten as

$$a < \frac{2(N+1)B}{A} \tag{5.100}$$

and (5.99) has the form

$$\frac{Na\gamma(N+1)}{2(N+1)B - aA} > -1,$$

which is equivalent to

$$a < \frac{2(N+1)B}{A - \gamma N(N+1)} = \frac{2(N+1)B}{A + cN}. \tag{5.101}$$

This inequality is stronger than (5.100), so if (5.101) holds, then the steady state is locally asymptotically stable, and if (5.101) is violated with strict inequality, then the steady state is unstable.

The stability results given above express sufficient conditions for the local asymptotic stability of the equilibrium, so they ensure the convergence of the adjustment process provided that the initial factors selected by the firms, $\varepsilon_k(0)$, are sufficiently close to the true slope $B$. This local analysis leaves open several questions. First of all, what are the necessary conditions for local stability, such that an exact delineation of the stability region in the space of the parameters can be obtained? What kinds of bifurcations occur when the boundaries of such stability regions are crossed? How does the steady state lose stability and what kind of disequilibrium asymptotic dynamics should be expected when the steady state is unstable? Finally, what are the extent and the shape of the basin of attraction of the steady state (when it is stable) or of other attractors when the steady state is unstable? To answer these questions is, in general, not easy if one considers a nonlinear $N$-dimensional dynamical system. Therefore, we will try to gain some insight into these problems by considering some simple situations, such as the symmetric case of an oligopoly with $N$ identical firms starting from identical initial guesses, and a duopoly with two heterogeneous firms that start from arbitrary initial guesses for the scale factors $\varepsilon_k(0)$, $k = 1, 2$.

*Example 5.12.* Consider first an oligopoly with identical firms such that $c_k = c$, $a_k = a$ for each $k$, and assume that they also have identical initial conditions,

$$\varepsilon_k(0) = \varepsilon(0).$$

So we have $\varepsilon_k(t) = \varepsilon(t)$ for each $t \geqslant 0$. The dynamics of $\varepsilon(t)$ are governed by the following one-dimensional difference equation

$$\varepsilon(t+1) = g(\varepsilon(t)) = \left(1 - \frac{Aa}{(N+1)B}\right)\varepsilon(t) + \frac{aBcN}{N+1}\frac{1}{\varepsilon(t)} + \frac{a(A-Nc)}{N+1}, \tag{5.102}$$

which can be derived easily from (5.91). At the unique positive equilibrium $\overline{\varepsilon} = B$ the derivative of the function $g$ becomes

$$g'(B) = 1 - \frac{aA}{(N+1)B} - \frac{acN}{(N+1)B},$$

and it is easy to realize that the condition for the local asymptotic stability $-1 < g'(B) < 1$ is fulfilled for

$$\frac{a(Nc+A)}{B(N+1)} < 2, \tag{5.103}$$

which can be rewritten as

$$a < \frac{2B(N+1)}{Nc+A}.$$

The stability condition (5.103) illustrates the stabilizing role of small values of the speed of adjustment $a$. The role of the number of firms is also clear, since the left hand side of the stability condition (5.103) is a decreasing function of $N$ if $c < A$ (note that the reservation price has to be larger than the unit cost in order to make production profitable). Hence, in our case a higher number of identical firms helps the firm to learn the true value of the slope of the market demand. If the expression on the left hand side of (5.103) is increased past the value 2, then the fixed point loses its stability via a flip (or period doubling) bifurcation, at which a stable cycle of period two is created around it.

However, in order to understand the global dynamical properties of the one-dimensional map (5.102), the graph of $g(\varepsilon)$ has to be examined for $\varepsilon > 0$. It is a hyperbola with a vertical asymptote at $\varepsilon = 0^+$, and for $\varepsilon \to +\infty$ it approaches the asymptote given by the equation

$$y = \left(1 - \frac{Aa}{(N+1)B}\right)\varepsilon + \frac{a(A - Nc)}{N+1}. \tag{5.104}$$

If $\frac{Aa}{B(N+1)} > 1$ then the map $g$ is decreasing, and for $\varepsilon \to +\infty$ it tends to $-\infty$ along the negatively sloped line (5.104). In this case, any positive trajectory converges to the steady state if the stability condition (5.103) is satisfied, whereas if (5.103) does not hold a stable cycle of period two may be the unique attractor: no other different kinds of attractors can exist for a decreasing map. On the other hand, if $\frac{Aa}{B(N+1)} < 1$ then the map $g$ is unimodal (see Fig. 5.5a). It decreases for $\varepsilon < \varepsilon_{\min}$, where

$$\varepsilon_{\min} = B\sqrt{\frac{Nac}{(N+1)B - Aa}},$$

and it increases for $\varepsilon > \varepsilon_{\min}$. As $\varepsilon \to +\infty$ it approaches the positively sloped line (5.104). This case may give rise to more complex dynamic properties. In fact, in this case the first period doubling bifurcation at which the steady state loses stability

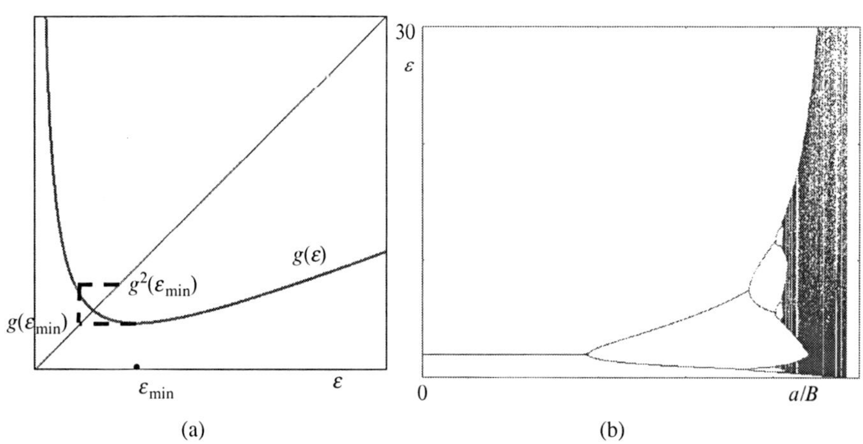


**Fig. 5.5** Oligopoly with $N$ identical firms starting from identical initial guesses on the scale factor. (**a**) The map $g$ and the trapping region of the dynamics. (**b**) Bifurcation diagram with respect to the parametric ratio $a/B$, where $a$ is the common speed of adjustment of the firms and $B$ is the slope of the demand function. Here $N = 3$, $A = 3$, $B = 2$ and $c = 2$

is followed by other period doublings and, in general, by the well-known period doubling cascade, that constitutes the typical route to chaotic behavior for smooth unimodal maps. So, complex dynamics, that include periodic cycles of any period and chaotic motion, can be obtained if the map is unimodal and the fixed point is unstable, that is if

$$\frac{2B(N+1)}{Nc+A} < a < \frac{B\,(N+1)}{A}. \tag{5.105}$$

This range is non-empty provided that $A < Nc$, that is if the reservation price is less than the firms' aggregated marginal costs. For example, if we consider the set of parameters $N = 3$, $A = 3$, $c = 2$, $B = 2$, the range given by (5.105) is $16/9 \simeq 1.78 < a < 8/3 \simeq 2.67$. This is confirmed by a numerical computation of the bifurcation diagram shown in Fig. 5.5b. The asymptotic dynamics are trapped inside the interval $[m, g(m)]$, where $m = g(\varepsilon_{\min}) > 0$ is the minimum value of the map (see Fig. 5.5a). For increasing values of the adjustment coefficient $a$ the minimum value $m$ decreases until it reaches the value $m = 0$ (for the set of parameters used to obtain the bifurcation diagram of Fig. 5.5b, this occurs at $a/B \simeq 1.1858$). This is the *final bifurcation*, after which the generic trajectory involves negative values. It is worth stressing that the same kind of bifurcation diagram, as the one shown in Fig. 5.5b can be obtained by increasing the reservation price $A$ or by increasing the marginal costs $c$. In cases where the sequence of scaling factors $\varepsilon(t)$ does not converge learning does not occur in the long run. ▼

*Example 5.13.* We now consider the case of a duopoly with heterogeneous players and we give a detailed study of the region of stability in the space of the parameters. For $N = 2$, the dynamic model (5.91) assumes the form of an iterated two-dimensional map $T : (\varepsilon_1(t), \varepsilon_2(t)) \to (\varepsilon_1(t+1), \varepsilon_2(t+1))$ defined by

$$\begin{aligned} \varepsilon_1(t+1) &= \varepsilon_1(t) + \frac{a_1}{3}\left[A\left(1 - \frac{\varepsilon_1(t)}{B}\right) + B\left(\frac{2c_1 - c_2}{\varepsilon_1(t)} + \frac{2c_2 - c_1}{\varepsilon_2(t)}\right) - (c_1 + c_2)\right], \\ \varepsilon_2(t+1) &= \varepsilon_2(t) + \frac{a_2}{3}\left[A\left(1 - \frac{\varepsilon_2(t)}{B}\right) + B\left(\frac{2c_1 - c_2}{\varepsilon_1(t)} + \frac{2c_2 - c_1}{\varepsilon_2(t)}\right) - (c_1 + c_2)\right]. \end{aligned} \tag{5.106}$$

In order to study the stability of the unique positive steady state $\bar{\varepsilon} = (\bar{\varepsilon}_1, \bar{\varepsilon}_2) = (B, B)$, we consider the Jacobian matrix computed at the equilibrium

$$\begin{pmatrix} 1 - \frac{a_1}{3B}(A + 2c_1 - c_2) & -a_1 \frac{2c_2 - c_1}{3B} \\ -a_2 \frac{2c_1 - c_2}{3B} & 1 - \frac{a_2}{3B}(A + 2c_2 - c_1) \end{pmatrix}, \tag{5.107}$$

from which the standard stability conditions are obtained (see Appendix F), namely

$$1 - Tr + Det > 0; \quad 1 + Tr + Det > 0; \quad Det < 1 \tag{5.108}$$

where $Tr$ and $Det$ are, respectively, the Trace and the Determinant of the Jacobian matrix (5.107). The first condition is always satisfied, hence the stability conditions, after some algebraic manipulations, reduce to

$$A\left(c_1+c_2+A\right)\frac{a_1a_2}{B^2}-6\left(2c_1-c_2+A\right)\frac{a_1}{B}-6\left(2c_2-c_1+A\right)\frac{a_2}{B}+36>0 \tag{5.109}$$

and

$$A\left(c_1+c_2+A\right)\frac{a_1a_2}{B^2}-3\left(2c_1-c_2+A\right)\frac{a_1}{B}-3\left(2c_2-c_1+A\right)\frac{a_2}{B}<0. \tag{5.110}$$

These two inequalities define a *region of stability* (we may also call it a *learning region*) in the space of the parameters. Moreover, the conditions (5.109) and (5.110) taken as equalities, define bifurcation hypersurfaces. This means that when one or more parameters are varied so that the equilibrium $\bar{\varepsilon}$ becomes unstable, if (1) the stability loss is due to a change of sign of the left hand side of (5.109), then a flip (or period doubling) bifurcation occurs, and if (2) the stability loss is due to a change of sign of the left hand side of (5.110), then a Neimark–Hopf bifurcation occurs. It is useful to represent the learning region by projecting it into the two-dimensional parameter plane $(a_1/B, a_2/B)$, where the bifurcation curves that bound the region of stability are equilateral hyperbolas (see Fig. 5.6, where $F$ denotes the positive branch of the hyperbola at which the flip bifurcation occurs, $H$ denotes the positive branch of the hyperbola at which the Neimark–Hopf bifurcation occurs, and the shaded area represents the learning region). If

$$c_1/2<c_2<2c_1, \tag{5.111}$$

then the two hyperbolas do not intersect, and the learning region is bounded only by the flip bifurcation curve (Fig. 5.6a), whereas if

$$2c_2<c_1<2c_2+A \quad \text{or} \quad 2c_1<c_2<2c_1+A \tag{5.112}$$

then the two hyperbolas intersect in the positive orthant of the plane $(a_1/B, a_2/B)$, so that the learning region is bounded by an arc of the Neimark–Hopf bifurcation curve and by two arcs of the flip bifurcation curve[1] (Fig. 5.6b).

If the parameters $a_1/B$ and/or $a_2/B$ are varied, so that they cross the boundary of the stability region along the portion of curve $F$, then the equilibrium point changes from a stable node to a saddle point via a supercritical flip bifurcation.[2]

---

[1] For $c_1 = 2c_2$ the curve $F$ degenerates into the pair of straight lines $a_1/B = 6/(3c_2 + A)$ and $a_2/B = 6/B$. For $c_2 = 2c_1$ the curve $F$ degenerates into the pair of straight lines $a_1 B = 6/A$ and $a_2/B = 6/(3c_1 + A)$.

[2] A rigorous proof of the supercritical nature of the flip bifurcation requires a center manifold reduction and the evaluation of higher order derivatives, up to the third order (see for example Guckenheimer and Holmes (1983)). This is a rather tedious calculation for a two-dimensional map,

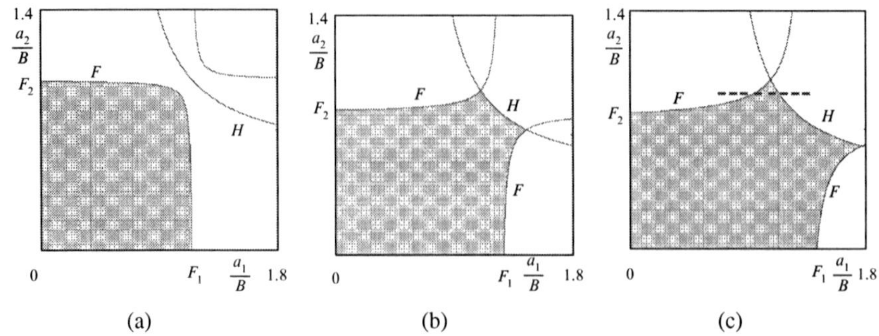


**Fig. 5.6** Duopoly with heterogeneous players, the bifurcation curves with respect to the parameters $a_1/B$ and $a_2/B$, where $a_i$ is the speed of adjustment of firm $i$ and $B$ is the slope of the demand curve. In all cases $A = 5$ and $c_1 = 0.5$. The *shaded area* is the learning region, the *curve* $F$ is the flip boundary and $H$ the Neimark–Hopf boundary. (**a**) Here $c_2 = 0.8$ so that $c_1/2 < c_2 < 2c_1$, the $F$ and $H$ curves do not intersect and the learning region is bounded only by the flip curve. (**b**) Now $c_2 = 1.3$ and the $F$ and $H$ curves intersect; the learning region is bounded by both curves. (**c**) Here $c_2 = 1.4$ and the portion of the arc of $H$ included in the boundary of the stability region increases due to increasing heterogeneity

This means that, just after the loss of stability of $\overline{\varepsilon}$, the long run evolution of the trajectories of (5.106) is characterized by the convergence to a periodic cycle of period two. So, if firms adopt the learning process introduced above, then they will never learn the true demand function. They will keep on underestimating/overestimating it, as the subjective scale factors continue to oscillate. If the cost parameters $c_1$ and $c_2$ are not too different, that is in the case of moderate heterogeneity in costs, then according to (5.111) the steady state $\overline{\varepsilon}$ can lose stability only via a period doubling bifurcation. This is particularly true if players are identical, as the analysis in Example 5.9 has already shown.

Let us now consider what happens if the parameters $a_1/B$ and $a_2/B$ are varied, so that they cross the boundary of the learning region along the portion of curve $H$. In this case, the equilibrium $\overline{\varepsilon}$ changes from a stable focus to an unstable focus via a supercritical Neimark–Hopf bifurcation.[3] This means that the long run evolution of the trajectories of (5.106) converges to a quasi-periodic motion around the steady state. Again, this implies that, on the basis of the adjustment process adopted, players will never learn the true demand function, as they will continue to over- and underestimate prices. This kind of route to instability can only occur if the two players are sufficiently heterogeneous with respect to cost parameters, according to

---

and we prefer to rely on numerical evidence as a stable 2-cycle close to the saddle $\overline{\varepsilon}$ is numerically detected whenever the parameters cross the bifurcation curve $F$.

[3] Also in this case, a rigorous proof of the supercritical nature of the Neimark–Hopf bifurcation requires a center manifold reduction and the evaluation of higher order derivatives, up to the third order (see for example Guckenheimer and Holmes (1983)). This is rather tedious in a two-dimensional map, and we prefer to rely on numerical evidence as a stable orbit surrounding the unstable focus $\overline{\varepsilon}$ is numerically detected whenever the parameters cross the bifurcation curve $H$.

(5.112). Moreover, if the difference between the cost parameters is increased, then the arc of the curve $H$ included in the boundary of the stability region becomes more extended (see Fig. 5.6c). Summarizing, on the basis of the results on the stability region in the plane $(a_1/B, a_2/B)$, we can say that the adjustment process can converge to the true demand function provided that both of the ratios $a_1/B$ and $a_2/B$ are sufficiently small. This means that for a given slope $B$, the speeds of adjustment $a_1$ and $a_2$ cannot be too large in order to ensure convergence of the adaptive learning process to the true demand. Indeed, increasing one or both speeds of adjustment may cause overshooting, characterized by oscillations of the scale factors that never settle to the true demand function. It is interesting to note that some bifurcation paths exist where an increase of one or both of the parameters $a_i/B_i$ may have both a stabilizing and a destabilizing effect. This occurs if (5.112) holds, so that the stability region has a shape like the one shown in Fig. 5.6b, c. One such bifurcation path is indicated by the dashed line in Fig. 5.6c. Along the first portion of this path an increase of $a_1/B$ and/or $a_2/B$ has a stabilizing effect: the equilibrium is first unstable, but becomes stable via a backward flip (or period halving) bifurcation. If we continue to increase $a_1/B$ and/or $a_2/B$ along the same path, we get a destabilizing effect because $\bar{\varepsilon}$ loses stability via a supercritical Neimark–Hopf bifurcation. Such a scenario can only happen if there is a sufficiently large degree of heterogeneity in costs because, as remarked above, the portion of the boundary of the learning region formed by the Neimark–Hopf bifurcation curve becomes smaller and smaller (until it finally disappears) as the heterogeneity in marginal costs is reduced.

It is also worth noting that the stability region shrinks as, *ceteris paribus*, the reservation price $A$ increases. In fact, the intersections $F_1$ and $F_2$ of the curve $F$ with the coordinate axes of the parameter plane $(a_1/B, a_2/B)$ are given by $F_1 = (6/(2c_1 - c_2 + A), 0)$ and $F_2 = (0, 6/(2c_2 - c_1 + A))$. Consequently, convergence of the learning process to the true demand is less likely to occur if reservation prices are higher. This confirms the results on local stability for the $N$-dimensional model given above. The stability analysis provided so far is only based on local stability and local bifurcations of the unique steady state. With the help of some numerical simulations we can explore what happens when the parameters are located far away from the boundaries of the stability region, and we can obtain some indication about the extent and the shape of the basin of attraction of the steady state or of the more complex attractors that replace the steady state if the parameters are outside the learning region. Let us consider, first, the following values of the parameters: $A = 5$, $B = 1$, $c_1 = 0.5$ and $c_2 = 0.6$. This gives a shape of the learning region similar to the one shown in Fig. 5.6a. In this case, when the parameters are inside the stability region the steady state is a stable node, as in Fig. 5.7a obtained with $a_1 = 0.9$, $a_2 = 1$. In this case, there are two real eigenvalues, one positive and one negative. This means that any trajectory of (5.106) starting close to the steady state $\bar{\varepsilon}$ converges to it through oscillations of decreasing amplitude. Note that in Fig. 5.7 the white region represents the set of points that generate feasible trajectories (in other words trajectories entirely included inside the positive orthant) and converging to the steady state, whereas the grey region represents the set of points that generate infeasible trajectories (which are trajectories involving negative values). Figure 5.7b

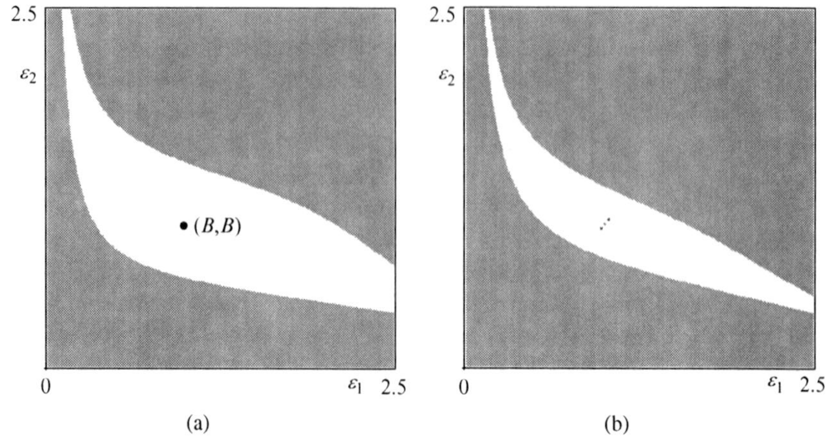


**Fig. 5.7** The duopoly with heterogeneous players. From initial values in the *white region* the system generates feasible trajectories, from initial values in the *grey region*, the trajectories become infeasible. Here $A = 5$, $B = 1$, $c_1 = 0.5$ and $c_2 = 0.6$. (**a**) The equilibrium is a stable node with one positive and one negative real eigenvalue. The speeds of adjustment are $a_1 = 0.9$ and $a_2 = 1$. (**b**) The steady state becomes a saddle point and a stable cycle of period 2 emerges. The speeds of adjustment are $a_1 = 0.9475$ and $a_2 = 1$

is obtained with a higher value of the speed of adjustment $a_1$, namely $a_1 = 0.9475$. In this case, as expected on the basis of the local stability analysis, the steady state is a saddle point, because a period doubling bifurcation has created a stable cycle of period 2, represented by the two small dots in Fig. 5.7b. This means that none of the two firms learns the demand and they keep on underestimating and overestimating it. As $a_1/B$ and/or $a_2/B$ are further moved away from the stability region, the periodic points move away from the unstable steady state, and so the amplitude of the oscillations increases. Moreover, other local bifurcations may occur, at which also the cycle of period two loses stability and more complex attractors may appear (for example, the 2-cycle may flip bifurcate to give rise to a stable cycle of period 4, and so on, until chaotic attractors appear after the well-known period-doubling cascade) or the attractor may have a contact with the boundary of its basin of attraction and disappear, after which the generic trajectory will be infeasible.

Let us now consider the case of a larger difference between the cost parameters $c_1$ and $c_2$, so that the condition (5.112) is satisfied and, consequently, the stability region is also bounded by a portion of the curve $H$ where a Neimark–Hopf bifurcation occurs. By setting $A = 5$, $B = 1$, $c_1 = 0.5$ and $c_2 = 1.3$, as in Fig. 5.6b, we consider a set of parameters inside the stability region, namely $a_1 = 1$, $a_2 = 0.9$. Hence, the equilibrium, shown in Fig. 5.8a with its feasible set of attraction, is a stable focus (complex conjugate eigenvalues of modulus less than 1). As expected, if we increase $a_1$ and/or $a_2$, so that $(a_1/B, a_2/B)$ crosses the boundary $H$ of the stability region, a supercritical Neimark–Hopf bifurcation occurs, at which the steady state is transformed into an unstable focus, and an attracting closed invariant curve is created around it (see Fig. 5.8b, obtained with $a_1 = 1.17$ and $a_2 = 0.9$).

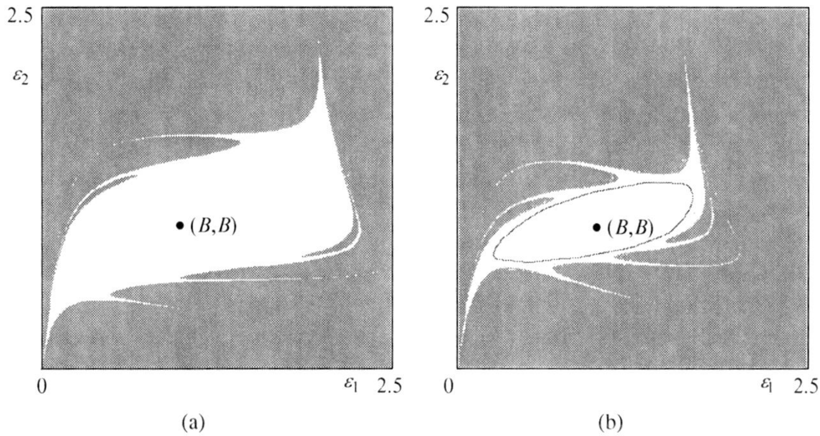


**Fig. 5.8** The duopoly with heterogeneous players. From initial values in the *white region* the system generates feasible trajectories, from initial values in the *grey region*, the trajectories become infeasible. Here $A = 5$, $B = 1$, $c_1 = 0.5$, $c_2 = 1.3$ and $a_2 = 0.9$. (**a**) The speed of adjustment of the first firm is $a_1 = 1$. The equilibrium is a stable focus. (**b**) The speed of adjustment of the first firm is $a_1 = 1.17$. After the value of $(a_1/B, a_2/B)$ crosses the boundary $H$ an attracting closed invariant curve is created around the (now) unstable equilibrium

As the parameters $a_1$ and/or $a_2$ are further increased, the size of the attracting closed orbit around the steady state increases, according to the Neimark–Hopf bifurcation theorem, and consequently the long-run oscillations of the scale factors $\varepsilon_k(t)$ will increase their amplitude until a contact between the boundaries of the attractor and the boundary of the feasible region occurs. This contact represents a global bifurcation (called *final bifurcation* in Mira et al. (1996), or *boundary crisis* in Grebogi et al. (1983)) that marks the disappearance of the attractor, because after the contact the generic trajectory is infeasible.

We do not analyze these dynamic properties of the model in greater detail, as here we are mainly interested in studying the conditions under which learning emerges. However, putting together the information gained by the two numerical simulations shown above, we can easily see what happens when the condition (5.112) holds (which implies a high degree of heterogeneity between the two firms) and the parameters $a_1/B$ and/or $a_2/B$ are gradually increased in such a way that we obtain two bifurcations which cause a transition between two different instability situations separated by a "window" of stability, like in the bifurcation path represented by the dashed line in Fig. 5.6c. Moving along that path by increasing the value of the parameter $a_1$, at first the equilibrium is unstable, with long-run dynamics characterized by oscillations of period 2. Then the learning process leads to a period halving (or backward flip) bifurcation, after which the equilibrium becomes stable. Then, by further increasing the speed of adjustment $a_1$ a supercritical Neimark–Hopf bifurcation occurs after which the equilibrium becomes unstable again, and the long-run dynamics of the learning process are characterized by quasi-periodic oscillations along a stable close invariant orbit around the unstable steady state. To conclude this example, it is worth making some remarks about the extent and the shape of the feasible region, which is represented by the white area in the figures shown above. If the parameters are inside the learning region, the extent of the feasible region provides important information about the robustness of the learning process. Detailed information about the feasible basin of the steady state $\bar{\varepsilon}$ provides an answer to the fundamental question: how far away from the true demand can the guesses of the players be in order to still guarantee the success of the learning process? First of all, it can be noticed that the maximum "distance" of a single subjective scale factor is not important, as the distance of all the scale factors must be considered. Even if one firm starts with an initial estimate for $\varepsilon_k$ very close to the true value $B$, the endogenous dynamics of the global learning process may not lead to convergence to the true demand in the long run due to the influence of its competitors. Although this remark may sound obvious ex post, we think that it is worth pointing this out. As a second and final remark we point out that the boundaries of the feasible region may be quite complicated. This can be clearly seen in Fig. 5.8b. A study of this kind of complexity requires an analysis of the global dynamic properties of the map (5.106). In particular, the creation of complicated topological structures may be related to the fact that the map (5.106) is noninvertible (as explained in Appendix C). ▼

Now we turn our attention to the continuous time system (5.92). Its Jacobian is the matrix $\boldsymbol{J}$ with eigenvalue equation

$$\prod_{k=1}^{N}\left(-\frac{a_k A}{(N+1)B}-\lambda\right)\left[1-\sum_{k=1}^{N}\frac{\frac{a_k B\gamma_k}{\varepsilon_k^2}}{\frac{a_k A}{(N+1)B}+\lambda}\right]=0, \tag{5.113}$$

which can be derived similarly to the discrete case. In this case we have the following stability theorem, which can be proved along the lines of the proof of Theorem 2.2.

**Theorem 5.5.** *Under assumption (5.97) the steady state of system (5.92) is always locally asymptotically stable.*

If condition (5.97) is violated, that is, when for at least one firm,

$$\gamma_k = \frac{1}{N+1}\sum_{l=1}^{N} c_l - c_k > 0,$$

then Theorems 5.4 and 5.5 no longer hold, and the asymptotic behavior of the dynamic systems becomes much more complicated.

### 5.3.2 Unknown Reservation Price with Known Slope

In this section we assume that all firms know the slope $B$ of the price function but the value of the reservation price $f(0) = A$ is unknown. In this case, the firms try to estimate and learn about the value of $A$. Let $\widetilde{f_k}(Q) = \varepsilon_k - BQ$ denote the believed price function of firm $k$, where the value of $\varepsilon_k$ will be repeatedly updated. The learning process will be similar to that introduced in the previous section.

Consider first the situation from the point of view of firm $k$. The profit of any firm $l$ (including itself) is given as

$$\widetilde{x_l}(\varepsilon_k - B\widetilde{Q_l} - B\widetilde{x_l}) - (c_l\widetilde{x_l} + d_l), \tag{5.114}$$

where the 'tilde' again indicates that we are dealing with quantities based on firm $k$'s estimate of the price function. So the best response of firm $l$ is

$$\widetilde{x_l} = \frac{\varepsilon_k - c_l}{2B} - \frac{\widetilde{Q_l}}{2},$$

implying that

$$\widetilde{x_l} = \frac{\varepsilon_k - c_l}{B} - \widetilde{Q}. \tag{5.115}$$

By summing these equations for all $l$, we find

$$\widetilde{Q} = \frac{N\varepsilon_k - \sum_{l=1}^{N} c_l}{B} - N\widetilde{Q},$$

so firm $k$ believes that the total output of the industry is

$$\widetilde{Q}^k = \frac{N\varepsilon_k - \sum_{l=1}^{N} c_l}{(N+1)B}.$$

Therefore firm $k$ will produce the output

$$x_k = \frac{\varepsilon_k - c_k}{B} - \widetilde{Q}^k = \frac{\varepsilon_k - (N+1)c_k + \sum_{l=1}^{N} c_l}{(N+1)B} \tag{5.116}$$

and expects the market price to be

$$\widetilde{p_k} = \widetilde{f_k}(\widetilde{Q}^k) = \frac{\varepsilon_k + \sum_{l=1}^{N} c_l}{N+1}. \tag{5.117}$$

In reality however each firm reasons independently in the same way, the expected price and produced amount depend on its estimated price function, so the actual total output of the industry becomes

$$Q = \sum_{k=1}^{N} x_k = \frac{\sum_{k=1}^{N} \varepsilon_k - \sum_{l=1}^{N} c_l}{(N+1)B},$$

with the corresponding equilibrium price

$$p = A - BQ = A - \frac{1}{N+1}\left(\sum_{l=1}^{N} \varepsilon_l - \sum_{l=1}^{N} c_l\right). \tag{5.118}$$

As in the previous section the firms adjust their beliefs about the price function based on the discrepancies

$$\Delta p_k = p - \widetilde{p}_k = \frac{1}{N+1}\left((N+1)A - \sum_{l=1}^{N} \varepsilon_l - \varepsilon_k\right). \tag{5.119}$$

They want to increase $\varepsilon_k$ if $\Delta p_k > 0$, and if $\Delta p_k < 0$ then they decrease the value of $\varepsilon_k$, and if $\Delta p_k = 0$, then they have no reason to change it. This adjustment concept can be again modeled by the discrete system (5.91) and its continuous counterpart (5.92). Notice that $\Delta p_k$ is a linear function of the state variables $\varepsilon_1, \ldots, \varepsilon_N$, therefore the corresponding dynamical systems are linear, and in this case local and global asymptotic stability are equivalent. Similarly to the previous case it is easy to show that both systems have a unique steady state, $\bar{\varepsilon}_k = A$ for all $k$ which corresponds to full knowledge of the price function. Notice first that if $\Delta p_k = 0$ for all $k$, then the $\varepsilon_k$ values are identical. If $\bar{\varepsilon}$ denotes their common value, then

$$0 = \frac{1}{N+1}((N+1)A - N\bar{\varepsilon} - \bar{\varepsilon})$$

implying that $\bar{\varepsilon} = A$.

Consider first the discrete case. The coefficient matrix has now the special structure

$$\boldsymbol{I} + \boldsymbol{J}$$

where

$$\boldsymbol{J} = \frac{1}{N+1}\begin{pmatrix} -2a_1 & -a_1 & \ldots & -a_1 \\ -a_2 & -2a_2 & \ldots & -a_2 \\ \vdots & \vdots & \ddots & \vdots \\ -a_N & -a_N & \ldots & -2a_N \end{pmatrix} = \boldsymbol{D} + \boldsymbol{a}\boldsymbol{b}^T \tag{5.120}$$

with

$$\boldsymbol{D} = diag\left(-\frac{a_1}{N+1}, \ldots, -\frac{a_N}{N+1}\right), \quad \boldsymbol{a} = \left(-\frac{a_1}{N+1}, \ldots, -\frac{a_N}{N+1}\right)^T,$$

and

$$b^T = (1, \ldots, 1).$$

By using again the results of Appendix E we can see that the eigenvalue equation of the coefficient matrix is

$$\prod_{k=1}^{N}\left(1-\frac{a_k}{N+1}-\lambda\right)\left[1+\sum_{k=1}^{N}\frac{-\frac{a_k}{N+1}}{1-\frac{a_k}{N+1}-\lambda}\right]=0. \tag{5.121}$$

By repeating the proof of Theorem 2.1 we can easily show the following result.

**Theorem 5.6.** *The steady state is globally asymptotically stable if and only if for all $k$,*

$$a_k < 2(N+1) \tag{5.122}$$

*and*

$$\sum_{k=1}^{N}\frac{a_k}{2(N+1)-a_k} < 1. \tag{5.123}$$

In the symmetric case $a_k \equiv a$, relations (5.122) and (5.123) reduce to

$$a < 2(N+1)$$

and

$$a < 2, \tag{5.124}$$

where the second inequality is the stronger of the two. Therefore the steady state is globally asymptotically stable if and only if (5.124) holds.

Consider next the continuous time model (5.92). Its coefficient matrix is $\boldsymbol{J}$ with eigenvalue equation

$$\prod_{k=1}^{N}\left(-\frac{a_k}{N+1}-\lambda\right)\left[1+\sum_{k=1}^{N}\frac{\frac{a_k}{N+1}}{\frac{a_k}{N+1}+\lambda}\right]=0. \tag{5.125}$$

By repeating the proof of Theorem 2.2 the following stability result is obtained.

**Theorem 5.7.** *The steady state is always globally asymptotically stable.*

Global asymptotic stability means that regardless of how inaccurate the initial estimations of parameter $A$ are, as $t \to \infty$, the estimates always converge to the true value of $A$.

We will next show that this nice stability property may be lost, if the firms obtain delayed price information. Assuming continuously distributed time lags and using the same weighting functions as in Sect. 2.6, the dynamical system (5.92) becomes

$$\dot{\varepsilon}_k(t) = \frac{a_k}{N+1}\left((N+1)A - \int_0^t w(t-s, T_k, m_k)\sum_{l=1}^{N}\varepsilon_l(s)\mathrm{d}s - \varepsilon_k(t)\right) \tag{5.126}$$

since at any time period $t$, firm $k$ uses the current estimate $\varepsilon_k(t)$ in computing $p_k(t)$ by (5.117), however it uses delayed actual price data, and therefore the computed discrepancy $\Delta p_k$ is based on continuously distributed lagged values of $\sum_{l=1}^{N}\varepsilon_l$.

Equation (5.126) constitutes a system of linear Volterra-type integro-differential equations. In order to compute the eigenvalues we seek the solution of the corresponding homogenous equations in the exponential form $\varepsilon_k(t) = v_k e^{\lambda t}$ $(k = 1, 2, \ldots, N)$. Substituting these into the homogenous equations implies that

$$\left(\lambda + \frac{a_k}{N+1}\right) v_k e^{\lambda t} + \frac{a_k}{N+1}\int_0^t w(t-s, T_k, m_k)\sum_{l=1}^{N} v_l e^{\lambda s}\mathrm{d}s = 0.$$

Letting $t \to \infty$ and using the limiting property of integral (D.3) we have

$$\left(\lambda + \frac{a_k}{N+1}\right) v_k + \frac{a_k}{N+1}\left(1 + \frac{\lambda T_k}{p_k}\right)^{-(m_k+1)} \sum_{l=1}^{N} v_l = 0$$

or

$$\left(\frac{N+1}{a_k}\lambda + 1\right)\left(1 + \frac{\lambda T_k}{p_k}\right)^{m_k+1} v_k + \sum_{l=1}^{N} v_l = 0, \tag{5.127}$$

where

$$p_k = \begin{cases} 1 & \text{if} \quad m_k = 0, \\ m_k & \text{if} \quad m_k > 0, \end{cases}$$

as before. Non-trivial solutions exist if and only if the determinant of the coefficient matrix is zero. This determinantal equation has the special structure

$$det \begin{pmatrix} A_1(\lambda) & B_1(\lambda) & \ldots & B_1(\lambda) \\ B_2(\lambda) & A_2(\lambda) & \ldots & B_2(\lambda) \\ \vdots & \vdots & \ddots & \vdots \\ B_N(\lambda) & B_N(\lambda) & \ldots & A_N(\lambda) \end{pmatrix},$$

with

$$A_k(\lambda) = \left(\frac{N+1}{a_k}\lambda + 1\right)\left(1 + \frac{\lambda T_k}{p_k}\right)^{m_k+1} + 1,$$

and

$$B_k(\lambda) = 1,$$

for all $k$. By applying relation (E.3) we see that the determinantal equation simplifies to

$$\prod_{k=1}^{N}\left\{\left(\frac{N+1}{a_k}\lambda+1\right)\left(1+\frac{\lambda T_k}{p_k}\right)^{m_k+1}\right\}\times$$

$$\times\left[1+\sum_{k=1}^{N}\frac{1}{\left(\frac{N+1}{a_k}\lambda+1\right)\left(1+\frac{\lambda T_k}{p_k}\right)^{m_k+1}}\right]=0. \tag{5.128}$$

The roots of the first part of the left hand side are

$$\lambda=-\frac{a_k}{N+1}\quad\text{and}\quad\lambda=-\frac{p_k}{T_k},$$

for $k = 1, 2, \ldots, N$, all are negative, therefore we should examine only the roots of the second part. So we turn our attention to the equation

$$1+\sum_{k=1}^{N}\frac{1}{\left(\frac{N+1}{a_k}\lambda+1\right)\left(1+\frac{\lambda T_k}{p_k}\right)^{m_k+1}}=0. \tag{5.129}$$

This is clearly equivalent to a polynomial equation, the roots of which can be determined by using computational methods in the general case. In order to obtain analytical results we will consider the case of symmetric firms, when the initial states are identical, $a_k = a, T_k = T, m_k = m$ and so $p_k \equiv p$. In this case (5.129) is reduced to

$$\left(\frac{N+1}{a}\lambda+1\right)\left(1+\frac{\lambda T}{p}\right)^{m+1}+N=0. \tag{5.130}$$

Assume first that $T = 0$, that is, there is no time delay. Then (5.130) assumes the linear form

$$\frac{N+1}{a}\lambda+1+N=0$$

with the only root $\lambda = -a$, so the steady state is globally asymptotically stable.

Note that this result is a special case of Theorem 5.7. Assume next that $T > 0$. For larger values of $m$ computational methods are needed to locate the eigenvalues and check stability. For $m = 0$ and $m = 1$, analytical methods are available, and the following theorem can be proved.

**Theorem 5.8.** *If $m = 0$, then the steady state is always globally asymptotically stable. In the case of $m = 1$ we have the following possibilities:*

*(i) If $N < 8$, then the steady state is always globally asymptotically stable,*

(ii) *If* $N = 8$, *then it is globally asymptotically stable if and only if* $LT \neq 1$, *where* $L = a/(N+1)$,

(iii) *If* $N \geq 9$, *then the steady state is globally asymptotically stable if and only if either* $LT < (LT)_1^*$ *or* $LT > (LT)_2^*$, *where* $(LT)_1^*$ *and* $(LT)_2^*$ *are given by (5.134). At these critical values Hopf bifurcations occur giving rise to the possibility of the birth of limit cycles.*

*Proof.*

Let $T > 0$ and $m = 0$. Then (5.130) becomes the quadratic equation

$$\lambda^2 \frac{T(N+1)}{a} + \lambda\left(T + \frac{N+1}{a}\right) + (N+1) = 0.$$

Since all coefficients are positive, all roots have negative real parts (see Lemma F.2 in Appendix F) implying global asymptotic stability.

Consider next the case of $T > 0$ and $m = 1$. Then (5.130) becomes a cubic equation, and with the notation $L = a/(N+1)$ it has the form

$$(\lambda + L)(1 + 2\lambda T + \lambda^2 T^2) + NL = 0,$$

that is,

$$T^2\lambda^3 + \lambda^2(2T + LT^2) + \lambda(1 + 2LT) + L(N+1) = 0. \tag{5.131}$$

Since all coefficients are positive, the Routh–Hurwitz stability condition implies that all eigenvalues have negative real parts if and only if

$$(2T + LT^2)(1 + 2LT) > T^2L(N+1), \tag{5.132}$$

which can be rewritten as a quadratic inequality in $LT$, namely

$$2(LT)^2 + (LT)(4 - N) + 2 > 0. \tag{5.133}$$

The discriminant of the left hand side is

$$(4 - N)^2 - 16 = N(N - 8).$$

Depending on the number of roots of the left hand side of (5.133) we have to consider the following cases.

Case 1. If $N < 8$, then the discriminant is negative, so (5.133) always holds and the steady state is always globally asymptotically stable.

Case 2. If $N = 8$, then there is a unique real root

$$LT = \frac{N-4}{4} = 1,$$

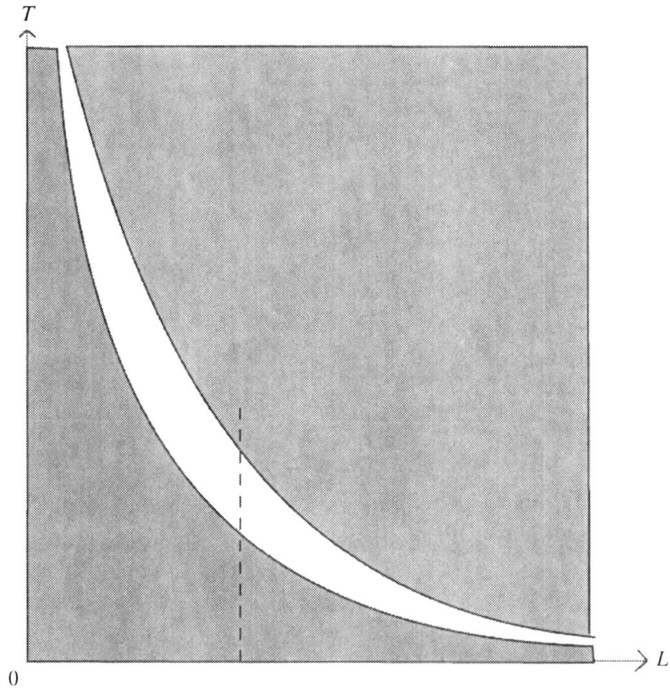


**Fig. 5.9** Stability region (*shaded*) in the $(L, T)$ plane

so the steady state is globally asymptotically stable if and only if $LT \neq 1$.

Case 3. If $N > 8$, then there are two real roots,

$$(LT)^*_{1,2} = \frac{N - 4 \pm \sqrt{N(N-8)}}{4}, \tag{5.134}$$

both are positive, and reciprocals of each other. So the steady state is globally asymptotically stable if and only if either $LT < (LT)^*_1$ or $LT > (LT)^*_2$, where we assume that $(LT)^*_1 < (LT)^*_2$.

The stability region in the $(L, T)$ plane is the shaded region shown in Fig. 5.9. The steady state is globally asymptotically stable under the lower hyperbola and above the upper hyperbola. The white region between the two hyperbolas shows where the steady state is unstable. If $N = 8$, then the hyperbolas coincide. In this case global asymptotic stability occurs outside the curve of the hyperbola. In Table 5.1 we show the values of $(LT)^*_1$ and $(LT)^*_2$ for several cases of $N \geq 8$.

Consider next $N \geq 9$ and a fixed value of $L$, and start gradually increasing the value of $T$ starting at a very small level. The steady state is globally asymptotically stable until we reach the lower hyperbola, and after crossing this hyperbola the steady state becomes unstable. Stability is regained after the upper hyperbola is

**Table 5.1** Values of $(LT)_1^*$ and $(LT)_2^*$

| $N$ | $(LT)_1^*$ | $(LT)_2^*$ |
|---|---|---|
| 8 | 1 | 1 |
| 9 | 0.5 | 2 |
| 10 | 0.382 | 2.618 |
| 11 | 0.314 | 3.186 |
| 15 | 0.188 | 5.312 |
| 20 | 0.127 | 7.873 |
| 30 | 0.0774 | 12.923 |
| 50 | 0.0436 | 22.956 |

crossed. We will prove that at these critical points Hopf bifurcations occur giving rise to the possibility of the birth of limit cycles around the steady state.

We select $T$ as the bifurcation parameter. At the critical points inequality (5.132) becomes equality, so (5.131) can be rewritten as

$$\lambda^3 T^2 + \lambda^2(2T + LT^2) + \lambda \frac{T^2 L(N+1)}{2T + LT^2} + L(N+1)$$

$$= (\lambda T^2 + (2T + LT^2))\left(\lambda^2 + \frac{L(N+1)}{2T + LT^2}\right) = 0,$$

showing that the roots are

$$\lambda_{1,2} = \pm i \sqrt{\frac{L(N+1)}{2T + LT^2}} \tag{5.135}$$

and

$$\lambda_3 = -\frac{2T + LT^2}{T^2} < 0.$$

Differentiating implicitly (5.131) with respect to $T$, we have

$$2T\lambda^3 + 3T^2\lambda^2\dot{\lambda} + 2\lambda\dot{\lambda}(2T + LT^2) + \lambda^2(2 + 2LT) + \dot{\lambda}(1 + 2LT) + 2\lambda L = 0,$$

where we use the notation $\dot{\lambda} = \frac{d\lambda}{dT}$, from which

$$\dot{\lambda} = \frac{-2T\lambda^3 - \lambda^2(2 + 2LT) - 2\lambda L}{3\lambda^2 T^2 + 2\lambda(2T + LT^2) + (1 + 2LT)}. \tag{5.136}$$

For the sake of simplifying the notation let

$$\alpha^2 = \frac{L(N+1)}{2T + LT^2} \left( = \frac{1+2LT}{T^2} \right),$$

then $\lambda_{1,2} = \pm i\alpha$ and at these values

$$\dot{\lambda} = \frac{\pm 2T\alpha^3 i + \alpha^2(2+2LT) \mp 2\alpha L i}{-3\alpha^2 T^2 \pm 2\alpha i(2T + LT^2) + (1+2LT)}$$

$$= \frac{\alpha^2(2+2LT) \pm 2i(T\alpha^3 - \alpha L)}{-2(1+2TL) \pm 2\alpha i(2T + LT^2)}$$

with real part

$$Re\dot{\lambda} = \frac{4\alpha^2(1-(LT)^2)}{4(1+2TL)^2 + 4\alpha^2(2T+LT^2)^2} \neq 0,$$

since it is easy to show that

$$(LT)_1^* < 1 < (LT)_2^*$$

based on the fact that $(LT)_1^*$, and $(LT)_2^*$ are reciprocals of each other. Hence all conditions of the Hopf bifurcation theorem are satisfied. ■

Larger values of $m$ lead to higher order polynomial equations and therefore the stability analysis becomes more complicated and requires the use of computational methods.

### 5.3.3 Unknown Slope with Known Reservation Price

In this section we assume that the firms know the value of the reservation price, $f(0) = A$, but they are uncertain about the slope $B$ of the price function. In this case firm $k$ believes that the price function is $\widetilde{f_k}(Q) = A - \varepsilon_k Q$ where the value of $\varepsilon_k$ is estimated and updated at each time period.

As in the previously discussed cases let us examine the way firm $k$ reasons in this situation. It believes that the profit of each firm (including its own) is

$$\widetilde{x_l}(A - \varepsilon_k \widetilde{Q_l} - \varepsilon_k x_l) - (c_l \widetilde{x_l} + d_l), \tag{5.137}$$

so the believed best response of firm $l$ is

$$\widetilde{x_l} = \frac{A - c_l}{2\varepsilon_k} - \frac{\widetilde{Q_l}}{2},$$

implying that

$$\widetilde{x}_l = \frac{A - c_l}{\varepsilon_k} - \widetilde{Q}. \tag{5.138}$$

By summing these equations for all values of $l$,

$$\widetilde{Q} = \frac{NA - \sum_{l=1}^{N} c_l}{\varepsilon_k} - N\widetilde{Q},$$

so firm $k$ believes that the total production level of the industry is

$$\widetilde{Q}^k = \frac{NA - \sum_{l=1}^{N} c_l}{(N+1)\varepsilon_k}. \tag{5.139}$$

Therefore firm $k$ will produce the output

$$x_k = \frac{A - c_k}{\varepsilon_k} - \widetilde{Q}^k = \frac{A - (N+1)c_k + \sum_{l=1}^{N} c_l}{(N+1)\varepsilon_k} \tag{5.140}$$

and the equilibrium price is believed to be

$$\widetilde{p}_k = \widetilde{f}_k(\widetilde{Q}^k) = \frac{A + \sum_{l=1}^{N} c_l}{N+1}, \tag{5.141}$$

as a consequence of (5.138) with $l = k$ and the particular form of the believed price function $\widetilde{f}_k$. Notice that $\widetilde{p}_k$ is the same for all firms, that is, the expected equilibrium prices are identical.

In reality the total production of the industry becomes

$$Q = \sum_{k=1}^{N} x_k = \frac{1}{N+1}\left((A + \sum_{l=1}^{N} c_l)\sum_{k=1}^{N}\frac{1}{\varepsilon_k} - (N+1)\sum_{k=1}^{N}\frac{c_k}{\varepsilon_k}\right),$$

with actual market price

$$p = A - BQ = A - \frac{B}{N+1}\left((A + \sum_{l=1}^{N} c_l)\sum_{k=1}^{N}\frac{1}{\varepsilon_k} - (N+1)\sum_{k=1}^{N}\frac{c_k}{\varepsilon_k}\right). \tag{5.142}$$

For firm $k$, the discrepancy between the actual and believed price is

$$\Delta p_k = p - \widetilde{p}_k = \frac{1}{N+1}\left(NA - B(A + \sum_{l=1}^{N} c_l)\sum_{k=1}^{N}\frac{1}{\varepsilon_k} + B(N+1)\sum_{k=1}^{N}\frac{c_k}{\varepsilon_k} - \sum_{l=1}^{N} c_l\right),$$

and since an increase in the value of $\varepsilon_k$ decreases the price estimate the dynamic processes (5.91) and (5.92) are now modified to become

$$\varepsilon_k(t+1) = \varepsilon_k(t) - a_k \Delta p_k \quad (k = 1, 2, \ldots, N) \tag{5.143}$$

and

$$\dot{\varepsilon}_k = -a_k \Delta p_k \quad (k = 1, 2, \ldots, N). \tag{5.144}$$

Notice also that $\Delta p_k$ is the same for all firms, so $\bar{\varepsilon}_1, \ldots, \bar{\varepsilon}_N$ is a steady state of the dynamical systems (5.143) or (5.144) if and only if

$$NA - B\left(A + \sum_{l=1}^{N} c_l\right) \sum_{k=1}^{N} \frac{1}{\varepsilon_k} + B(N+1) \sum_{k=1}^{N} \frac{c_k}{\varepsilon_k} - \sum_{l=1}^{N} c_l = 0.$$

Clearly $\bar{\varepsilon}_1 = \cdots = \bar{\varepsilon}_N = B$ satisfies this equation, so the full knowledge of the price function is a steady state. However, this is single linear equality in the variables $1/\varepsilon_1, \ldots, 1/\varepsilon_N$, therefore there are infinitely many positive steady states. That is, at any other steady state there is no discrepancy between expected and actual prices, so all firms believe that their price functions are correct, but they are not. Therefore no learning is possible in this case (see Fudenberg and Levine (1998), Marimon (1997), or Kirman and Salmon (1995)).

## 5.4 Uncertain Price Functions

In this section we assume that the firms face an uncertain price function, so firm $k$ believes that the price function is $\widetilde{f}_k(Q) + \eta_k$, where $\widetilde{f}_k(Q)$ is the estimate of $f(Q)$ and $\eta_k$ is a random error. It is also assumed that $E(\eta_k) = 0$ and $Var(\eta_k) = \sigma_k^2$ for all $k$. The believed profit of firm $k$,

$$\widetilde{\varphi}_k = x_k(\widetilde{f}_k(Q) + \eta_k) - C_k(x_k), \tag{5.145}$$

is therefore also random with expectation

$$E(\widetilde{\varphi}_k) = x_k \widetilde{f}_k(Q) - C_k(x_k) \tag{5.146}$$

and variance

$$Var(\widetilde{\varphi}_k) = x_k^2 \sigma_k^2. \tag{5.147}$$

For the sake of simplicity no externalities are considered. The firms want to maximize their expected profits and at the same time to ensure as low variance of the profit as possible. Therefore at each time period each firm faces a multi-objective optimization problem where $E(\widetilde{\varphi}_k)$ is maximized and $Var(\widetilde{\varphi}_k)$ is minimized. Assume that $\gamma_k$ measures the relative importance of lowering $Var(\widetilde{\varphi}_k)$ compared to the increase of $E(\widetilde{\varphi}_k)$ by firm $k$, then this firm might decide to maximize the utility function

$$\begin{aligned}\Psi_k(x_1,\dots,x_N) &= E(\widetilde{\varphi}_k) - \gamma_k Var(\widetilde{\varphi}_k) \\ &= x_k \widetilde{f}_k(Q) - (C_k(x_k) + \sigma_k^2 \gamma_k x_k^2).\end{aligned} \tag{5.148}$$

Notice that this function has the same form as the profit of firm $k$ with believed price function $\widetilde{f}_k$ and cost function $C_k(x_k) + \sigma_k^2 \gamma_k x_k^2$. Therefore all results of Sect. 5.1 can be applied to this case by assuming that the firms do not update their information about the distributions of the random variables $\eta_k$ based on the repeated price observations during the dynamic process. In reality however at each time period a new sample element for each random variable $\eta_k$ becomes available, so each firm is able to update $E(\eta_k)$ and $Var(\eta_k)$ by using Bayesian methodology. If we now use the notation $E(\eta_k(t)) = e_k(t)$ and $Var(\eta_k(t)) = \sigma_k^2(t)$ for the updated expectation and standard deviation of $\eta_k$ at time period $t$, then at the next period the utility function of firm $k$ becomes

$$x_k(\widetilde{f}_k(x_k + Q_k^E(t+1)) + e_k(t)) - (C_k(x_k) + \sigma_k^2(t)\gamma_k x_k^2),$$

so the best response of firm $k$ as well as the resulting dynamic models with both discrete and continuous time scales become time variant. The details of Bayesian updating of the distributions of $\eta_k$ as well as an examination of the asymptotic behavior of the corresponding time variant dynamical systems are beyond the scope of this book. The interested reader may consult Cyert and DeGroot (1971, 1973, 1987) to find out more about the relevant methodology.

# Chapter 6
# Overview and Directions for Future Research

In Chap. 1 we introduced the classical Cournot model and after setting up the general framework we focused on a number of specific examples involving combinations of linear and hyperbolic price functions and linear and quadratic cost functions, also taking careful account of capacity constraints. These examples illustrated the variety of reaction functions that can occur and the various types of equilibria (possibly multiple) both in the interior of the domain of interest and on its boundaries. We then went on to introduce the various types of adjustment processes that underpin the dynamic processes, the study of the local and global dynamics of which has occupied much of the space in this book. In particular we considered discrete time and continuous time versions of partial adjustment towards the best response with naive expectations and adaptive expectations as well as the gradient adjustment process. We then introduced some of the basic tools for the analysis of global dynamics via some examples involving duopoly or symmetric and semi-symmetric oligopolies. We introduced the important concept of basins of attraction of different equilibria and the important tool of the critical curve and the concept of border collision bifurcations. Already with the simple examples considered we see the types of complexity that can arise in oligopoly models under the type of dynamic adjustment processes we consider here.

In the second chapter we considered the widely studied class of concave oligopolies. We first obtained the properties of the reaction function both with and without cost externalities, and then used these to study the local and global dynamics of discrete time and continuous time concave oligopolies under the various best response processes of Chap. 1. We made use of the determinantal relation in Appendix E to obtain results on local stability. The full array of the tools for the analysis of the global dynamics were brought to bear to obtain interesting results in a number of special cases of price and cost functions, as well as on the nature of the oligopoly, such as whether it is a duopoly or semi-symmetric. We saw in particular the important role of border collision bifurcations in determining the global dynamics and how the number of firms in the oligopoly, the capacity constraints of firms and their speeds of adjustment are all important bifurcation parameters. The chapter concluded with a study of the local dynamics of continuous time oligopolies with continuously distributed informational time lags, and we saw how such time lags can have a strong influence of local bifurcation behavior.

G.I. Bischi et al., *Nonlinear Oligopolies*, DOI 10.1007/978-3-642-02106-0_6,

Chapter 3 considered general oligopolies and started with an analysis of the case of isoelastic price functions under both continuous time and discrete time adjustment processes. Again the results of Appendix E were invoked to analyze the local stability and we saw that for semi-symmetric oligopolies in both the discrete time and continuous time cases this is determined qualitatively by the same set of graphs, though of course quantitatively the two cases differ. With regard to the global analysis of the discrete time model we saw the richness of the local bifurcations with respect to the number of firms, the cost ratio and speeds of adjustment in the semi-symmetric case. We found also that the speeds of adjustment played a role in the generation of border collisions and hence global bifurcations. The remainder of the chapter considered the role of cost externalities that are captured by the assumption of a certain type of non-monotonic reaction function. Here we focused on the duopoly case and saw that such models can generate situations of several coexisting equilibria that are locally stable, each having its own basin of attraction. In the case of identical speeds of adjustment we were able to analyze and understand in some detail the way in which the different equilibria can be born and the way in which the structure of their basins of attraction change with key parameters, due mainly to the occurrence of contact bifurcations. Some numerical examples of the non-identical speed of adjustment situation illustrate how the basins can become even more complex in this case. The disconnected nature of the basins of attraction means that the outcome (in the sense of to which equilibrium the game converges) of oligopolies with cost externalities is highly path dependent. These examples convey in a very clear way the important distinction between local bifurcations and global bifurcations.

In Chap. 4 we apply the analysis of the first three chapters to a number of models that are an extension of the basic oligopoly set-up or are dynamic economic games that essentially reduce to classical oligopolies. These are market share attraction games, labor-managed oligopolies, oligopolies with intertemporal demand interaction, oligopolies with production adjustment costs and oligopolies with partial cooperation amongst the firms. Such extensions of the basic oligopoly model and dynamic economic games exhibit the range of behaviors observed in the basic oligopolies of the previous chapters.

Finally in Chap. 5 we considered learning behavior under incomplete knowledge of the demand relationship. We started by considering oligopolies in which firms have misspecified price functions but otherwise we still use the adjustment processes of Chap. 1. Now the possibility of subjective equilibria arises, the local and global dynamics of which are studied through some examples that illustrate how such subjective equilibria are born, their local stability properties and the (sometimes complicated) nature of their basins of attraction. Next we assume that firms use some kind of approximate learning procedure to resolve their incomplete knowledge of the price function. Using a number of specific examples we study the local stability of the equilibria and how its loss can give rise to fluctuating attractors as various parameters change. Global analysis indicates how the learning scheme can affect the basin of attraction of a stable equilibrium. Next we study other types of learning schemes by firms as they try to determine the true shape of the price function. We focus on the case of linear price and cost functions and consider three scenarios in which firms have different types of partial information about some parameters of the price function and seek to learn about the remaining parameters by some adjustment process. Again via specific examples we see that these learning schemes can generate the type of local and global bifurcations seen in the previous chapters. Finally we conclude this chapter with a brief discussion of uncertain price functions which brings us to the edge of the field of statistical learning, which presents a whole different field of research.

From the point of view of nonlinear dynamical systems the book has introduced the still relatively new (at least for economists) concept of border collisions and illustrated its use in a number of examples. The examples have emphasized how there are in fact two types of complexity of importance in dynamic economic models. The first is the familiar one arising as a result of local bifurcations, which frequently occur when equilibria lose local stability via Hopf or flip bifurcations and local stability of an equilibrium gives way to some sort of fluctuation around it. The other, less familiar one, arises when a border collision occurs, and basins of attraction of different equilibria undergo a change in their structure. Also there may be co-existing attractors within the same basin of attraction. A typical result of such bifurcations is that the outcome of the economic adjustment process under consideration may be highly sensitive to initial conditions. Future research in economic applications in this area will probably focus on the systematic description of the different sources of such bifurcations, the elaboration of the types of examples where such border collision bifurcations can occur and the typical sorts of behavior that can emerge from them. It would also be useful to try to understand the economic origins of the different types of such bifurcations.

With regard to the specific models we have studied in this book, a number of issues are likely to occupy the attention of researchers in the years ahead.

Considering first the case of concave oligopolies, we have derived most of our results under the assumptions (A)–(C) in Sect. 2.1 (or their modifications in different model types) which we recall placed restraints on the inverse demand function and cost functions so as to guarantee the concavity of the profit function and in the concave case the monotonicity of the best response functions. An important task for future research will be to study the implications of relaxing any one of these assumptions. We have seen in Example 1.2 that just by relaxing the condition (C) how more complicated equilibrium situations can arise.

A number of our examples involved the isoelastic price function, which is widely used in the literature on oligopoly because it affords a lot of analytical tractability. However this price function has the disadvantage that it has no reservation price (or rather the reservation price is infinite) and this is rather unrealistic. Future research should try to introduce a reservation price into this price function, either by using a translated hyperbola, or by truncating the function at some (presumably high) price. The isoelastic price function has also been used frequently in conjunction with a convex cost function, but what would happen if we were to allow a certain amount of concavity into the cost function? This could arise for instance if there were an increasing return to scales effect at low outputs and a decreasing returns to scale effect at higher levels of output. Uniqueness of the equilibrium could be lost in such situations or there may be no equilibrium at all. Also many of the local and global stability results we have derived rely heavily on the special analytical properties of the cost functions, so we might expect to see a much richer set of dynamic outcomes.

With regard to the modified and extended oligopolies of Chap. 4 a number of extensions can be envisaged. The market share attraction games are equivalent to oligopolies with isoelastic price functions, so all the remarks of the previous paragraph apply to this class of model as well. In our analysis of labor-managed oligopolies we assumed very special forms for the labor demand functions, but both equilibrium results as well as the dynamic analysis will change if we consider more general forms for these demand functions. The models with intertemporal demand interaction were analyzed under the assumptions of the concave oligopolies so again the relaxation of the assumptions on the price and cost functions will lead to a richer set of outcomes for the equilibria and the dynamics. In the models with production adjustment costs we have assumed that this additional cost component depends on the output change from the previous period. A more realistic assumption might be to make this cost depend on a state variable related to the capacity limit that adjusts dynamically in such a way that the firm increases it if output needs to go beyond it. The analysis of oligopolies under partial cooperation also relies very much on the concavity assumptions being satisfied by the profit functions. Here also it would be of interest to study the situations in which these concavity conditions are relaxed. It would also be interesting to include partial cooperation into some of the extensions described earlier in this chapter.

In the learning schemes in the models with misspecified and uncertain cost functions we have adopted various assumptions on the learning behavior of the firms, from remaining statically with the same misspecification over every time period, to updating their estimate of it based on the most recently observed price. There is now a vast literature on learning in dynamic economic models, see for example Fudenberg and Levine (1998), and many of these ideas could be brought into the problems considered in Chap. 5. Many of these schemes are probabilistic in nature so this strand of research will involve the analysis of economic models evolving dynamically under random influences, this is an area into which research has barely begun as it involves bringing together the theory of dynamical systems and the theory of stochastic processes.

# Appendix A
# Elements of Lyapunov Theory

Consider a time-invariant nonlinear dynamical system

$$x(t+1) = g(x(t)) \tag{A.1}$$

or

$$\dot{x}(t) = g(x(t)), \tag{A.2}$$

where $g : \mathbb{D} \mapsto \mathbb{R}^n$ with $\mathbb{D}$ being a set in $\mathbb{R}^n$. It is also assumed that $g$ is continuous on $\mathbb{D}$, and starting with arbitrary initial state $x(0) \in \mathbb{D}$, the unique solution of (A.1) or (A.2) exists for all $t > 0$ and remains in $\mathbb{D}$. A vector $\bar{x} \in \mathbb{D}$ is an equilibrium of system (A.1) if and only if $\bar{x} = g(\bar{x})$, and it is an equilibrium of system (A.2) if and only if $g(\bar{x}) = \mathbf{0}$. If in any time period $x(t)$ becomes $\bar{x}$, then the state remains at the equilibrium for all future time periods. Therefore equilibria of a dynamical system are sometimes called the steady states of the system. If $x(0)$ is selected nearby an equilibrium, then the state might go away from the equilibrium, it might stay close to the equilibrium for all future times or it even might converge to the equilibrium as $t \to \infty$. In all cases distances between state vectors have to be defined in order to decide if a state vector is close to the equilibrium or not. The distance between any two vectors is usually defined as the *norm* of their difference. The norm is a mathematical way to characterize the lengths of real vectors.

A norm $\|.\|$ in the $n$-dimensional vector space $\mathbb{R}^n$ is a real valued function defined on all $n$-element vectors such that the following conditions are satisfied:

1. $\|v\| \geq 0$ for all $v \in \mathbb{R}^n$, and $\|v\| = 0$ if and only if $v = \mathbf{0}$
2. $\|\alpha v\| = |\alpha| \cdot \|v\|$ for all $v \in \mathbb{R}^n$ and real numbers $\alpha$
3. $\|v + w\| \leq \|v\| + \|w\|$ for all $v, w \in \mathbb{R}^n$.

It is easy to prove that all vector norms are continuous functions, that is, $\|v\|$ is continuous in $v$. The proof is based on conditions (2) and (3), since

$$\|v\| \leq \|v - w\| + \|w\|$$

and

$$\|w\| \leq \|w - v\| + \|v\| = \|v - w\| + \|v\|,$$

implying that

$$\|\boldsymbol{v}\| - \|\boldsymbol{w}\| \le \|\boldsymbol{v} - \boldsymbol{w}\|$$

and

$$\|\boldsymbol{w}\| - \|\boldsymbol{v}\| \le \|\boldsymbol{v} - \boldsymbol{w}\|.$$

Therefore

$$\big|\|\boldsymbol{v}\| - \|\boldsymbol{w}\|\big| \le \|\boldsymbol{v} - \boldsymbol{w}\|,$$

showing that the distance between $\|\boldsymbol{v}\|$ and $\|\boldsymbol{w}\|$ cannot exceed the distance between $\boldsymbol{v}$ and $\boldsymbol{w}$.

We can also show that with any vector norm the circular neighborhoods of any vector $\boldsymbol{u} \in \mathbb{R}^n$, denoted

$$U = \{\boldsymbol{v} | \boldsymbol{v} \in \mathbb{R}^n, \quad \|\boldsymbol{v} - \boldsymbol{u}\| < \varepsilon\},$$

are convex sets. In order to prove this property assume that $\boldsymbol{x}$ and $\boldsymbol{y}$ are in $U$, then both $\|\boldsymbol{x} - \boldsymbol{u}\|$ and $\|\boldsymbol{y} - \boldsymbol{u}\|$ are less than $\varepsilon$. With any vector

$$\boldsymbol{z} = \alpha\boldsymbol{x} + (1-\alpha)\boldsymbol{y} \quad (0 \le \alpha \le 1),$$

we have

$$\begin{aligned}\|\boldsymbol{z} - \boldsymbol{u}\| &= \|\alpha(\boldsymbol{x} - \boldsymbol{u}) + (1-\alpha)(\boldsymbol{y} - \boldsymbol{u})\| \\ &\le |\alpha| \cdot \|\boldsymbol{x} - \boldsymbol{u}\| + |1-\alpha| \cdot \|\boldsymbol{y} - \boldsymbol{u}\| < \alpha\varepsilon + (1-\alpha)\varepsilon = \varepsilon,\end{aligned}$$

which proves the assertion.

Consider a linear segment $\boldsymbol{u} + t(\boldsymbol{v} - \boldsymbol{u})$ $(0 \le t \le 1)$ connecting points $\boldsymbol{u}$ and $\boldsymbol{v}$ in $\mathbb{R}^n$. Let $0 = t_0 < t_1 < \cdots < t_k = 1$ and $\boldsymbol{u}_l = \boldsymbol{u} + t_l(\boldsymbol{v} - \boldsymbol{u})$ for $l = 0, 1, \ldots, k$. Then

$$\|\boldsymbol{v} - \boldsymbol{u}\| = \sum_{l=1}^{k} \|\boldsymbol{u}_l - \boldsymbol{u}_{l-1}\|, \tag{A.3}$$

since

$$\begin{aligned}\sum_{l=1}^{k} \|\boldsymbol{u}_l - \boldsymbol{u}_{l-1}\| &= \sum_{l=1}^{k} \|(\boldsymbol{u} + t_l(\boldsymbol{v} - \boldsymbol{u})) - (\boldsymbol{u} + t_{l-1}(\boldsymbol{v} - \boldsymbol{u}))\| \\ &= \sum_{l=1}^{k} \|(t_l - t_{l-1})(\boldsymbol{v} - \boldsymbol{u})\| = \sum_{l=1}^{k} (t_l - t_{l-1})\|\boldsymbol{v} - \boldsymbol{u}\| \\ &= (t_k - t_0)\|\boldsymbol{v} - \boldsymbol{u}\| = \|\boldsymbol{v} - \boldsymbol{u}\|.\end{aligned}$$

Assume next that $\boldsymbol{f} : [a, b] \longmapsto \mathbb{R}^n$ is a continuous function. We can easily prove that

$$\left\| \int_a^b \boldsymbol{f}(t)\mathrm{d}t \right\| \le \int_a^b \left\|\boldsymbol{f}(t)\right\| \mathrm{d}t. \tag{A.4}$$

By using the definition of the Riemann integral we have

$$\left\| \sum_{i=1}^{N} f(\tau_i)(t_i - t_{i-1}) \right\| \le \sum_{i=1}^{N} \left\| f(\tau_i) \right\| (t_i - t_{i-1})$$

where $a = t_0 < t_1 < \cdots < t_N = b$ and $\tau_i \in [t_{i-1}, t_i]$ for all $i$. By letting $N \to \infty$ we conclude (A.4).

In practical applications three particular vector norms are usually used, namely

$$\|\boldsymbol{v}\|_\infty = \max\{|v_1|, |v_2|, \ldots, |v_n|\},$$

$$\|\boldsymbol{v}\|_1 = |v_1| + |v_2| + \cdots + |v_n|$$

and

$$\|\boldsymbol{v}\|_2 = \sqrt{|v_1|^2 + |v_2|^2 + \cdots + |v_n|^2},$$

where $v_i$ is the $i$th element of $\boldsymbol{v}$ for $i = 1, 2, \ldots, n$. The norm $\|.\|_2$ is usually called the *Euclidean* norm.

All three norms satisfy conditions (1)–(3). Notice that $\|.\|_2$ is the $n$-dimensional generalization of the well known definition of the lengths of 2 and 3 dimensional real vectors.

Let $\boldsymbol{T}$ be an invertible $n \times n$ matrix, and $\|.\|$ a given vector norm. Then a new vector norm can be defined as

$$\|\boldsymbol{u}\|_T = \|\boldsymbol{T}\boldsymbol{u}\|.$$

Clearly this norm also satisfies conditions (1)–(3).

Assume now that $\boldsymbol{x}(0)$ is the initial state of a system. Then the following stability types can be considered.

**Definition A.1.** An equilibrium $\bar{\boldsymbol{x}}$ is called stable (or marginally stable) if for all $\varepsilon_1 > 0$ there exists an $\varepsilon > 0$ such that $\|\boldsymbol{x}(0) - \bar{\boldsymbol{x}}\| < \varepsilon$ implies that for all $t > 0$, $\|\boldsymbol{x}(t) - \bar{\boldsymbol{x}}\| < \varepsilon_1$.

**Definition A.2.** An equilibrium $\bar{\boldsymbol{x}}$ is asymptotically stable (or locally asymptotically stable) if it is stable and there is an $\varepsilon > 0$ such that $\|\boldsymbol{x}(0) - \bar{\boldsymbol{x}}\| < \varepsilon$ implies that $\boldsymbol{x}(t)$ converges to $\bar{\boldsymbol{x}}$ as $t \to \infty$.

**Definition A.3.** An equilibrium $\bar{\boldsymbol{x}}$ is globally asymptotically stable in $\mathbb{D}$ if it is stable and for arbitrary $\boldsymbol{x}(0) \in \mathbb{D}$, $\boldsymbol{x}(t)$ converges to $\bar{\boldsymbol{x}}$ as $t \to \infty$.

The (marginal) stability of an equilibrium means that the entire state trajectory remains close to the equilibrium if the initial state is selected close enough to the equilibrium. If in addition the state trajectory converges to the equilibrium as $t \to \infty$,

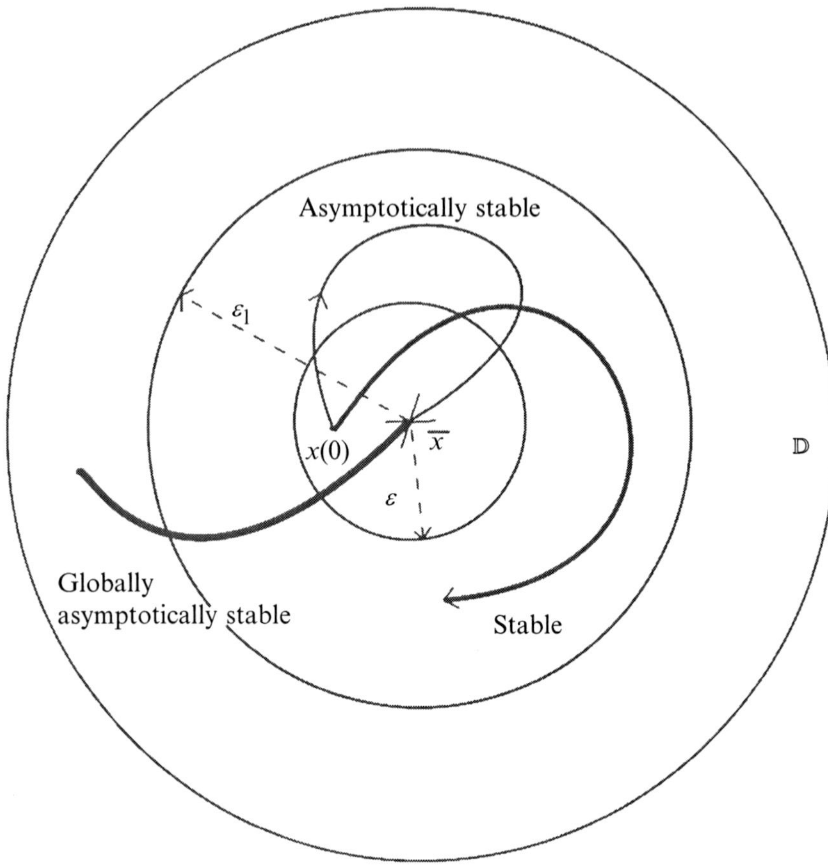


**Fig. A.1** Stability concepts

then the equilibrium is (locally) asymptotically stable. Global asymptotic stability occurs if with any initial state $\boldsymbol{x}(0) \in \mathbb{D}$, the state trajectory converges to the equilibrium as $t \to \infty$. Figure A.1 illustrates these concepts.

Assume now that $\bar{\boldsymbol{x}}$ is an equilibrium of the system (A.1) or (A.2) and let $\Omega$ be a subset of $\mathbb{D}$ such that $\bar{\boldsymbol{x}} \in \Omega$.

**Definition A.4.** A real valued function $\mathbb{V}$ defined in $\Omega$ is called a Lyapunov function if is satisfies the following conditions:

(a) $\mathbb{V}$ is continuous on $\Omega$;
(b) The global minimum of $\mathbb{V}$ on $\Omega$ occurs at the equilibrium $\bar{\boldsymbol{x}}$;
(c) For any state trajectory $\boldsymbol{x}(t)$ contained in $\Omega$, $\mathbb{V}(\boldsymbol{x}(t))$ is non-increasing in $t$.

Notice that the Lyapunov function concept is a straightforward generalization of that of the energy function in mechanical systems. Condition (a) requires that $\mathbb{V}$ has no discontinuities, (b) means that $\mathbb{V}$ has its smallest value at the equilibrium, and condition (c) generalizes the well-known property of mechanical systems that the energy of a free mechanical system never increases and with friction it always decreases.

Assume next that $\Omega$ is a spherical region

$$\Omega = \{x \mid \|x - \bar{x}\| \leq r_0\} \subseteq D. \tag{A.5}$$

**Theorem A.1.** *If there exists a Lyapunov function on $\Omega$, then $\bar{x}$ is (marginally) stable.*

*Proof.* The proof for the discrete time and continuous time cases are very similar, therefore we present it here only for the discrete case.

Select an $\varepsilon_1 > 0$ and assume that $\varepsilon_1 \leq r_0$. Notice first that from the continuity of $g$ we know the existence of a $\delta \in (0, \varepsilon_1)$ such that $\|x - \bar{x}\| < \delta$ implies that $\|g(x) - \bar{x}\| < r_0$, since $\bar{x} = g(\bar{x})$ and $g$ is continuous. Therefore if $\|x(t) - \bar{x}\| < \delta$ with some $t \geq 0$, then $\|x(t+1) - \bar{x}\| < r_0$ showing that $x(t+1) \in \Omega$.

Define next

$$m = \min\{\mathbb{V}(x) \mid \delta \leq \|x - \bar{x}\| \leq r_0\}, \tag{A.6}$$

which exists since the defining set is compact and $\mathbb{V}$ is continuous. Since the defining set does not contain $\bar{x}$ and $\bar{x}$ is the only global minimizer of the Lyapunov function, $\mathbb{V}(\bar{x}) < m$. The continuity of $\mathbb{V}$ implies the existence of an $\varepsilon \in (0, \delta)$ such that $\mathbb{V}(x) < m$ as $\|x - \bar{x}\| < \varepsilon$.

Finally we prove that this $\varepsilon$ satisfies the condition of Definition A.1. Select an $x(0)$ such that $\|x(0) - \bar{x}\| < \varepsilon$. Then $\mathbb{V}(x(0)) < m$, and the non-increasing property of the Lyapunov function implies that $\mathbb{V}(x(t)) < m$ for all $t \geq 0$. The first part of the proof guarantees that $x(t) \in \Omega$. The definition of $m$ implies that $\|x(t) - \bar{x}\| < \delta < \epsilon_1$ which completes the proof. ∎

**Theorem A.2.** *In addition to the conditions of the previous theorem assume that $\mathbb{V}(x(t))$ strictly decreases in $t$ unless $x(t) = \bar{x}$. Then $\bar{x}$ is (locally) asymptotically stable.*

*Proof.* Only the discrete time case is shown, the proof in the continuous time case is similar.

Select $\varepsilon > 0$ as in the proof of the previous theorem. We shall show that $\|x(0) - \bar{x}\| < \varepsilon$ implies that $x(t) \to \bar{x}$ as $t \to \infty$. Assume that this limit relation does not hold. Since the sequence $\{x(t)\}$ is bounded it has a convergent subsequence such that $x(t_k) \to x^* \neq \bar{x}$ as $k \to \infty$. The sequence $\{x(t_k + 1)\}$ is also bounded, so it also must have a convergent subsequence $x(t_{k_i} + 1) \to x^{**}$ as $i \to \infty$. From the strict monotonicity of the Lyapunov function we see that for all $i \geq 0$,

$$\mathbb{V}(x(t_{k_{i+1}})) \leq \mathbb{V}(x(t_{k_i} + 1) < \mathbb{V}(x(t_{k_i}),$$

and by letting $i \to \infty$ we have

$$\mathbb{V}(x^*) \leq \mathbb{V}(x^{**}) \leq \mathbb{V}(x^*),$$

therefore

$$\mathbb{V}(\boldsymbol{x}^*) = \mathbb{V}(\boldsymbol{x}^{**}). \tag{A.7}$$

The continuity of the function $\boldsymbol{g}$ implies that

$$\begin{aligned}\boldsymbol{x}^{**} &= \lim_{i\to\infty} \boldsymbol{x}(t_{k_i} + 1) = \lim_{i\to\infty} \boldsymbol{g}(\boldsymbol{x}(t_{k_i})) \\ &= \boldsymbol{g}(\lim_{i\to\infty} \boldsymbol{x}(t_{k_i})) = \boldsymbol{g}(\boldsymbol{x}^*),\end{aligned}$$

which contradicts relation (A.7) and the strict monotonicity of the Lyapunov function. ■

**Theorem A.3.** *Assume that a Lyapunov function is defined on the entire state space* $\mathbb{D}$*. Assume also that* $\mathbb{V}(\boldsymbol{x}(t))$ *strictly decreases in* $t$ *unless* $\boldsymbol{x}(t) = \bar{\boldsymbol{x}}$*, furthermore* $\mathbb{V}(\boldsymbol{x}) \to \infty$ *as* $\|\boldsymbol{x}\| \to \infty$*. Then* $\bar{\boldsymbol{x}}$ *is globally asymptotically stable.*

*Proof.* We show again only the discrete time case. Let $\boldsymbol{x}(0) \in \mathbb{D}$ be arbitrary, then for all $t \geq 0$, $\mathbb{V}(\boldsymbol{x}(t)) \leq \mathbb{V}(\boldsymbol{x}(0))$. Therefore the sequence $\{\boldsymbol{x}(t)\}$ is bounded, and the proof can continue along the lines of the proof of the previous theorem. ■

The particular choice of the Lyapunov function depends on the special properties of the dynamical system being examined. The most popular choice is

$$\mathbb{V}(\boldsymbol{x}) = \|\boldsymbol{x} - \bar{\boldsymbol{x}}\|^2,$$

where the Euclidean norm is selected. This function clearly satisfies properties (a) and (b) of Definition A.4, so only the monotonicity condition has to be established. In the discrete case we have to prove that

$$\|\boldsymbol{x}(t+1) - \bar{\boldsymbol{x}}\| \leq \|\boldsymbol{x}(t) - \bar{\boldsymbol{x}}\|$$

for marginal stability and the corresponding strict inequality for asymptotic stability. Notice that the additional condition of Theorem A.3 is also satisfied. In the continuous case we have to show the monotonicity of the function

$$\mathbb{V}(\boldsymbol{x}(t)) = (\boldsymbol{x}(t) - \bar{\boldsymbol{x}})^T(\boldsymbol{x}(t) - \bar{\boldsymbol{x}})$$

by showing that its derivative is non-positive or negative. It is easy to see that this derivative can be expressed as

$$\begin{aligned}\frac{d}{dt}\mathbb{V}(\boldsymbol{x}(t)) &= \dot{\boldsymbol{x}}(t)^T(\boldsymbol{x}(t) - \bar{\boldsymbol{x}}) + (\boldsymbol{x}(t) - \bar{\boldsymbol{x}})^T\dot{\boldsymbol{x}}(t) \\ &= 2(\boldsymbol{x}(t) - \bar{\boldsymbol{x}})^T\boldsymbol{g}(\boldsymbol{x}(t)).\end{aligned}$$

In proving that this expression is non-positive or negative, the particular form of the function $\boldsymbol{g}$ has to be used. Unfortunately this function is not always monotonic, and even if it is, then the actual proof is different for different cases.

# Appendix B
# Local Linearization

Consider a time-invariant nonlinear dynamical system

$$x(t+1) = g(x(t)) \tag{B.1}$$

in discrete time or

$$\dot{x}(t) = g(x(t)), \tag{B.2}$$

in continuous time where $g : \mathbb{D} \mapsto \mathbb{R}^n$, with $\mathbb{D} \subset \mathbb{R}^n$ being a set in $\mathbb{R}^n$. A vector $\bar{x} \in \mathbb{D}$ is an equilibrium of system (B.1) if and only if $\bar{x} = g(\bar{x})$, and it is an equilibrium of system (B.2) if and only if $g(\bar{x}) = \mathbf{0}$. Let $x \in \mathbb{D}$ be an arbitrary point. If $x$ is interior, then we assume that $g$ is differentiable at $x$, and if $x$ is on the boundary, then we assume that $g$ can be extended outside $\mathbb{D}$ to an open neighborhood of $x$, and this extension is differentiable at $x$. The Jacobian of $g$ at the point $x$ is defined as the matrix

$$J(x) = \begin{pmatrix} \frac{\partial g_1}{\partial x_1}(x) & \dots & \frac{\partial g_1}{\partial x_n}(x) \\ \vdots & & \\ \frac{\partial g_n}{\partial x_1}(x) & \dots & \frac{\partial g_n}{\partial x_n}(x) \end{pmatrix}.$$

In this Appendix the relation between the local and global asymptotic stability of the equilibrium and some properties of the Jacobian will be summarized. Some of the conditions will be based on the computation of matrix norms. Let $\mathbb{R}^{n\times n}$ denote the set of all $n \times n$ real matrices. A real valued function $A \longmapsto \|A\|$, defined for all $A \in \mathbb{R}^{n\times n}$, is a matrix norm if it satisfies the following conditions:

1. $\|A\| \geq 0$ for all $A \in \mathbb{R}^{n\times n}$, and $\|A\| = 0$ if and only if $A$ is the zero matrix with all elements being equal to zero
2. $\|\alpha A\| = |\alpha| \cdot \|A\|$ for all $A \in \mathbb{R}^{n\times n}$ and all real numbers $\alpha$
3. $\|A + B\| \leq \|A\| + \|B\|$ for all $A, B \in \mathbb{R}^{n\times n}$.

Similarly to the case of vector norms it is easy to prove that matrix norms are also continuous matrix functions.

There are many particular vector norms which are used in practical applications. A large class of matrix norms can be generated from vector norms in the following way. Let $\|\cdot\|$ be a given vector norm in $\mathbb{R}^n$ (see Appendix A), and for any $\boldsymbol{A} \in \mathbb{R}^{n\times n}$ define

$$\|\boldsymbol{A}\| = \max_{\boldsymbol{v}\neq \boldsymbol{0}} \frac{\|\boldsymbol{A}\boldsymbol{v}\|}{\|\boldsymbol{v}\|} = \max_{\|\boldsymbol{v}\|=1} \|\boldsymbol{A}\boldsymbol{v}\|,$$

where the numerator and the denominator use the same given vector norm. It is easy to prove that matrix norms generated by vectors norms always satisfy conditions (1)–(3) and in addition, for all $\boldsymbol{A}$ and $\boldsymbol{B} \in \mathbb{R}^{n\times n}$,

4. $\|\boldsymbol{A}\boldsymbol{B}\| \leq \|\boldsymbol{A}\| \cdot \|\boldsymbol{B}\|$.

Furthermore it is an additional important fact, that for all $\boldsymbol{v} \in \mathbb{R}^n$ and $\boldsymbol{A} \in \mathbb{R}^{n\times n}$,

5. $\|\boldsymbol{A}\boldsymbol{v}\| \leq \|\boldsymbol{A}\| \cdot \|\boldsymbol{v}\|$, if the matrix norm $\|\boldsymbol{A}\|$ is generated from the vector norm which is used to compute both $\|\boldsymbol{v}\|$ and $\|\boldsymbol{A}\boldsymbol{v}\|$.

If property (5) holds for a given vector norm and a particular matrix norm, then we say that the two norms are *compatible*.

The matrix norms generated from the vector norms $\|\boldsymbol{v}\|_\infty$, $\|\boldsymbol{v}\|_1$ and $\|\boldsymbol{v}\|_2$ are given as follows. Let $a_{ij}$ denote the $(i, j)$ elements of matrix $\boldsymbol{A}$, then

$$\|\boldsymbol{A}\|_\infty = \max_i \left\{ \sum_{j=1}^{n} |a_{ij}| \right\} \quad \text{(row norm)}$$

$$\|\boldsymbol{A}\|_1 = \max_j \left\{ \sum_{i=1}^{n} |a_{ij}| \right\} \quad \text{(column norm)}$$

and

$$\|\boldsymbol{A}\|_2 = \sqrt{\max_i \lambda_i(\boldsymbol{A}^T\boldsymbol{A})} \quad \text{(Euclidean norm)}$$

where $\lambda_i(\boldsymbol{A}^T\boldsymbol{A})$ $(i = 1, 2, \ldots, n)$ are the eigenvalues of the product $\boldsymbol{A}^T\boldsymbol{A}$, and $\boldsymbol{A}^T$ is the transpose of $\boldsymbol{A}$. It can be proved that all eigenvalues of $\boldsymbol{A}^T\boldsymbol{A}$ are real and nonnegative.

Notice that conditions (1)–(3) for matrix norms do not imply that there is a vector norm which is compatible with it. For example, consider the matrix norm $\|\boldsymbol{A}\| = \frac{1}{2}\|\boldsymbol{A}\|_\infty$, then $\|\boldsymbol{I}\| = \frac{1}{2}$, so with $\boldsymbol{a} \neq \boldsymbol{0}$ and any vector norm,

$$\|\boldsymbol{I}\boldsymbol{x}\| = \|\boldsymbol{x}\| > \frac{1}{2}\|\boldsymbol{x}\| = \|\boldsymbol{I}\| \cdot \|\boldsymbol{x}\|.$$

Let $A$ be a real $n \times n$ matrix and $\lambda$ an eigenvalue of $A$. If $\|\cdot\|$ is a matrix norm which is compatible with a vector norm, then

$$|\lambda| \le \|A\|.$$

This relation is a simple consequence of the eigenvalue equation of matrix $A$,

$$Av = \lambda v,$$

where $v \neq \mathbf{0}$ is an associated eigenvector to $\lambda$. Then

$$\|Av\| \le \|A\| \cdot \|v\|$$

and

$$\|\lambda v\| = |\lambda| \cdot \|v\|,$$

so

$$|\lambda| \cdot \|v\| \le \|A\| \cdot \|v\|.$$

The assertion is obtained by dividing both sides by $\|v\| > 0$.

Consider first the discrete time system (B.1).

**Theorem B.1.** *Let $\overline{x} \in \mathbb{D}$ be an equilibrium and assume that $J(x)$ exists in an open neighborhood of $\overline{x}$, $\|J(\overline{x})\| < 1$ with some matrix norm and $J$ is continuous at $\overline{x}$. Then $\overline{x}$ is locally asymptotically stable.*

*Proof.* Since $J(x)$ is continuous at $\bar{x}$, there is an $\varepsilon > 0$ such that $J(x)$ exists for all

$$x \in \mathbb{U} = \{x \mid \|x - \bar{x}\| < \varepsilon\}$$

and $\|J(x)\| \le q$ with some $0 \le q < 1$. Then with any $x \in \mathbb{U}$,

$$g(x) - \bar{x} = g(x) - g(\bar{x}) = \int_0^1 J(\bar{x} + t(x - \bar{x}))(x - \bar{x})\mathrm{d}t.$$

Therefore

$$\|g(x) - \bar{x}\| \le \int_0^1 \|J(x + t(x - \bar{x}))\| \cdot \|x - \bar{x}\|\mathrm{d}t \le q \cdot \|x - \bar{x}\|. \tag{B.3}$$

Starting with arbitrary initial state $x(0) \in \mathbb{U}$, the entire state sequence generated by (B.1) remains in $\mathbb{U}$, furthermore for all $t \ge 0$,

$$\|x(t+1) - \bar{x}\| = \|g(x(t)) - \bar{x}\| \le q \cdot \|x(t) - \bar{x}\|.$$

Consequently for all $t \geq 1$,

$$\|x(t) - \bar{x}\| \leq q^t \|x(0) - \bar{x}\| \tag{B.4}$$

showing that $x(t) \to \bar{x}$ as $t \to \infty$. ■

A slight modification of the above proof can be used to show the following sufficient condition for global asymptotic stability of the equilibrium.

**Theorem B.2.** *Assume that $\mathbb{D}$ is convex, and $g$ is continuously differentiable on $\mathbb{D}$. If with some matrix norm, $\|J(x)\| \leq q < 1$ for all $x \in \mathbb{D}$, where $q$ is a constant, then $\bar{x}$ is globally asymptotically stable.*

The conditions of this theorem can be relaxed to cases when $g$ is a continuous and piece-wise differentiable function. Assume now that $\mathbb{D}$ is convex and is the union of the closed sets $\mathbb{D}^{(1)}, \mathbb{D}^{(2)}, \ldots$, with mutually exclusive interiors. The restriction of $g$ to the region $\mathbb{D}^{(k)}$ is denoted by $g^{(k)}$ and we assume that it can be extended to an open set containing $\mathbb{D}^{(k)}$ and that it is differentiable there. Let $J^{(k)}(x)$ denote the Jacobian of $g^{(k)}$ and assume, for all $k$ and $x \in \mathbb{D}^{(k)}$, that $\|J^{(k)}(x)\| \leq q < 1$ where $\|.\|$ is a matrix norm that is compatible with some vector norm and $q$ is a scaler. Assume in addition that for the linear segment between $\bar{x}$ and any $x \in \mathbb{D}$ there are finitely many values[1] $0 = t_0 < t_1 < \cdots < t_{K(x)} = 1$ such that for each entire subsegment,

$$[\bar{x} + t_l(x - \bar{x}), \bar{x} + t_{l+1}(x - \bar{x})] \subset \mathbb{D}^{(k_l)}$$

with some $k_l$.

**Theorem B.3.** *Under the above conditions $\bar{x}$ is globally asymptotically stable in $\mathbb{D}$.*

*Proof.* Notice that with any $x \in \mathbb{D}$,

$$\begin{aligned}
\|g(x) - \bar{x}\| = \|g(x) - g(\bar{x})\| &= \left\| \sum_{l=0}^{K(x)-1} (g(\bar{x} + t_{l+1}(x - \bar{x})) - g(\bar{x} + t_l(x - \bar{x}))) \right\| \\
&\leq \sum_{l=0}^{K(x)-1} \left\| \int_{t_l}^{t_{l+1}} J^{(k_l)}(\bar{x} + t(x - \bar{x}))(x - \bar{x})\mathrm{d}t \right\| \\
&\leq \sum_{l=0}^{K(x)-1} \int_{t_l}^{t_{l+1}} \|J^{(k_l)}(\bar{x} + t(x - \bar{x}))\|.\|x - \bar{x}\|\mathrm{d}t \\
&\leq q\|x - \bar{x}\| \sum_{l=0}^{K(x)-1} (t_{l+1} - t_l) = q \cdot \|x - \bar{x}\|,
\end{aligned}$$

and then the proof can follow the lines of the proof of Theorem B.1. ■

[1] $K(x)$ is the number of subregions that a linear segment goes through between $x$ and the equilibrium.

The assumption that $\boldsymbol{g}^{(k)}$ can be extended to an open set containing $\mathbb{D}^{(k)}$ can be replaced by the following. For the linear segment between $\bar{\boldsymbol{x}}$ and any $\boldsymbol{x} \in \mathbb{D}$ there are finitely many values $0 = t_0 < t_1 < \cdots < t_{K(\boldsymbol{x})} = 1$ such that

(a) Subsegments $[\bar{\boldsymbol{x}} + t_l(\boldsymbol{x} - \bar{\boldsymbol{x}}),\ \bar{\boldsymbol{x}} + t_{l+1}(\boldsymbol{x} - \bar{\boldsymbol{x}})] \subset \mathbb{D}^{(k_l)}$ for all $l$
(b) For each such subsegment there are sequences $\{\boldsymbol{u}_k\}$ and $\{\boldsymbol{v}_k\}$ such that $\boldsymbol{u}_k \to \bar{\boldsymbol{x}} + t_l(\boldsymbol{x} - \bar{\boldsymbol{x}})$, $\boldsymbol{v}_k \to \bar{\boldsymbol{x}} + t_{l+1}(\boldsymbol{x} - \bar{\boldsymbol{x}})$ and the entire linear segment $[\boldsymbol{u}_k, \boldsymbol{v}_k]$ is in the interior of $\mathbb{D}^{(k_l)}$.

Since there are infinitely many matrix norms, and the condition of the theorem might hold with one matrix norm and not with others, the above conditions are difficult to check in practical applications.

For example, in the cases of matrices

$$\boldsymbol{A}_1 = \begin{pmatrix} 0.8 & 0 \\ 0.8 & 0 \end{pmatrix}, \quad \boldsymbol{A}_2 = \begin{pmatrix} 0.8 & 0.8 \\ 0 & 0 \end{pmatrix} \quad \text{and} \quad \boldsymbol{A}_3 = \begin{pmatrix} 0.51 & 0.51 \\ 0.51 & 0 \end{pmatrix},$$

$$\|\boldsymbol{A}_1\|_\infty = 0.8 < 1, \quad \|\boldsymbol{A}_1\|_1 = 1.6 > 1, \quad \|\boldsymbol{A}_1\|_2 = \sqrt{1.28} \simeq 1.13 > 1,$$
$$\|\boldsymbol{A}_2\|_\infty = 1.6 > 1, \quad \|\boldsymbol{A}_2\|_1 = 0.8 < 1, \quad \|\boldsymbol{A}_2\|_2 = \sqrt{1.28} \simeq 1.13 > 1$$

and

$$\|\boldsymbol{A}_3\|_\infty = 1.02 > 1, \quad \|\boldsymbol{A}_3\|_1 = 1.02 > 1,$$
$$\|\boldsymbol{A}_3\|_2 = \sqrt{\frac{3+\sqrt{5}}{2}}(0.51) \simeq 0.825 < 1.$$

That is, only one of the most popular matrix norms is below one, the other two norms are greater than one. It is well-known that if all eigenvalues of a matrix are inside the unit circle, then there is a matrix norm such that the norm of this matrix is below one, (see for example, Ortega and Rheinholdt (1970)). Therefore we can reformulate Theorem B.1 as follows:

**Theorem B.4.** *Let $\bar{\boldsymbol{x}} \in \mathbb{D}$ be an equilibrium and assume that $\boldsymbol{J}(\boldsymbol{x})$ exists in an open neighborhood of $\bar{\boldsymbol{x}}$, and is continuous at $\bar{\boldsymbol{x}}$. Assume furthermore that all eigenvalues of $\boldsymbol{J}(\bar{\boldsymbol{x}})$are inside the unit circle. Then $\bar{\boldsymbol{x}}$ is locally asymptotically stable.*

Unfortunately this eigenvalue criterion cannot be extended to prove global asymptotic stability. That is, the assumption that for all $\boldsymbol{x} \in \mathbb{D}$ the eigenvalues of $\boldsymbol{J}(\bar{\boldsymbol{x}})$ are inside the unit circle does not necessarily imply the $\bar{\boldsymbol{x}}$ is globally asymptotically stable. In fact Cima et al. (1997, 1999) present counterexamples for $n = 2$ and $n \geq 3$.

Instability of equilibria cannot be proved by showing that a particular matrix norm of the Jacobian at $\bar{\boldsymbol{x}}$ is larger than one, since there is the possibility that another norm of the Jacobian is less than one. However it is well known that if at least one eigenvalue of $\boldsymbol{J}(\bar{\boldsymbol{x}})$ is outside the unit circle, then $\bar{\boldsymbol{x}}$ is unstable. For an elementary proof see Li and Szidarovszky (1999*a*). If for all eigenvalues $\lambda_i$ of $\boldsymbol{J}(\bar{\boldsymbol{x}})$, $|\lambda_i| \leq 1$ and at least one eigenvalue is located on the unit circle, then no conclusion can be given, since $\bar{\boldsymbol{x}}$ can be unstable, marginally stable, and even locally (or globally) asymptotically stable. Such examples are given next.

*Example B.1.* Consider first the two-dimensional linear system

$$\boldsymbol{x}(t+1) = \begin{pmatrix} 1 & 1 \\ 0 & 1 \end{pmatrix} \boldsymbol{x}(t),$$

with both eigenvalues of the coefficient matrix (which is also the Jacobian) being on the unit circle. It is easy to show by finite induction that for all $t \geq 1$,

$$\begin{pmatrix} 1 & 1 \\ 0 & 1 \end{pmatrix}^t = \begin{pmatrix} 1 & t \\ 0 & 1 \end{pmatrix}.$$

Consequently, with any $\boldsymbol{x}(0) = (x_1(0), x_2(0))^T$,

$$\boldsymbol{x}(t) = \begin{pmatrix} 1 & t \\ 0 & 1 \end{pmatrix} \boldsymbol{x}(0), \tag{B.5}$$

and when $x_2(0) > 0$, $x_1(t) \to \infty$. Therefore the zero equilibrium is unstable.

*Example B.2.* Consider next the single dimensional system

$$x(t+1) = -x(t)$$

with the unique equilibrium $\bar{x} = 0$. Notice that the Jacobian of the right hand side is $-1$ with unit absolute value. Clearly, for all $t \geq 0$,

$$x(t) = (-1)^t x(0)$$

showing that the zero equilibrium is only marginally stable.

*Example B.3.* Consider now the simple nonlinear system

$$x(t+1) = x(t)e^{-x(t)^2}.$$

Notice first that $x(0) = 0$ implies that $x(t) = 0$ for all $t \geq 1$, if $x(0) > 0$, then $x(t) > 0$ and if $x(0) < 0$, then $x(t) < 0$ for all $t \geq 0$. Furthermore if $x(0) \neq 0$, then

$$| x(t+1) | < | x(t) |,$$

showing that the positive state sequence is strictly decreasing and the negative state sequence is strictly increasing. Therefore in both cases the state sequence is bounded by zero, so convergent. Let $x^*$ be the limit. Letting $t \to \infty$ in the defining difference equation we have

$$x^* = x^* e^{-(x^*)^2},$$

showing that $x^* = 0$, which is the unique equilibrium of the system. Hence the zero equilibrium is globally asymptotically stable. In this case the Jacobian is the derivative of the right hand side:

$$\frac{d}{dx}(xe^{-x^2}) = e^{-x^2} + xe^{-x^2}(-2x)$$

which equals 1 at $x = 0$.

We next turn our attention to continuous time systems (B.2), Theorem B.4 can be modified as follows (see for example, Bellman (1969)).

**Theorem B.5.** *Assume that all eigenvalues of $\boldsymbol{J}(\bar{\boldsymbol{x}})$ have negative real parts, then $\bar{\boldsymbol{x}}$ is locally asymptotically stable.*

The global stability version of this theorem is not valid in general. Cima et al. (1997) provide a counterexample for all $n \geq 3$ where all eigenvalues of the Jacobian of $\boldsymbol{g}$ have negative real parts for all $\boldsymbol{x} \in \mathbb{R}^n$, but the equilibrium of the continuous time system is unstable. There has been intensive research on this problem for the case of $n = 2$. Many authors proved the global asymptotic stability with different additional conditions, and finally Gutierrez (1995) proved the global asymptotic stability of the equilibrium without additional assumptions in the case of continuously differentiable functions. This result was extended without requiring the continuity of the Jacobian by Fernandes et al. (2004). Their main result is the following:

**Theorem B.6.** *Assume $\mathbb{D} = \mathbb{R}^2$, $\boldsymbol{g}(\bar{\boldsymbol{x}}) = \boldsymbol{0}$ and $\boldsymbol{g}$ is differentiable everywhere. Assume furthermore that all eigenvalues of the Jacobian of $\boldsymbol{g}$ have negative real parts on $\mathbb{D}$. Then $\bar{\boldsymbol{x}}$ is the of system* (B.2) *and it is globally asymptotically stable.*

The eigenvalues of the Jacobian at the equilibrium might also indicate the instability of the equilibrium, since similarly to the discrete time case we can show that if at least one eigenvalue of $\boldsymbol{J}(\bar{\boldsymbol{x}})$ has positive real part, then $\bar{\boldsymbol{x}}$ is unstable. If all eigenvalues of $\boldsymbol{J}(\bar{\boldsymbol{x}})$ have non-positive real parts and at least one eigenvalue is zero or pure complex, then $\bar{\boldsymbol{x}}$ may be unstable, marginally stable, or even asymptotically stable. Such examples are given next.

*Example B.4.* Consider first the two-dimensional linear system

$$\dot{\boldsymbol{x}} = \begin{pmatrix} 0 & 1 \\ 0 & 0 \end{pmatrix} \boldsymbol{x}$$

which has both eigenvalues equal to zero. It is easy to show that the fundamental matrix (which is the matrix exponential) is given as

$$\begin{pmatrix} 1 & t \\ 0 & 1 \end{pmatrix},$$

thus showing the instability of the zero equilibrium.

*Example B.5.* In the case of the system $\dot{x} = 0$, the Jacobian is zero with zero eigenvalue. All solutions are constant and all real numbers are equilibria. Clearly all equilibria are marginally stable.

*Example B.6.* Consider finally the system driven by the single dimensional differential equation

$$\dot{x} = -x^3.$$

Here $\bar{x} = 0$ is the only equilibrium, and the Jacobian is the derivative

$$\frac{\mathrm{d}}{\mathrm{d}x}(-x^3) = -3x^2,$$

giving zero value at the equilibrium. Since the equation is separable, one can easily find the state trajectories

$$x(t) = \frac{x(0)}{\sqrt{1 + 2tx(0)^2}}.$$

Clearly $x(t) \to 0$ as $t \to \infty$ showing the global asymptotic stability of the equilibrium.

▼

In the case of linear systems local and global asymptotic stability are equivalent. A discrete time invariant linear system is asymptotically stable if and only if all eigenvalues of the coefficient matrix are inside the unit circle, and a time invariant continuous linear system is asymptotically stable if and only if all eigenvalues have negative real parts.

Continuous systems based on certain adjustment principles can often be written as

$$\dot{\boldsymbol{x}} = \boldsymbol{K}(\boldsymbol{g}(\boldsymbol{x})) \tag{B.6}$$

where $\boldsymbol{K} : \mathbb{R}^n \to \mathbb{R}^n$ is an adjustment function with sign preserving components. If $\bar{\boldsymbol{x}}$ is an equilibrium of this system, then the Jacobian of the right hand side has the special form

$$\boldsymbol{J}(\bar{\boldsymbol{x}}) = \boldsymbol{J}_K(\boldsymbol{g}(\bar{\boldsymbol{x}}))\boldsymbol{J}_g(\bar{\boldsymbol{x}}) = \boldsymbol{J}_K(\boldsymbol{0})\boldsymbol{J}_g(\bar{\boldsymbol{x}}),$$

where $\boldsymbol{J}_K$ and $\boldsymbol{J}_g$ are the Jacobians of $\boldsymbol{K}$ and $\boldsymbol{g}$, respectively. In analyzing the asymptotic behavior of this system the following result can be applied.

**Theorem B.7.** *Assume $\boldsymbol{J}_K(\boldsymbol{0})$is positive definite and $\boldsymbol{J}_g(\bar{\boldsymbol{x}}) + \boldsymbol{J}_g(\bar{\boldsymbol{x}})^T$ is negative definite. Then all eigenvalues of $\boldsymbol{J}(\bar{\boldsymbol{x}})$ have negative real parts implying the (local) asymptotic stability of the equilibrium of system (B.6).*

*Proof.* Consider the time invariant linear continuous system

$$\dot{z} = \boldsymbol{J}_K(\boldsymbol{0})\boldsymbol{J}_g(\bar{\boldsymbol{x}})z \tag{B.7}$$

and notice that $\bar{z} = \mathbf{0}$ is an equilibrium. Select the Lyapunov function

$$\mathbb{V}(z) = z^T \boldsymbol{J}_K(\mathbf{0})^{-1} z. \tag{B.8}$$

Since $\boldsymbol{J}_K(\mathbf{0})^{-1}$ is also positive definite, conditions $(a)$ and $(b)$ of Definition A.4 are satisfied. Let $z(t)$ be a trajectory of system (B.7), then

$$\begin{aligned}
\frac{\mathrm{d}}{\mathrm{d}t}\mathbb{V}(z(t)) &= \dot{z}(t)^T \boldsymbol{J}_K(\mathbf{0})^{-1} z(t) + z(t)^T \boldsymbol{J}_K(\mathbf{0})^{-1} \dot{z}(t) \\
&= 2z(t)^T \boldsymbol{J}_K(\mathbf{0})^{-1} \dot{z}(t) \\
&= 2z(t)^T \boldsymbol{J}_K(\mathbf{0})^{-1} \boldsymbol{J}_K(\mathbf{0}) \boldsymbol{J}_g(\bar{x}) z(t) \\
&= 2z(t)^T \boldsymbol{J}_g(\bar{x}) z(t) \\
&= z(t)^T \left( \boldsymbol{J}_g(\bar{x}) + \boldsymbol{J}_g(\bar{x})^T \right) z(t) < 0,
\end{aligned}$$

unless $z(t) = \mathbf{0}$. Hence the zero equilibrium of system (B.7) is asymptotically stable implying that all eigenvalues of its coefficient matrix have negative real parts. ■

# Appendix C
# Noninvertible Maps and Critical Sets

In this appendix we give some definitions, properties and simple examples of discrete dynamical systems represented by the iteration of noninvertible maps.

## C.1 Definitions and Simple Examples

A map $T : S \to S$, $S \subseteq \mathbb{R}^n$, defined by $\mathbf{x}' = T(\mathbf{x})$, transforms a point $\mathbf{x} \in S$ into a unique point $\mathbf{x}' \in S$. The point $\mathbf{x}'$ is called the *rank-1 image* of $\mathbf{x}$, and a point $\mathbf{x}$ such that $T(\mathbf{x}) = \mathbf{x}'$ is a *rank-1 preimage* of $\mathbf{x}'$.

If $\mathbf{x} \neq \mathbf{y}$ implies $T(\mathbf{x}) \neq T(\mathbf{y})$ for each $\mathbf{x}, \mathbf{y}$ in $S$, then $T$ is an *invertible map* in $S$, because the inverse mapping $\mathbf{x} = T^{-1}(\mathbf{x}')$ is uniquely defined. Otherwise $T$ is said to be a *noninvertible map*, because points $\mathbf{x}$ exist that have several rank-1 preimages, i.e., the inverse relation $\mathbf{x} = T^{-1}(\mathbf{x}')$ is multivalued. So, noninvertible means "many-to-one", that is, distinct points $\mathbf{x} \neq \mathbf{y}$ may have the same image, $T(\mathbf{x}) = T(\mathbf{y}) = \mathbf{x}'$.

Geometrically, the action of a noninvertible map can be thought of as "folding and pleating" the space $S$, so that distinct points are mapped into the same point. This is equivalently stated by saying that several inverses are defined in some points of $S$, and these inverses "unfold" $S$.

For a noninvertible map, $S$ can be subdivided into regions $Z_k$, $k \geq 0$, whose points have $k$ distinct rank-1 preimages. Generally, for a continuous map, as the point $\mathbf{x}'$ varies in $\mathbb{R}^n$, pairs of preimages appear or disappear as this point crosses the boundaries separating different regions. Hence, such boundaries are characterized by the presence of at least two coincident (merging) preimages. This leads us to the definition of the *critical sets*, one of the distinguishing features of noninvertible maps (see Gumowski and Mira (1980), Mira et al. (1996)):

**Definition C.1.** The *critical set* $CS$ of a continuous map $T$ is defined as the locus of points having at least two coincident $rank-1$ preimages, located on a set $CS_{-1}$, called *the set of merging preimages*.

The critical set $CS$ is generally formed by $(n-1)$-dimensional hypersurfaces of $\mathbb{R}^n$, and portions of $CS$ separate regions $Z_k$ of the phase space characterized by a different number of $rank - 1$ preimages, for example $Z_k$ and $Z_{k+2}$ (this is the standard occurrence for continuous maps). The critical set $CS$ is the $n$-dimensional generalization of the notion of local minimum or local maximum of a one-dimensional map, and of the notion of *critical curve* $LC$ of a noninvertible two-dimensional map. This terminology, and notation, originates from the notion of critical point as it is used in the classical works of Julia and Fatou. The set $CS_{-1}$ is the generalization of local extremum point of a one-dimensional map, and of the *fold curve* $LC_{-1}$ of a two-dimensional noninvertible map.

As an illustration, we consider the one-dimensional quadratic map (logistic map)

$$x' = f(x) = \mu x(1 - x). \tag{C.1}$$

This map has a unique critical point $c = \mu/4$, which separates the real line into the two subsets: $Z_0 = (c, +\infty)$, where no inverses are defined, and $Z_2 = (-\infty, c)$, whose points have two rank-1 preimages (Fig. C.1a). These preimages can be computed by the two inverses

$$x_1 = f_1^{-1}(x') = \frac{1}{2} - \frac{\sqrt{\mu\,(\mu - 4x')}}{2\mu}; \quad x_2 = f_2^{-1}(x') = \frac{1}{2} + \frac{\sqrt{\mu\,(\mu - 4x')}}{2\mu}. \tag{C.2}$$

If $x' \in Z_2$, its two rank-1 preimages, computed according to (C.2), are located symmetrically with respect to the point $c_{-1} = 1/2 = f_1^{-1}(\mu/4) = f_2^{-1}(\mu/4)$. Hence, $c_{-1}$ is the point where the two merging preimages of $c$ are located. As the map (C.1) is differentiable, at $c_{-1}$ the first derivative vanishes.

We remark that in general the condition of vanishing derivative is not sufficient to define the critical points of rank-0 since such a condition may be also satisfied by points which are not local extrema (for example the inflection points with horizontal tangent). Moreover, for continuous and piecewise differentiable maps the condition of vanishing derivative is not necessary as well, because such maps may have the property that the images of points where the map is not differentiable are critical points, according to the definition given above. This occurs whenever such points are local maxima or minima, like in the cases shown in Figs. C.2a, b. In Fig. C.2a,

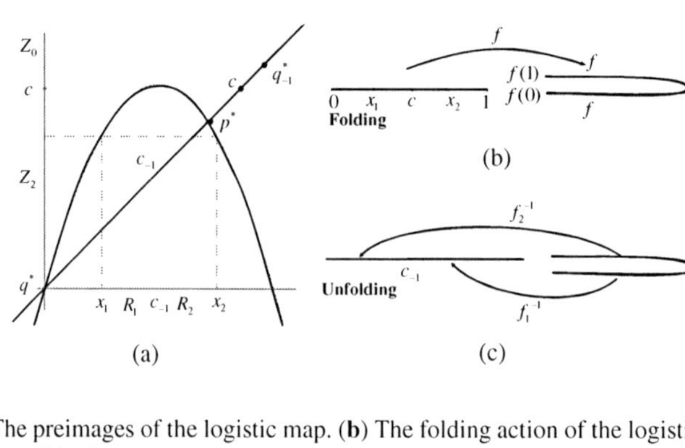


**Fig. C.1** (**a**) The preimages of the logistic map. (**b**) The folding action of the logistic map. (**c**) The unfolding action of the inverses

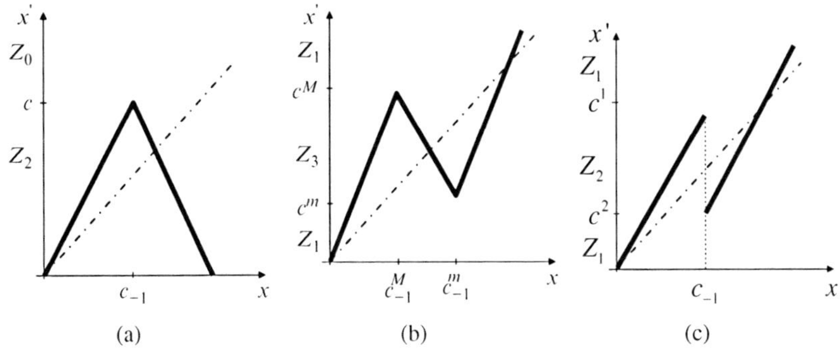


**Fig. C.2** The preimage regions of certain maps. (**a**) The tent-map. (**b**) A bimodal piecewise linear map. (**c**) A discontinuous map. Notice that in (**a**) and (**b**) the number of preimages in adjacent regions differ by 2, whereas in (**c**) they differ by 1

a typical $Z_0 - Z_2$ tent map is shown, where the kink point behaves like the critical point of the logistic map even if it is not obtained as the image of a point with vanishing derivative. The same reasoning applies to the "bimodal" $Z_1 - Z_3 - Z_1$ piecewise linear function shown in Fig. C.2b.

Up to now we have considered continuous maps, but the properties of critical points can easily be extended also to piecewise continuous maps $T$. In this case a point of discontinuity may behave as a critical point of $T$, even if the definition in terms of merging preimages cannot be applied. This happens when the ranges of the map on the two sides of the discontinuity have an overlapping zone, so that at least one of the two limiting values of the function at the discontinuity separates regions having a different number of rank-1 preimages (see for example the map shown in Fig. C.2c). The difference with respect to the case of a continuous map is that now the number of distinct rank-1 preimages through a critical point differs generally by one (instead of two), that is, a critical value $c$ (in general the critical set $CS$) separates regions $Z_k$ and $Z_{k+1}$. A one-dimensional example is shown in Fig. C.2c, where the point of discontinuity is a critical point $c_{-1}$, and both the two limiting values of the function in $c_{-1}$ are critical points, say $c^1$ and $c^2$, associated with $c_{-1}$, as both $c^1$ and $c^2$ separate regions $Z_1$ and $Z_2$. Notice that now the critical points have no merging rank-1 preimages. More on the properties and bifurcations of discontinuous maps of the plane can be found in Mira et al. (1996).

In order to explain the geometric action of a critical point in a continuous map, let us consider, again, the logistic map, and note that as $x$ moves from 0 to 1 the corresponding image $f(x)$ spans the interval $[0, c]$ twice, the critical point $c$ being the turning point. In other words, if we consider how the segment $\gamma = [0, 1]$ is transformed by the map $f$, we can say that it is *folded and pleated* to obtain the image $\gamma' = [0, c]$. Such folding gives a geometric reason why two distinct points of $\gamma$, say $x_1$ and $x_2$, located symmetrically with respect to the point $c_{-1} = 1/2$, are mapped into the same point $x' \in \gamma'$ due to the folding action of $f$ (see Fig. C.1b). The same conclusions can be obtained by looking at the two inverse mappings $f_1^{-1}$ and $f_2^{-1}$

defined in $(-\infty, a/4]$ according to (C.2). We can consider the range of the map $f$ formed by the superposition of two half-lines $(-\infty, a/4]$, joined at the critical point $c = a/4$ (Fig. C.1c), and on each of these half-lines a different inverse is defined. In other words, instead of saying that two distinct maps are defined on the same half-line we say that the range is formed by two distinct half lines on each of which a unique inverse map is defined. This point of view gives a geometric visualization of the critical point $c$ as the point in which two distinct inverses merge. The action of the inverses, say $f^{-1} = f_1^{-1} \cup f_2^{-1}$, causes an unfolding of the range by mapping $c$ into $c_{-1}$ and by opening the two half-lines one on the right and one on the left of $c_{-1}$, so that the whole real line $\mathbb{R}$ is covered. So, the map $f$ folds the real line, the two inverses unfold it.

Another interpretation of the folding action of a critical point is the following. Since $f(x)$ is increasing for $x \in [0, 1/2)$ and decreasing for $x \in (1/2, 1]$, its application to a segment $\gamma_1 \subset [0, 1/2)$ is orientation preserving, whereas its application to a segment $\gamma_2 \subset (1/2, 1]$ is orientation reversing. This suggests that an application of $f$ to a segment $\gamma_3 = [a, b]$ including the point $c_{-1} = 1/2$ preserves the orientation of the portion $[a, c_{-1}]$, that is $f([a, c_{-1}]) = [f(a), c]$, whereas it reverses the portion $[c_{-1}, b]$, so that $f([c_{-1}, b]) = [f(b), c]$, so that $\gamma_3' = f(\gamma_3)$ is folded, the folding point being the critical point $c$.

Let us now consider the case of a continuous two-dimensional map $T : S \to S$, $S \subseteq \mathbb{R}^2$, defined by

$$T : \begin{cases} x_1' = T_1(x_1, x_2), \\ x_2' = T_2(x_1, x_2). \end{cases} \tag{C.3}$$

If we solve the system of the two (C.3) with respect to the unknowns $x_1$ and $x_2$, then, for a given $(x_1', x_2')$, we may have several solutions, representing rank-1 preimages (or backward iterates) of $(x_1', x_2')$, say $(x_1, x_2) = T^{-1}(x_1', x_2')$, where $T^{-1}$ is in general a multivalued relation. In this case we say that $T$ is noninvertible, and the critical set (formed by critical curves, denoted by $LC$ from the French "Ligne Critique") constitutes the set of boundaries that separate regions of the plane characterized by a different number of rank-1 preimages. According to the definition, along $LC$ at least two inverses give merging preimages, located on $LC_{-1}$ (following the notations of Gumowski and Mira (1980), Mira et al. (1996)).

For a continuous and (at least piecewise) differentiable noninvertible map of the plane, the set $LC_{-1}$ is included in the set where $det\boldsymbol{J}(x_1, x_2)$ changes sign, with $\boldsymbol{J}$ being the Jacobian matrix of $T$, since $T$ is locally an orientation preserving map near points $(x_1, x_2)$ such that $det\boldsymbol{J}(x_1, x_2) > 0$ and orientation reversing if $det\boldsymbol{J}(x_1, x_2) < 0$. In order to explain this point, let us recall that when an affine transformation $\mathbf{x}' = \boldsymbol{A}\mathbf{x} + \mathbf{b}$, where $\boldsymbol{A} = \{a_{ij}\}$ is a $2 \times 2$ matrix and $\mathbf{b} \in \mathbb{R}^2$, is applied to a plane figure, then the area of the transformed figure grows, or shrinks, by a factor $\rho = |det\boldsymbol{A}|$, and if $det\boldsymbol{A} > 0$ then the orientation of the figure is preserved, whereas if $det\boldsymbol{A} < 0$ then the orientation is reversed. This property also holds for the linear approximation of (C.3) in a neighborhood of a point $\mathbf{p} = (x_1, x_2)$, given by an affine map with $\boldsymbol{A} = \boldsymbol{J}$, $\boldsymbol{J}$ being the Jacobian matrix evaluated at the point $\mathbf{p}$

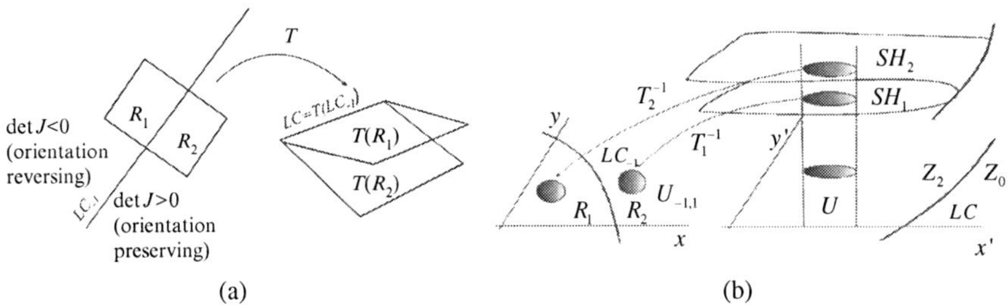


**Fig. C.3** (**a**) A qualitative visualization of a map of the plane, and how the folding relates to the sign of the Jacobian matrix. (**b**) Visualizing a Riemann foliation of the plane, in the case of a $Z_0 - Z_2$ noninvertible map

$$\boldsymbol{J}(\mathbf{p}) = \begin{pmatrix} \partial T_1/\partial x_1 & \partial T_1/\partial x_2 \\ \partial T_2/\partial x_1 & \partial T_2/\partial x_2 \end{pmatrix}. \tag{C.4}$$

A qualitative visualization is given in Figs. C.3a, b. Of course, if the map is continuously differentiable then the change of the sign of $det(\boldsymbol{J})$ occurs along points where $det(\boldsymbol{J})$ vanishes, thus giving the characterization of the fold line $LC_{-1}$ as the locus where the Jacobian vanishes.

In order to give a geometrical interpretation of the action of a multi-valued inverse relation $T^{-1}$, it is useful to consider a region $Z_k$ as the superposition of $k$ sheets, each associated with a different inverse. Such a representation is known as *Riemann foliation* of the plane (see for example Mira et al. (1996)). Different sheets are connected by folds joining two sheets, and the projections of such folds on the phase plane are arcs of $LC$. This is shown in the qualitative sketch of Fig. C.3b, where the case of a $Z_0 - Z_2$ noninvertible map is considered. This graphical representation of the unfolding action of the inverses also gives an intuitive idea of the mechanism which causes the creation of disconnected basins for noninvertible maps of the plane.

To give an example, let us again consider a quadratic map $T : (x, y) \to (x', y')$, extensively studied in Mira et al. (1996) and Abraham et al. (1997), defined by

$$T : \begin{cases} x' = ax + y, \\ y' = b + x^2. \end{cases} \tag{C.5}$$

Given $x'$ and $y'$, if we try to solve the algebraic system with respect to the unknowns $x$ and $y$ we get two solutions, given by

$$T_1^{-1} : \begin{cases} x = -\sqrt{y' - b}, \\ y = x' + a\sqrt{y' - b}, \end{cases} ; \qquad T_2^{-1} : \begin{cases} x = \sqrt{y' - b}, \\ y = x' - a\sqrt{y' - b}, \end{cases} \tag{C.6}$$

if $y' \geq b$, and no solutions if $y' < b$. So, (C.5) is a $Z_0 - Z_2$ noninvertible map, where $Z_0$ (the region whose points have no preimages) is the half plane $Z_0 = \{(x, y) \,|\, y < b\}$ and $Z_2$ (region whose points have two distinct rank-1 preimages)

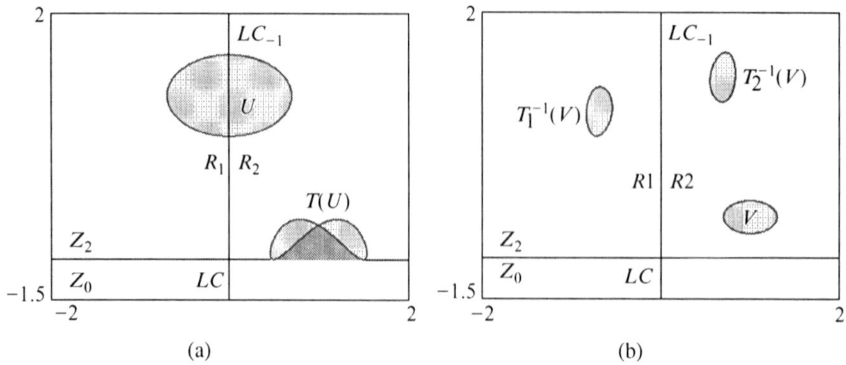


**Fig. C.4** A quadratic map example. Here $a = -0.3$ and $b = -1$. (**a**) The folding of the ball $U$ by the map along the critical line $LC$. (**b**) The unfolding action of the inverses of the map

is the half plane $Z_2 = \{(x, y) \,|\, y > b\}$. The line $y = b$, which separates these two regions, is $LC$, that is the locus of points having two merging rank-1 preimages, located on the line $x = 0$, that represents $LC_{-1}$. Since (C.5) is a continuously differentiable map, the points of $LC_{-1}$ necessarily belong to the set of points at which the Jacobian determinant vanishes, in other words $LC_{-1} \subseteq J_0$, where $J_0 = \{(x, y) \,|det\boldsymbol{J}(x, y) = -2x = 0\}$. In this case $LC_{-1}$ coincides with $J_0$ (the vertical axis $x = 0$) and the critical curve $LC$ is the image of $LC_{-1}$, that is $LC = T(LC_{-1}) = T(\{x = 0\}) = \{(x, y) \,|\, y = b\}$.

In order to show the folding action related to the presence of the critical lines, we consider a plane figure (a circle) $U$ separated by $LC_{-1}$ into two portions, say $U_1 \in R_1$ and $U_2 \in R_2$ (Fig. C.4a) and we apply the map (C.5) to the points of $U$. The image $T(U_1) \cap T(U_2)$ is a non-empty set included in the region $Z_{k+2}$, which is the region whose points $p'$ have rank-1 preimages $p_1 = T_1^{-1}(p') \in U_1$ and $p_2 = T_2^{-1}(p') \in U_2$. This means that two points $p_1 \in U_1$ and $p_2 \in U_2$, located at opposite sides with respect to $LC_{-1}$, are mapped in the same side with respect to $LC$, in the region $Z_{k+2}$. This is also expressed by saying that the ball $U$ is "folded" by $T$ along $LC$ on the side with more preimages (see Fig. C.4a). The same concept can be equivalently expressed by stressing the "unfolding" action of $T^{-1}$, obtained by the application of the two distinct inverses in $Z_{k+2}$ which merge along $LC$. Indeed, if we consider a ball $V \subset Z_{k+2}$, then the set of its $rank - 1$ preimages $T_1^{-1}(V)$ and $T_2^{-1}(V)$ is made up of two balls $T_1^{-1}(V) \in R_1$ and $T_2^{-1}(V) \in R_2$. These balls are disjoint if $V \cap LC = \emptyset$ (Fig. C.4b).

Many of the considerations made above, for one-dimensional and two-dimensional noninvertible maps, can be generalized to $n$-dimensional ones, even if their visualization becomes more difficult. First of all, from the definition of critical set it is clear that the relation $CS = T(CS_{-1})$ holds in any case. Moreover, the points of $CS_{-1}$ where the map is continuously differentiable are necessarily points where the Jacobian determinant vanishes, so that

$$CS_{-1} \subseteq J_0 = \{p \in \mathbb{R}^n | det J(p) = 0\} \tag{C.7}$$

In fact, in any neighborhood of a point of $CS_{-1}$ there are at least two distinct points which are mapped by $T$ in the same point. Accordingly, the map is not locally invertible in points of $CS_{-1}$, and (C.7) follows from the implicit function theorem. This property provides an easy method to compute the critical set for continuously differentiable maps – from the expression of the Jacobian determinant one computes the locus of points at which it vanishes, then the set obtained after an application of the map to these points is the critical set $CS$.

Also the geometric properties illustrated above for the two-dimensional noninvertible map (C.5) can be easily generalized to the case of the critical set of an $n$-dimensional noninvertible map. It is worth noting that, in general, for piecewise differentiable maps the set of points where the map is not differentiable may belong to $CS_{-1}$, that is the images by $T$ of such points may separate regions characterized by a different number of rank-1 preimages (see for example Mira (1987)). Moreover, piecewise continuous maps may have points of $CS_{-1}$ at the discontinuities and, differently from the case of continuous maps, the corresponding portions of $CS$ may separate regions that differ by an odd number of preimages (see Mira (1987)). In any case, the importance of the set $CS$ lies in the fact that its points separate regions $Z_k$ characterized by a different number of preimages. This property may also be shared by points where some inverses are not defined due to a vanishing denominator, as shown in Bischi et al. (1999, 2001a, 2003a).

## C.2 Discrete Time Dynamical Systems as Iterated Maps

A *discrete-time dynamical system*, defined by the difference equation

$$\mathbf{x}(t+1) = T(\mathbf{x}(t)), \tag{C.8}$$

can be viewed as the result of the repeated application (or *iteration*) of a map $T$. Indeed, the point $\mathbf{x}$ represents the state of a system, and $T$ represents the "unit time advancement operator" $T : \mathbf{x}(t) \to \mathbf{x}(t+1)$. Starting from an *initial condition* $\mathbf{x}_0 \in S$, the iteration of $T$ inductively defines a unique *trajectory*

$$\tau(\mathbf{x}_0) = \left\{\mathbf{x}(t) = T^t(\mathbf{x}_0), t = 0, 1, 2, \ldots\right\}, \tag{C.9}$$

where $T^0$ is the identity map and $T^t = T(T^{t-1})$. As $t \to +\infty$, a trajectory may diverge, or it may converge to a fixed point of the map $T$, which is a point $\overline{\mathbf{x}}$ such that $T(\overline{\mathbf{x}}) = \overline{\mathbf{x}}$. It may also asymptotically approach another kind of invariant set, such as a periodic cycle , or a closed invariant curve or a more complex attractor, for example a so called chaotic attractor (see for example Devaney (1989), Guckenheimer and Holmes (1983) and Medio and Lines (2001)). We recall that a set $A \subset \mathbb{R}^n$ is *invariant* for the map $T$ if it is mapped onto itself, $T(A) = A$.

This means that if $x \in A$ then $T(x) \in A$, so that A is *trapping,* and every point of $A$ is an image of some point of $A$. A closed invariant set $A$ is an *attractor* if (1) it is *Lyapunov stable,* that is for every neighborhood $W$ of $A$ there exists a neighborhood $V$ of $A$ such that $T^t(V) \subset W \ \forall t \geq 0$; (2) a neighborhood $U$ of $A$ exists such that $T^t(\mathbf{x}) \to A$ as $t \to +\infty$ for each $\mathbf{x} \in U$.

The *basin* of an attractor $A$ is the set of all points that generate trajectories converging to $A$

$$\mathcal{B}(A) = \left\{ \mathbf{x} | T^t(\mathbf{x}) \to A \quad \text{as } t \to +\infty \right\}. \tag{C.10}$$

Let $U(A)$ be a neighborhood of an attractor $A$ whose points converge to $A$. Of course $U(A) \subseteq \mathcal{B}(A)$, and also the points that are mapped into $U$ after a finite number of iterations belong to $\mathcal{B}(A)$. Hence, the basin of $A$ is given by

$$\mathcal{B}(A) = \bigcup_{n=0}^{\infty} T^{-n}(U(A)), \tag{C.11}$$

where $T^{-n}(x)$ represents the set of the rank-$n$ preimages of $x$ (the points mapped into $x$ after $n$ applications of $T$).

Let $\mathcal{B}$ be a basin of attraction and $\partial\mathcal{B}$ its boundary. From the definition it follows that $\mathcal{B}$ is trapping with respect to the forward iteration of the map $T$ and invariant with respect to the backward iteration of all the inverses $T^{-1}$. Points belonging to $\partial\mathcal{B}$ are mapped into $\partial\mathcal{B}$ both under forward and backward iteration of $T$. This implies that if an unstable fixed point or cycle belongs to $\partial\mathcal{B}$ then $\partial\mathcal{B}$ must also contain all of its preimages of any rank. In particular, if a saddle point, or a saddle cycle, belongs to $\partial\mathcal{B}$, then $\partial\mathcal{B}$ must also contain the whole stable set (see Gumowski and Mira (1980), Mira et al. (1996)).

A problem that often arises in the study of nonlinear dynamical systems concerns the existence of several attracting sets, each with its own basin of attraction. In this case the dynamic process becomes path dependent, which means that the kind of long-run dynamics that characterizes the system depends on the starting condition. Another important problem in the study of applied dynamical systems is the delineation of a bounded region of the state space in which the system dynamics are ultimately trapped, despite the complexity of the long-run time patterns. This is useful information, even more useful than a detailed description of the step-by-step time evolution.

Both of these questions require an analysis of the global properties of the dynamical system, that is, an analysis which is not based on the linear approximation of the map. When the map $T$ is noninvertible, its global dynamical properties can be usefully characterized by using the formalism of critical sets, by which the folding action associated with the application of the map, as well as the "unfolding" associated with the action of the inverses, can be described. Loosely speaking, the repeated application of a noninvertible map repeatedly folds the state space along the critical sets and their images, and often this allows one to define a bounded region in which asymptotic dynamics are trapped. As some parameter is varied, global bifurcations that cause sudden qualitative changes in the properties of the attracting sets can be detected by observing contacts of critical curves with invariant sets. The repeated application of the inverses "repeatedly unfolds" the state space, so that a neighborhood of an attractor may have preimages far from it, thus giving rise to complicated topological structures of the basins, that may be formed by the union of several (even infinitely many) disconnected portions. In fact, from (C.11) it follows that in order to study the extension of a basin and the structure of its boundaries one has to consider the properties of the inverse relation $T^{-1}$. The route to more and more complex basin boundaries, as some parameter is varied, is characterized by global bifurcations, also called contact bifurcations, due to contacts between the critical set and the invariant sets that form the boundaries of the basins of attraction.

## C.3 Critical Sets and the Delineation of Trapping Regions

Portions of the critical set $CS$ and its images $CS_k = T^k(CS)$ can be used to obtain the boundaries of trapping regions to which the asymptotic dynamics of the iterated points of a noninvertible map are confined. This can be easily explained for a one-dimensional noninvertible map, for example the quadratic map (C.1). In fact, it is quite evident that if we iterate the logistic map for $3 < \mu < 4$ starting from an initial condition inside the interval $[c_1, c]$, with $c_1 = f(c)$, no images can be obtained out of this interval (see Fig. C.5), that is the interval along the 45° line formed by the critical point $c$ and its rank-1 image $c_1$ is trapping. Moreover, any trajectory

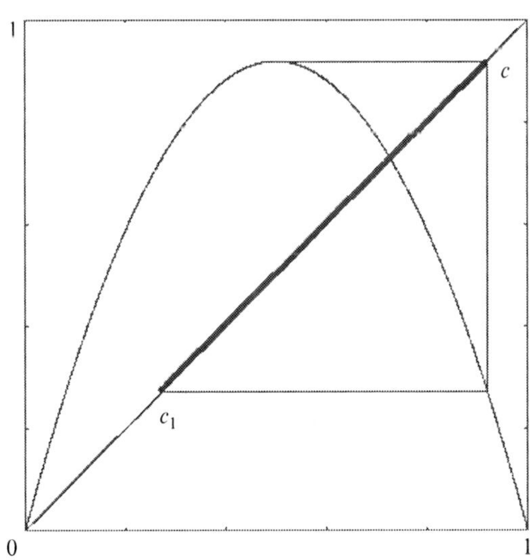


**Fig. C.5** The trapping region of the quadratic map. Trajectories starting from any point in (0, 1), will enter the trapping region after a finite number of iterations

generated from an initial condition in (0, 1), enters $[c_1, c]$ after a finite number of iterations. Following the terminology introduced in Mira et al. (1996), the interval $[c_1, c]$ is called *absorbing*.

In general, for an $n$-dimensional map, an *absorbing region* $\mathcal{A}$ *(intervals in* $\mathbb{R}$*, areas in* $\mathbb{R}^2$*, volumes in* $\mathbb{R}^3, \ldots$*)* is defined as a bounded set whose boundary is given by portions of the critical set $CS$ and its images of increasing order $CS_k = T^k(CS)$, such that a neighborhood $U \supset \mathcal{A}$ exists whose points enter $\mathcal{A}$ after a finite number of iterations and then never escape it, since $T(\mathcal{A}) \subseteq \mathcal{A}$, which is to say that $\mathcal{A}$ is trapping (see for example Mira et al. (1996) for more details).

Loosely speaking, we can say that the iterated application of a noninvertible map, folding and folding again the space, defines trapping regions bounded by critical sets of increasing order.

Sometimes, smaller absorbing regions are nested inside a bigger one. This can be illustrated, again, for the logistic map (C.1), as shown in Fig. C.6a, where inside the absorbing interval $[c_1, c]$ a trapping subset is obtained by higher rank images of the critical point, given by $\mathcal{A} = [c_1, c_3] \cup [c_2, c]$. In Fig. C.6b it is shown that, for the same parameter value $\mu = 3.61$ as in Fig. C.6a, the numerical iteration of the logistic map gives points that are trapped inside the two-cyclic interval $\mathcal{A}$.

Inside an absorbing region one or more attractors may exist. However, if a chaotic attractor exists which fills up a whole absorbing region then the boundary of the chaotic attractor is formed by portions of critical sets. This is the situation shown in Fig. C.6a, b, where the absorbing interval $\mathcal{A} = [c_1, c_3] \cup [c_2, c]$ is invariant and filled up by a chaotic trajectory, as shown in Fig. C.6b.

To better illustrate the foregoing point, we also give a two-dimensional example, obtained by using the map (C.5). In Fig. C.7a, a chaotic trajectory is shown, and in Fig. C.7b its outer boundary is obtained by the union of a segment of $LC$ and three iterates $LC_i = T^i(LC)$, $i = 1, 2, 3$.

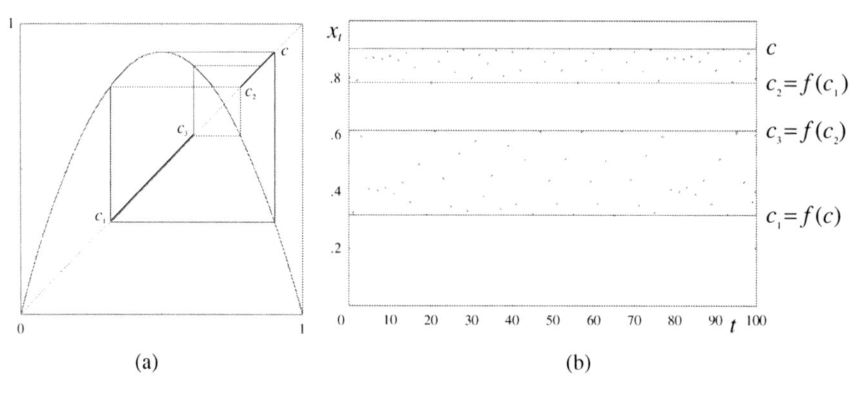


**Fig. C.6** Illustrating a trapping subset inside the absorbing set of Fig. C.5 for the quadratic map with $\mu = 3.61$. (**a**) The delineation of the trapping subset $[c_1, c_3] \cup [c_2, c]$. (**b**) The iterates of the map remain trapped inside the two cyclic interval

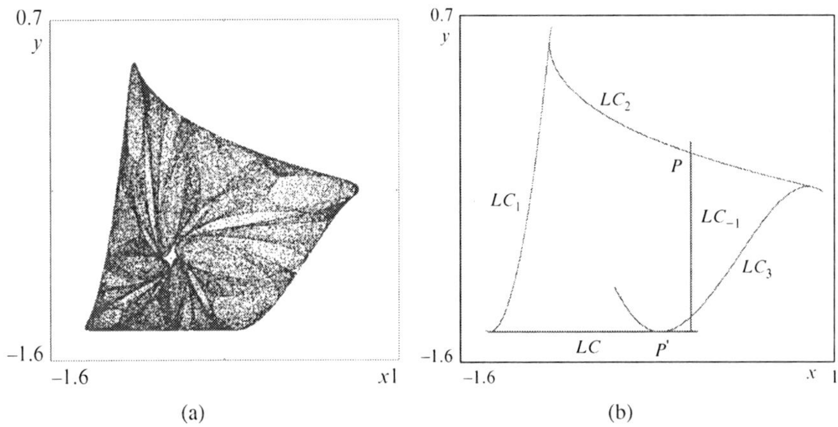


**Fig. C.7** Delineating the absorbing area of the two-dimensional map given in (C.5). Here $a = -0.3$ and $b = -1.4$. (**a**) A chaotic trajectory of the map. (**b**) The boundary of the absorbing area formed by the critical line and three of its iterates. The location of the starting line $LC_{-1}$ is discussed in the text

Indeed, following Mira et al. (1996) (see also Bischi and Gardini (1998)) a practical procedure can be outlined for obtaining the boundary of an absorbing area (although it is difficult to give a general method). Starting from a portion of $LC_{-1}$, approximately located in the region occupied by the area of interest, compute its images under $T$ of increasing rank until a closed region is obtained. When such a region is mapped into itself, then it is an absorbing area $\mathcal{A}$. The length of the initial segment is to be set, in general, by a trial and error method, although several suggestions are given in the books referenced above. Once an absorbing area $\mathcal{A}$ is found, in order to see if it is invariant or not the same procedure must be repeated by taking only the portion

$$\gamma = \mathcal{A} \cap LC_{-1} \tag{C.12}$$

as the starting segment. Then one of the following two cases occurs:

Case 1. The union of $m$ iterates of $\gamma$ (for a suitable $m$) covers the whole boundary of $\mathcal{A}$; in which case $\mathcal{A}$ is an invariant absorbing area, and

$$\partial\mathcal{A} \subset \bigcup_{k=1}^{m} T^k(\gamma) \tag{C.13}$$

Case 2. No natural $m$ exists such that $\bigcup_{k=1}^{m} T^k(\gamma)$ covers the whole boundary of $\mathcal{A}$; in which case $\mathcal{A}$ is not invariant but strictly mapped into itself. An invariant absorbing area is obtained by $\cap_{n>0} T^n(\mathcal{A})$ (and may be obtained by a finite number of images of $\mathcal{A}$).

The application of this procedure to the problem of the delineation of the chaotic area of Fig. C.7a by portions of critical curves suggests, on the basis of Fig. C.7b,

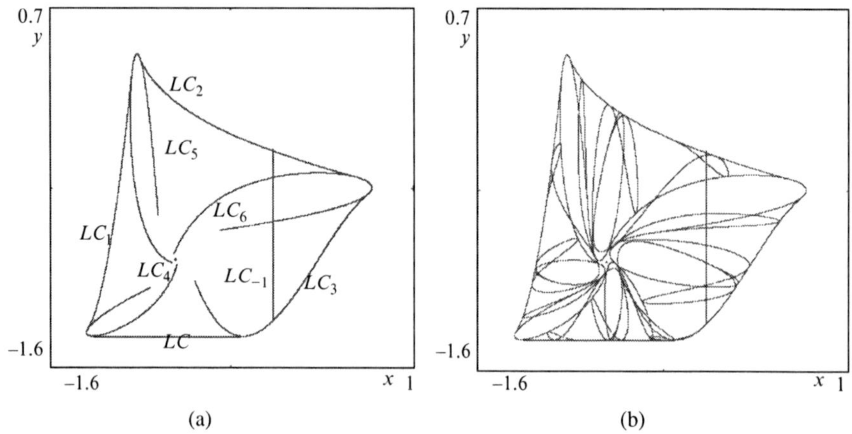


**Fig. C.8** Delineating more precisely the structure of the absorbing area of the quadratic map given by (C.5). (**a**) Higher order iterates of the boundary curves. (**b**) After a sufficient number of further iterates the inner boundaries of the chaotic area emerge. These should be compared with the frequently visited areas of the chaotic trajectory in Fig. C.7a

that we take a smaller segment $\gamma$ and that we take a higher number of iterates in order to also obtain the inner boundary. The result is shown in Fig. C.8a, where after four iterates we get the outer boundary. After a few more iterates the inner boundary of the chaotic area is also obtained, as shown in Fig. C.8b. As can be clearly seen, and as clearly expressed by the strict inclusion in (C.13), the union of the images also include several arcs internal to the invariant area $\mathcal{A}$. Indeed, the images of the critical arcs which are mapped inside the area play a particular role, because these curves represent the "foldings" of the plane under forward iterations of the map, and this is the reason why these inner curves often denote the portions of the region which are more frequently visited by a generic trajectory inside it (compare Fig. C.7a and C.8b). This is due to the fact that points close to a critical arc $LC_i$, $i \geq 0$, are more frequently visited, because there are several distinct parts of the invariant area which are mapped into the same region (close to $LC_i$) in $i+1$ iterations. Many similar examples are given in the literature on noninvertible maps, see for example Mira et al. (1996).

Examples of applications in dynamic economic modeling are given in Bischi and Naimzada (1999), Bischi et al. (2000a), Puu (2003), Agliari et al. (2000a), Agliari et al. (2000b), Agliari et al. (2002b), Agliari et al. (2004), Chiarella et al. (2001), Chiarella et al. (2002), and Sushko et al. (2003).

## C.4 Critical Sets and the Creation of Disconnected Basins

From (C.11) it is clear that the properties of the inverses are important in order to understand the structure of the basins and the main bifurcations that change their qualitative properties. In the case of noninvertible maps, the multiplicity of preimages may lead to basins with complex structures, such as multiply connected or disconnected sets, sometimes formed by infinitely many disconnected portions (see Mira et al. (1994), Mira and Rauzy (1995), Mira et al. (1996), Chap. 5 and Abraham et al. (1997), Chap. 5). In the context of noninvertible maps it is useful to define the *immediate basin* $\mathcal{B}_0(A)$, of an attracting set $A$, as the largest connected component of the basin that contains $A$. Then the total basin can be expressed as

$$\mathcal{B}(A) = \bigcup_{n=0}^{\infty} T^{-n}(\mathcal{B}_0(A))$$

where $T^{-n}(x)$ represents the set of all the rank-$n$ preimages of $x$, in other words the set of points which are mapped into $x$ after $n$ iterations of the map $T$. The backward iteration of a noninvertible map *repeatedly unfolds* the phase space, and this implies that the basins may be disconnected, that is they are formed by several disjoint portions. Also in this case, we first illustrate this property by using a one-dimensional map based on an evolutionary game proposed in Bischi et al. (2003b). In Fig. C.9a, b the graph of a $Z_1 - Z_3 - Z_1$ noninvertible map is shown, where $Z_3$ is the portion of the co-domain bounded by the relative minimum value $c_{\min}$ and the relative maximum value $c_{\max}$. In the situation shown in Fig. C.9a we have three attractors: the fixed point $z^*$, with $\mathcal{B}(z^*) = (-\infty, q^*)$, the attractor $A$ around $x^*$, with basin $\mathcal{B}(A) = (q^*, r^*)$ bounded by two unstable fixed points, and $+\infty$ (attracting positively diverging trajectories) with basin $\mathcal{B}(+\infty) = (r^*, +\infty)$. In this case all the basins are immediate basins, each being given by an open interval. In the situation shown in Figure C.9(a), both basin boundaries $q^*$ and $r^*$ are in $Z_1$,

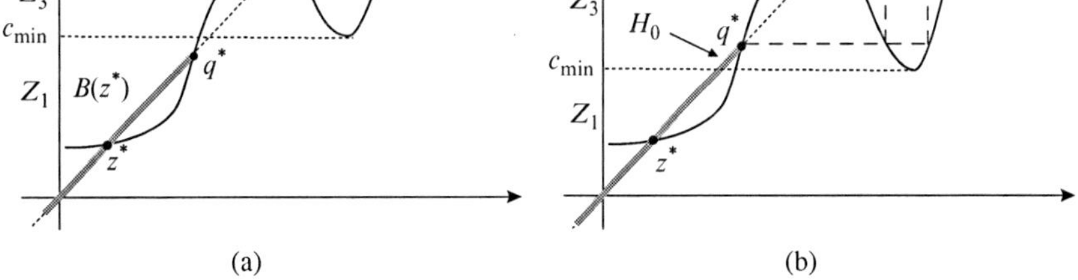


**Fig. C.9** The global bifurcation of a one-dimensional noninvertible $Z_1 - Z_3 - Z_1$, map. (**a**) The attractors of the map are $z^*$, $x^*$ and $+\infty$, and their basins are $(-\infty, q^*)$, $(q^*, r^*)$ and $(r^*, \infty)$ respectively. Note that $c_{\min}$ is above $q^*$. (**b**) After a parametric change $c_{\min}$ moves below $q^*$ and a global bifurcation has occurred. Now the basin of $z^*$ includes the (countably infinite number of) disconnected portions, $H_{-1}$, $H_{-2}$ etc. on $(x^*, r^*)$. These are the preimages of the portion $(c_{\min}, q^*)$

so they have only themselves a unique preimage (as for an invertible map). However, the situation drastically changes if, for example, some parameter change causes the minimum value $c_{\min}$ to move downwards sufficiently that it goes below $q^*$ (as in Fig. C.9b). After the global bifurcation, which occurs when $c_{\min} = q^*$, the portion $(c_{\min}, q^*)$ enters $Z_3$, so new preimages $f^{-k}(c_{\min}, q^*)$ appear with $k \geq 1$. These preimages constitute an infinite (countable) set of disconnected portions of $\mathcal{B}(z^*)$ nested inside $\mathcal{B}(A)$, represented by the thick portions of the diagonal in Fig. C.9b, bounded by the infinitely many preimages of any rank, say $q^*_{-k}$, $k \in \mathbb{N}$, of $q^*$, that accumulate in a left neighborhood of the fixed point $r^*$. In fact, as $r^*$ is a repelling fixed point for the forward iteration of $f$, it is an attracting fixed point for the backward iteration of the same map. So, the contact between the critical point $c_{\min}$ and the basin boundary $q^*$ marks the transition from simple connected to disconnected basins. Similar global bifurcations, due to contacts between critical sets and basin boundaries, also occur in higher dimensional maps.

Also in higher dimensional cases, the global bifurcations which give rise to complex topological structures of the basins, like those formed by disconnected sets, can be explained in terms of contacts of basin boundaries and critical sets. In fact, if a parameter variation causes a crossing between a basin boundary and a critical set which separates different regions $Z_k$ so that a portion of a basin enters a region where an higher number of inverses is defined, then new components of the basin may suddenly appear at the contact. However, for maps of dimension greater than 1, such kinds of bifurcations can be very rarely studied by analytical methods, since the analytical equations of such singularities are not known in general. Hence such studies are mainly performed by geometric and numerical methods.

Several examples of two-dimensional noninvertible maps that have disconnected basins can be found in this book. See also Agliari et al. (2000a, b), Agliari et al. (2002b), Agliari et al. (2004), Bischi and Kopel (2001), Bischi and Kopel (2003*a*), Bischi and Naimzada (1999), Bischi et al. (2000a), Bischi et al. (2003b), Puu (2003). Examples in three dimensions are given in Agliari et al. (2000b) and Bischi et al. (2001b).

# Appendix D
# Continuously Distributed Time Lags

Continuous time dynamical systems with continuously distributed time lags are frequently modeled with Volterra-type integro-differential equations, when some or all state variables in the usual differential equation model are replaced by certain averages of past values. If $x(t)$ is such a variable then its weighted average is

$$\bar{x}(t) = \int_0^t w(t-s, T, m)x(s)ds, \tag{D.1}$$

where the weighting function is of the form

$$w(t-s, T, m) = \begin{cases} \frac{1}{T}e^{-\frac{t-s}{T}} & \text{if } m = 0 \\ \frac{1}{m!}\left(\frac{m}{T}\right)^{m+1}(t-s)^m e^{-\frac{m(t-s)}{T}} & \text{if } m \geq 1. \end{cases} \tag{D.2}$$

Here $m$ is a non-negative integer and $T$ is a positive real parameter.

First we will examine some fundamental properties of this special weighting function.

(a) The area under the weighting function converges to 1 as $t \to \infty$.
For $m = 0$ we have

$$\int_0^t \frac{1}{T}e^{-\frac{t-s}{T}}ds = \left[\frac{1}{T}\cdot\frac{e^{-\frac{t-s}{T}}}{\frac{1}{T}}\right]_0^t = 1 - e^{-\frac{t}{T}},$$

and for $m \geq 1$ by introducing the new integration variable $x = m(t-s)/T$ we have

$$\begin{aligned}\int_0^t \frac{1}{m!}\left(\frac{m}{T}\right)^{m+1}(t-s)^m e^{-\frac{m(t-s)}{T}}ds &= \int_0^{\frac{mt}{T}} \frac{1}{m!}\left(\frac{m}{T}\right)^{m+1}\left(\frac{Tx}{m}\right)^m e^{-x}\frac{Tdx}{m} \\ &= \frac{1}{m!}\int_0^{\frac{mt}{T}} x^m e^{-x}dx.\end{aligned}$$

Notice that the integral converges to $\Gamma(m+1) = m!$, so the right hand side converges to 1.

(b) For $m = 0$, weights are exponentially declining with the most weight given to the most current data. For $m \geq 1$, zero weight is given to the most current data, rising to maximum at $s = t - T$ and declining exponentially thereafter.
For $m = 0$, the weighting function is a declining exponential function of $(t-s)$. For $m \geq 1$,

$$\begin{aligned}\frac{d}{ds}w(t-s,T,m) &= \frac{1}{m!}\left(\frac{m}{T}\right)^{m+1} e^{-\frac{m(t-s)}{T}}\left(-m(t-s)^{m-1} + (t-s)^m\frac{m}{T}\right)\\ &= \frac{1}{(m-1)!}\left(\frac{m}{T}\right)^{m+1} e^{-\frac{m(t-s)}{T}}(t-s)^{m-1}\left(\frac{t-s}{T}-1\right)\end{aligned}$$

which is negative for $t - s < T$, positive for $t - s > T$, and zero if $t - s = T$.

(c) As $m$ increases, the weighting function becomes more peaked around $t-s = T$, and as $m \to \infty$, the weighting function converges to the Dirac delta function centered at $s = t - T$.
This property can be easily proved by examining the ratio

$$w(t-s,T,m+1)/w(t-s,T,m)$$

with fixed $T$ and $t - s$.

(d) As $T \to 0$, the weighting function tends to the Dirac delta function with all $m \geq 0$.
Notice that the weighting function is the product of a polynomial and a decreasing exponential function of $\frac{1}{T}$ unless $t - s = T$.

Figures D.1 and D.2 show the plots of the weighting function with changing values of $m$ and $T$.

In analyzing continuous time systems with continuously distributed time lags, integrals of the form

$$\frac{1}{e^{\lambda t}}\int_0^t w(t-s,T,m)e^{\lambda s}ds \tag{D.3}$$

often arise. Notice first that by introducing the new variable $x = t - s$, (D.3) can be simplified as

$$\frac{1}{e^{\lambda t}}\int_0^t w(x,T,m)e^{(t-x)\lambda}dx = \int_0^t w(x,T,m)e^{-\lambda x}dx.$$

If $m = 0$, then we have

$$\int_0^t \frac{1}{T}e^{-\frac{x}{T}}e^{-\lambda x}dx = \frac{1}{T}\left[\frac{-e^{-x(\lambda+\frac{1}{T})}}{\lambda+\frac{1}{T}}\right]_0^t = (1+\lambda T)^{-1}(1-e^{-t(\lambda+\frac{1}{T})}),$$

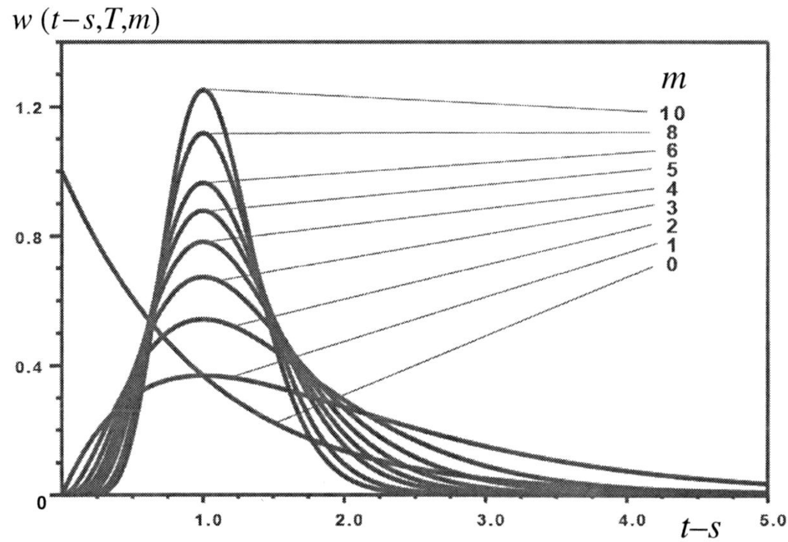


**Fig. D.1** Dependence of $w$ on $m$ is the case of $T = 1$

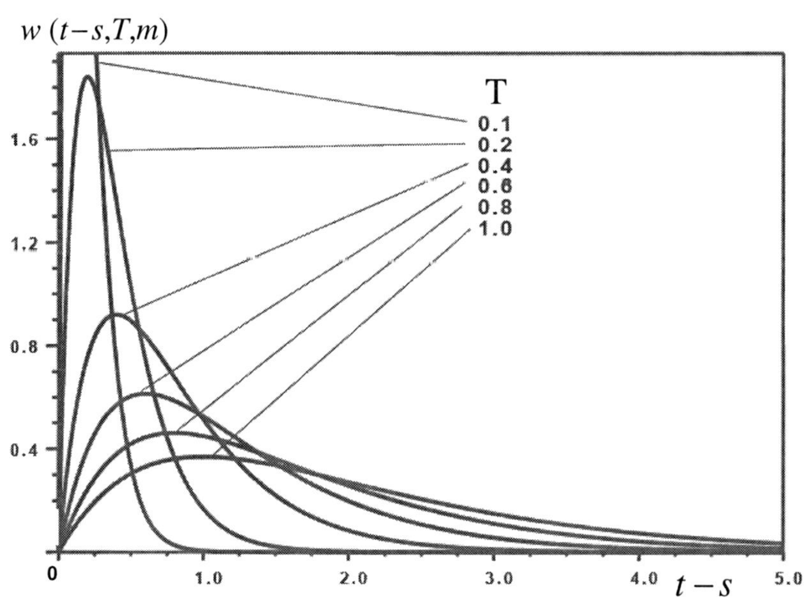


**Fig. D.2** Dependence of $w$ on $T$ is the case of $m = 1$

which converges to $(1+\lambda T)^{-1}$ if we assume that $\mathrm{Re}\lambda + \frac{1}{T} > 0$. If $m > 1$, then the integral becomes

$$\int_0^t \frac{1}{m!}\left(\frac{m}{T}\right)^{m+1} x^m e^{-\frac{mx}{T}} e^{-\lambda x}dx = \int_0^{(\lambda+\frac{m}{T})t} \frac{1}{m!}\left(\frac{m}{T}\right)^{m+1}\left(\frac{z}{\lambda+\frac{m}{T}}\right)^m e^{-z}\frac{dz}{\lambda+\frac{m}{T}}$$
$$= \frac{1}{m!}\left(\int_0^{(\lambda+\frac{m}{T})t} z^m e^{-z}dz\right)\left(1+\frac{\lambda T}{m}\right)^{-(m+1)},$$

where we have introduced the new variable $z = (1+\frac{m}{T})x$. If $\mathrm{Re}\lambda + \frac{m}{T} > 0$, then the integral term always converges to $\Gamma(m+1) = m!$, so the entire expression tends to $(1+\frac{\lambda T}{m})^{-(m+1)}$.

Finally we will demonstrate that Volterra-type integro-differential equations with (D.1)-type integral terms can be rewritten as systems of ordinary differential equations by introducing additional state variables. Therefore all tools known from the stability theory of ordinary differential equations can be used to analyze the asymptotic behavior of the equilibrium of such dynamical systems.

Consider first the case of $m = 0$. Introduce the new state variable

$$X_0(t) = \int_0^t \frac{1}{T}e^{-\frac{(t-s)}{T}}x(s)ds, \tag{D.4}$$

then simple differentiation shows that

$$\dot{X}_0(t) = \frac{1}{T}[x(t) - X_0(t)]. \tag{D.5}$$

By replacing the integral of the form (D.1) with $X_0(t)$ in the integro-differential equation and adding the additional equation (D.4) this integral term disappears from the equations describing the dynamical system.

Assume next that $m \geq 1$. Then introduce the new state variables

$$X_k^{(m)}(t) = \int_0^t \frac{1}{k!}\left(\frac{m}{T}\right)^{k+1}(t-s)^k e^{-\frac{m(t-s)}{T}}x(s)ds \qquad (0 \leq k \leq m). \tag{D.6}$$

Then simple differentiation shows that for $k = 1, 2, \ldots, m$,

$$\dot{X}_k^{(m)}(t) = \frac{m}{T}[X_{k-1}^{(m)}(t) - X_k^{(m)}(t)] \tag{D.7}$$

and

$$\dot{X}_0^{(m)}(t) = \frac{m}{T}[x(t) - X_0^{(m)}(t)]. \tag{D.8}$$

Therefore by replacing integral (D.1) with the new state variable $X_m^{(m)}(t)$ in the integro-differential equation, adding the new variables $X_0^{(m)}(t), \ldots, X_m^{(m)}(t)$ and

(D.7) and (D.8) to the system, the integral term disappears again from the equations describing the dynamical system.

After we repeat the above procedure for all integral terms, a larger system of ordinary differential equations is obtained which is clearly equivalent to the original system of Volterra-type integro-differential equations.

# Appendix E
# A Determinantal Identity

In analysing the local asymptotic stability of discrete dynamic oligopolies the eigenvalue equation of the associated Jacobians have to be determined. The Jacobians have similar special structure which allows us to give a simple representation of their characteristic polynomials.

Our method is based on the following simple identity.

**Lemma E.1.** *Let $\boldsymbol{a}, \boldsymbol{b}, \in \mathbb{R}^N$ be two real column vectors, then*

$$det(\boldsymbol{I} + \boldsymbol{a}\boldsymbol{b}^T) = 1 + \boldsymbol{a}\boldsymbol{b}^T. \tag{E.1}$$

*Proof.* Let $D_N$ denote this determinant. We will use finite induction with respect to $N$ to prove identity (E.1). If $N = 1$, then

$$D_1 = det(1 + a_1 b_1) = 1 + a_1 b_1$$

so (E.1) clearly holds. If $N > 1$, then with the notation $\boldsymbol{a} = (a_i)$ and $\boldsymbol{b} = b_i$ we have

$$D_N = det \begin{pmatrix} 1 + a_1 b_1 & a_1 b_2 & \dots & a_1 b_N \\ a_2 b_1 & 1 + a_2 b_2 & \dots & a_2 b_N \\ \vdots & \vdots & & \vdots \\ a_N b_1 & a_N b_2 & \dots & 1 + a_N b_N \end{pmatrix}.$$

Subtract the $a_N/a_{N-1}$–multiple of row $N-1$ from the last row, then the $a_{N-1}/a_{N-2}$ –multiple of row $N-2$ from row $N-1$, and so on, and finally subtract the $a_2/a_1$-multiple of row 2 from the first row. Then the value of the determinant remains the same, so

$$D_N = det \begin{pmatrix} 1 + a_1 b_1 & a_1 b_2 & \dots & a_1 b_{N-1} & a_1 b_N \\ -\frac{a_2}{a_1} & 1 & & & \\ & -\frac{a_3}{a_2} & & & \\ & & \ddots & & \\ & & & 1 & \\ & & & -\frac{a_N}{a_{N-1}} & 1 \end{pmatrix},$$

where all other matrix elements are zeros. Expanding this determinant with respect to its last column we obtain a recursive relation

$$D_N = D_{N-1} \cdot 1 + (-1)^{N-1} a_1 b_N \left(-\frac{a_2}{a_1}\right)\left(-\frac{a_3}{a_2}\right) \dots \left(-\frac{a_N}{a_{N-1}}\right) \quad = D_{N-1} + a_N b_N$$

completing the proof. $\square$

Two particular applications of identity (E.1) will be shown next.
Consider first the determinant with the simple structure

$$Det = det \begin{pmatrix} A_1(\lambda) & B_1(\lambda) & \dots & B_1(\lambda) \\ B_2(\lambda) & A_2(\lambda) & \dots & B_2(\lambda) \\ \vdots & \vdots & & \vdots \\ & & \ddots & \\ B_N(\lambda) & B_N(\lambda) & \dots & A_N(\lambda) \end{pmatrix}. \tag{E.2}$$

By introducing vectors $\boldsymbol{b}(\lambda) = (B_1(\lambda), \dots, B_N(\lambda))^T, \mathbf{1}^T = (1, \dots, 1)$ and the diagonal matrix $\boldsymbol{D}(\lambda) = diag(A_1(\lambda) - B_1(\lambda), \dots, A_N(\lambda) - B_N(\lambda))$ we can rewrite the determinant as

$$det(\boldsymbol{D}(\lambda) + \boldsymbol{b}(\lambda) \cdot \mathbf{1}^T) = det(\boldsymbol{D}(\lambda) \cdot det(\boldsymbol{I} + \boldsymbol{D}^{-1}(\lambda)\boldsymbol{b}(\lambda)\mathbf{1}^T).$$

The first determinant is diagonal, the second has the special structure of (E.1) with $\boldsymbol{a} = \boldsymbol{D}^{-1}(\lambda)\boldsymbol{b}(\lambda)$ and $\boldsymbol{b}^T = \mathbf{1}^T$. So by using identity (E.1) we have

$$Det = \prod_{k=1}^{N} (A_k(\lambda) - B_k(\lambda)) \cdot \left[ 1 + \sum_{k=1}^{N} \frac{B_k(\lambda)}{A_k(\lambda) - B_k(\lambda)} \right]. \tag{E.3}$$

Consider next a special matrix

$$\boldsymbol{A} = \begin{pmatrix} a_1 & b_1 & \dots & b_1 \\ b_2 & a_2 & \dots & b_2 \\ \vdots & \vdots & & \vdots \\ b_N & b_N & \dots & a_N \end{pmatrix}. \tag{E.4}$$

The characteristic polynomial of this matrix can be determined by using relation (E.3). Notice that

$$det(\boldsymbol{A} - \lambda \boldsymbol{I}) = det \begin{pmatrix} a_1 - \lambda & b_1 & \dots & b_1 \\ b_2 & a_2 - \lambda & \dots & b_2 \\ \vdots & \vdots & & \vdots \\ b_N & b_N & \dots & a_N - \lambda \end{pmatrix},$$

which is the special case of (E.2) by selecting $A_k(\lambda) = a_k - \lambda$ and $B_k(\lambda) = b_k$.

Therefore relation (E.3) gives the characteristic polynomial of matrix $\boldsymbol{A}$ as

$$det(\boldsymbol{A} - \lambda \boldsymbol{I}) = \prod_{k=1}^{N} (a_k - b_k - \lambda) \cdot \left[ 1 + \sum_{k=1}^{N} \frac{b_k}{a_k - b_k - \lambda} \right]. \tag{E.5}$$

# Appendix F
# Stable Quadratic Polynomials

Consider the quadratic polynomial

$$\lambda^2 + p\lambda + q = 0 \tag{F.1}$$

with real coefficients.

In examining the asymptotic stability of two-dimensional dynamical systems the following result can be used.

**Lemma F.1.** *All roots of (F.1) are inside the unit circle if and only if*

$$1 + p + q > 0$$
$$1 - p + q > 0$$

*and*

$$q < 1.$$

**Proof.** The roots are

$$\lambda_{1,2} = \frac{-p \pm \sqrt{p^2 - 4q}}{2}. \tag{F.2}$$

Assume first complex roots. This is the case when $q > p^2/4$, then

$$\lambda_{1,2} = -\frac{p}{2} \pm i\frac{\sqrt{4q - p^2}}{2},$$

So $|\lambda_{1,2}| < 1$ if and only if

$$\frac{p^2}{4} + \frac{4q - p^2}{4} = q < 1.$$

Assume next that the roots are real. Then $q \leq p^2/4$, and $|\lambda_{1,2}| < 1$ if and only if

$$-2 + p < \pm\sqrt{p^2 - 4q} < 2 + p. \tag{F.3}$$

Clearly $-2 < p < 2$, otherwise this relation cannot hold for both signs of the square root. Notice that (F.3) is equivalent to

$$\sqrt{p^2 - 4q} < \min\{2 + p, 2 - p\},$$

or equivalently

$$p^2 - 4q < \min\{4 + 4p + p^2, 4 - 4p + p^2\},$$

which can be rewritten as

$$1 + p + q > 0 \quad \text{and} \quad 1 - p + q > 0.$$

The cases of real and complex roots are shown in Figure F.1. The assertion can be obtained by combining the two cases. ■

Consider next a real matrix

$$\begin{pmatrix} a & b \\ c & d \end{pmatrix},$$

the characteristic polynomial of which is

$$\varphi(\lambda) = (a - \lambda)(d - \lambda) - bc = \lambda^2 - \lambda(a + d) + (ad - bc).$$

Let $Tr = a + d$ denote the trace and $Det = ad - bc$ the determinant of this matrix, then

$$\varphi(\lambda) = \lambda^2 - \lambda Tr + Det.$$

The eigenvalues of this matrix are inside the unit circle if and only if

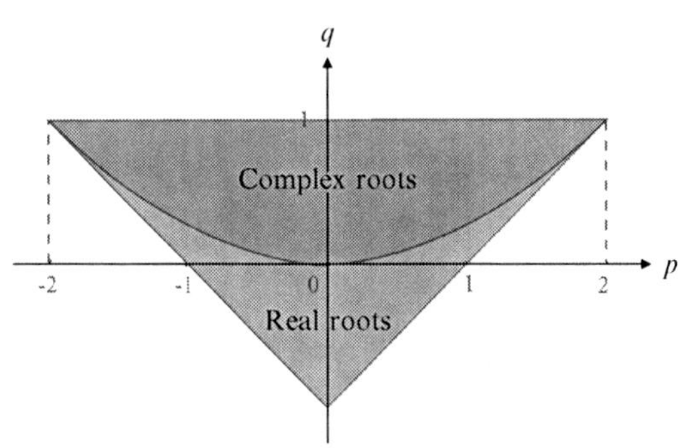


**Fig. F.1** The stability region (shaded) of the quadratic polynomial (F.1) in the $(p, q)$ plane. It shows the bounding lines $1 + p + q > 0$, $1 - p + q > 0$ and $q < 1$. Also shown are the regions where the roots of (F.1) are real and where they are complex, with the boundary between the two regions being the parabola $q = p^2/4$.

$$
\begin{aligned}
1 + Tr + Det &> 0, \\
1 - Tr + Det &> 0, \\
Det &< 1.
\end{aligned}
$$

The continuous time counterpart of Lemma F.1 can be formulated in the following way.

**Lemma F.2.** *The roots of (F.1) have negative real parts if and only if $p, q > 0$.*

**Proof.** Assume first that the roots are complex, $\lambda_{1,2} = a \pm ib$ with $a < 0$. Then $p = -(\lambda_1 + \lambda_2) = -2a > 0$ and $q = \lambda_1\lambda_2 = a^2 + b^2 > 0$. If the roots are real and negative, then $p = -(\lambda_1 + \lambda_2) > 0$ and $q = \lambda_1\lambda_2 > 0$.

Assume next that $p, q > 0$. If the roots are complex, then $Re\lambda_{1,2} = -p/2 < 0$. If the roots are real, then from (F.2), both roots are negative, since $\sqrt{p^2 - 4q} < p$.

■

# References


Abraham, R., Gardini, L., & Mira, C. (1997). *Chaos in discrete dynamical systems (A visual introduction in two dimension)*. Berlin: Springer.

Agliari, A. (2006). Homoclinic connections and subcritical Neimark bifurcation in a duopoly model with adaptively adjusted productions. *Chaos, Solitons & Fractals*, *29*, 739–755.

Agliari, A., Bischi, G. I., & Gardini, L. (2002a). Some methods for the global analysis of dynamic games represented by noninvertible maps. *In* T. Puu & I. Sushko (Eds.), *Oligopoly dynamics: Models and tools* (pp. 31–83). Berlin: Springer. Chapter 3.

Agliari, A., Chiarella, C., & Gardini, L. (2004). A stability analysis of the perfect foresight map in nonlinear models of monetary dynamics. *Chaos, Solitons & Fractals*, *21*(2), 371–386.

Agliari, A., Gardini, L., Delli Gatti, D., & Gallegati, M. (2000a). Global dynamics in a nonlinear model for the equity ratio. *Chaos, Solitons & Fractals*, *11*, 961–985.

Agliari, A., Gardini, L., & Puu, T. (2000b). The dynamics of a triopoly Cournot game. *Chaos, Solitons & Fractals*, *11*, 2531–2560.

Agliari, A., Gardini, L., & Puu, T. (2002b). Global bifurcations of basins in a triopoly game. *International Journal of Bifurcation and Chaos*, *12*(10), 2175–2207.

Agliari, A., Gardini, L., & Puu, T. (2005). Some global bifurcations related to the appearance of closed invariant curves. *Mathematics and Computers in Simulation* , *68*(3), 201–219.

Agliari, A., Gardini, L., & Puu, T. (2006). Global bifurcation in duopoly when the Cournot point is destabilized via a subcritical Neimark bifurcation. *International Game Theory Review*, *8*, 1–20.

Al-Nowaihi, A., & Levine, P. (1985). The stability of the Cournot oligopoly model: A reassessment. *Journal of Economic Theory*, *35*, 307–321.

Amir, R., & Lambson, V. (2000). On the effects of entry in Cournot markets. *Review of Economic Studies*, *67*, 235–254.

Arnold, V., Varchenko, A., & Goussein-Zadé, S. (1986). *Singularités des applications différentiables*. Moscow: Editions MIR.

Avrutin, V., & Schanz, M. (2006). Multi-parametric bifurcations in a scalar piecewise-linear map. *Nonlinearity*, *19*, 531–552.

Avrutin, V., Schanz, M., & Banerjee, S. (2006). Multi-parametric bifurcations in a piecewise-linear discontinuous map. *Nonlinearity*, *19*, 1875–1906.

Balch, M. (1971). Oligopolies, advertising, and non-cooperative games. *In* H. Kuhn & G. Szego (Eds.), *Differential games and related topics* (pp. 301–311). New York: North-Holland.

Banerjee, S., Karthik, M., Yuan, G., & Yorke, J. (2000a). Bifurcations in one-dimensional piecewise smooth maps – theory and applications in switching circuits. *IEEE Transactions on Circuits and Systems – I: Fundamental Theory and Applications*, *47*(3), 389–394.

Banerjee, S., Ranjan, P., & Grebogi, C. (2000b). Bifurcations in two-dimensional piecewise smooth maps – theory and applications in switching circuits. *IEEE Transactions on Circuits and Systems – I: Fundamental Theory and Applications*, *47*(5), 633–643.

Baumol, W., & Quandt, R. (1964). Rules of thumb and optimally imperfect decisions. *The American Economic Review*, *54*, 23–46.

Bellman, R. (1969). *Stability theory of differential equations*. New York: Dover.

Bischi, G. I., Carini, R., Gardini, L., & Tenti, P. (2004a). *Sulle orme del caos. Comportamenti complessi in modelli matematici semplici*. Torino: Bruno Mondadori Editore.

Bischi, G. I., Chiarella, C., & Kopel, M. (2004b). The long run outcomes and global dynamics of a duopoly game with misspecified demand functions. *International Game Theory Review*, *6*(3), 343–380.

Bischi, G. I., Dawid, H., & Kopel, M. (2003b). Spillover effects and the evolution of firm clusters. *Journal of Economic Behavior and Organization*, *50*, 47–75.

Bischi, G. I., & Gardini, L. (1998). Role of invariant and minimal absorbing areas in chaos synchronization. *Physical Review E*, *58*, 5710–5719.

Bischi, G. I., Gardini, L., & Kopel, M. (2000a). Analysis of global bifurcations in a market share attraction model. *Journal of Economic Dynamics and Control*, *24*, 855–879.

Bischi, G. I., Gardini, L., & Mira, C. (1999). Plane maps with denominator. Part I: Some generic properties. *International Journal of Bifurcation and Chaos* , *9*(1), 119–153.

Bischi, G. I., Gardini, L., & Mira, C. (2001a). Maps with a vanishing denominator. A survey of some results. *Nonlinear Analysis*, *47*(4), 2171–2185.

Bischi, G. I., Gardini, L., & Mira, C. (2003a). Plane maps with denominator. Part II: Noninvertible maps with simple focal points. *International Journal of Bifurcation and Chaos*, *13*(8), 2253–2277.

Bischi, G. I., & Kopel, M. (2001). Equilibrium selection in a nonlinear duopoly game with adaptive expectations. *Journal of Economic Behavior and Organization*, *46*(1), 73–100.

Bischi, G. I., & Kopel, M. (2003*a*). Long run evolution, path dependence and global properties of dynamic games: A tutorial. *CUBO Mathematical Journal* , *5*(3), 437–468.

Bischi, G. I., & Kopel, M. (2003*b*). Multistability and path dependence in a dynamic brand competition model. *Chaos, Solitons & Fractals*, *18*, 561–576.

Bischi, G. I., Kopel, M., & Szidarovszky, F. (2005). Expectation-stock dynamics in multi-agent fisheries. *In Contributions to the Theory of Games* (Vol. 137, pp. 299–329) of *Annals of Operations Research*.

Bischi, G. I., & Lamantia, F. (2002). Nonlinear duopoly games with positive cost externalities due to spillover effects. *Chaos, Solitons & Fractals*, *13*, 808–822.

Bischi, G. I., Mammana, C., & Gardini, L. (2000b). Multistability and cyclic attractors in duopoly games. *Chaos, Solitons & Fractals*, *11*, 543–564.

Bischi, G. I., Mroz, L., & Hauser, H. (2001b). Studying basin bifurcations in nonlinear triopoly games by using 3D visualization. *Nonlinear Analysis*, *47*(8), 5325–5341. (Special issue WCNA 2000).

Bischi, G. I., & Naimzada, A. (1999). Global analysis of a duopoly game with bounded rationality. *In* J. A. Filar, V. Gaitsgory & M. Koichi (Eds.), *Advances in dynamic games and applications* (Vol. 5, pp. 361–385). Boston: Birkhauser.

Bischi, G. I., Naimzada, A., & Sbragia, L. (2007). Oligopoly games with local monopolistic approximation. *Journal of Economic Behavior and Organization*, *62*, 371–388.

Bischi, G. I., Sbragia, L., & Szidarovszky, F. (2008). Learning the demand function in a repeated Cournot oligopoly game. *International Journal of Systems Science*, *39*(4), 403–419.

Bonanno, G., & Zeeman, C. (1985). Limited knowledge of demand and oligopoly equilibria. *Journal of Economic Theory*, *35*, 276–283.

Bresnahan, T. F. (1986). Quantifying the competitive effects of production joint ventures. *International Journal of Industrial Organization*, *4*, 155–175.

Brock, W., & Hommes, C. (1997). A rational route to randomness. *Econometrica*, *65*(5), 1059–1095.

Brousseau, V., & Kirman, A. (1993). The dynamics of learning in N-person games with the wrong N. *In* K. Binmore, A. Kirman, & P. Tani (Eds.), *Frontiers of game theory* (pp. 71–93). Cambridge, Massachusetts: MIT.

Bulow, J., Geanakoplos, J., & Klemperer, P. (1985*a*). Holding idle capacity to deter entry. *The Economic Journal*, *95*, 178–182.

Bulow, J., Geanakoplos, J., & Klemperer, P. (1985*b*). Multimarket oligopoly: Strategic substitutes and complements. *Journal of Political Economy*, *93*, 488–511.

Chiarella, C., Dieci, R., & Gardini, L. (2001). Asset price dynamics in a financial market with fundamentalists and chartists. *Discrete Dynamics in Nature and Society*, *6*, 69–99.

Chiarella, C., Dieci, R., & Gardini, L. (2002). Speculative behavior and complex asset price dynamics: A global analysis. *Journal of Economic Behavior and Organization, 49*(2), 173–197.

Chiarella, C., & Khomin, A. (1996). An analysis of the complex dynamic behaviour of nonlinear oligopoly models with time lags. *Chaos, Solitons & Fractals, 7*(12), 2049–2065.

Chiarella, C., & Szidarovszky, F. (2001*a*). The birth of limit cycles in nonlinear oligopolies with continuously distributed information lags. *In* M. Dror, P. L'Ecyer, & F. Szidarovszky (Eds.), *Modelling Uncertainty* (pp. 249–268). Dordrecht: Kluwer.

Chiarella, C., & Szidarovszky, F. (2001*b*). The nonlinear Cournot model under uncertainty with continuously distributed time lags. *Central European Journal of Operations Research, 9*(3), 183–196.

Chiarella, C., & Szidarovszky, F. (2002). The asymptotic behavior of dynamic rent-seeking games. *Computers and Mathematics with Applications, 43*, 169–178.

Chiarella, C., & Szidarovszky, F. (2004). Dynamic oligopolies without full information and with continuously distributed time lags. *Journal of Economic Behavior and Organization, 54*(4), 495–511.

Chiarella, C., & Szidarovszky, F. (2005). The complex asymptotic behavior of dynamic oligopolies with partially cooperating firms. *Pure Mathematics and Application, 16*(4), 365–375.

Chiarella, C., & Szidarovszky, F. (2008a). Dynamic oligopolies with production adjustment costs. *Scientia Iranica, 15*(1), 120–124.

Chiarella, C., & Szidarovszky, F. (2008b). Discrete dynamic oligopolies with intertemporal demand interactions. *Mathematica Pannonica, 19*(1), 107–115.

Cima, A., Gasull, A., & Mañosas, F. (1999). The discrete Markus-Yamabe problem. *Nonlinear Analysis, 35*, 343–354.

Cima, A., van der Essen, A., Gasull, A., Hubbers, E., & Mañosas, F. (1997). A polynomial counterexample to the Markus-Yamabe conjecture. *Advances in Mathematics, 131*, 453–457.

Clark, C. (1990). *Mathematical Bioeconomics* (2nd ed.). New York: Wiley.

Clayton, M. J., & Jorgensen, B. (2005). Optimal cross holdings with externalities and strategic interactions. *Journal of Business, 78*(4), 1505–1522.

Collet, P., & Eckmann, J. P. (1980). *Iterated maps on the interval as dynamical systems*. Boston, MA: Birkhauser.

Cooper, L. G., & Nakanishi, M. (1988). *Market-share analysis*, Dordrecht: Kluwer.

Cournot, A. (1838). *Recherches sur les Principes Mathématiques de la Théorie des Richessess*, Hachette, Paris. (English translation (1960): *Researches into the Mathematical Principles of the Theory of Wealth*, Kelley, New York).

Cox, J., & Walker, M (1998). Learning to play Cournot duopoly strategies. *Journal of Economic Behavior and Organization, 36*, 141–161.

Cushing, J. (1977). *Integro-differential equations and delay models in population dynamics*. Berlin: Springer.

Cyert, R. M., & DeGroot, M. H. (1971). Interfirm learning and the kinked demand curve. *Journal of Economic Theory, 3*(1), 272–287.

Cyert, R. M., & DeGroot, M. H. (1973). An analysis of cooperation and learning in a duopoly context. *American Economic Review, 63*(1), 24–37.

Cyert, R. M., & DeGroot, M. H. (1987). *Bayesian analysis and uncertainty in economic theory*. Lanham, MD: Rowman and Littlefield.

Dana, R.-A., & Montrucchio, L. (1986). Dynamic complexity in duopoly games. *Journal of Economic Theory, 40*, 40–56.

Dana, R.-A., & Montrucchio, L. (1987). On rational dynamic strategies in infinite horizon models where agents discount the future. *Journal of Economic Behavior and Organization, 8*, 497–511.

Devaney, R. (1989). *An introduction to chaotic dynamical systems*. Menlo Park, CA: The Benjamin/Cummings.

Di Bernardo, M., Feigen, M., Hogan, S., & Homer, M. (1999). Local analysis of c-bifurcations in *n*-dimensional piecewise smooth dynamical systems. *Chaos, Solitons & Fractals, 10*(11), 1881–1908.

Dixit, A. (1986). Comparative statics of oligopoly. *International Econonomic Review, 27*, 107–121.

Dockner, E., Jorgensen, S., Long, N., & Sorger, G. (2000). *Differential games in economics and management science*. Cambridge University Press, Cambridge, UK.

Fernandes, A., Gutierrez, C., & Rabanal, R. (2004). Global asymptotic stability for differentiable vector fields of $\mathbb{R}^2$. *Journal of Differential Equations* , *206*(2), 470–482.

Fisher, F. (1961). The stability of the Cournot oligopoly solution: The effect of speeds of adjustment and increasing marginal costs. *Review of Economic Studies*, *28*, 125–135.

Flam, S. (1993). Oligopolistic competition: From stability to chaos. *In* L. F.Gori and M.Galeotti (Eds.), *Nonlinear dynamics in economics and social sciences. Lecture notes in economics and mathematical systems* (Vol. 399). Berlin: Springer.

Forgo, F., Szep, J., & Szidarovszky, F. (1999). *Introduction to the theory of games*. Dordrecht: Kluwer.

Frank, C. (1965). Entry in a Cournot market. *Review of Economic Studies*, *32*, 245–250.

Friedman, J. W. (1977). *Oligopoly and the theory of games*. Amsterdam: North Holland.

Friedman, J. W. (1982). Oligopoly theory. *In* K. Arrow and M. Intriligator (Eds.), *Handbook of mathematical economics* (Vol. II, pp. 491–534). Amsterdam: North-Holland.

Friedman, L. (1958). Game-theory models in the allocation of advertising expenditures. *Operations Research*, *6*, 699–709.

Fudenberg, D., & Levine, D. (1998). *The theory of learning in games*. Cambridge, MA: MIT.

Furth, D. (1986). Stability and instability in oligopoly. *Journal of Economic Theory*, *40*, 197–228.

Furth, D. (2009). Anything goes with heterogeneous but not always with homogeneous oligopoly. *Journal of Economic Dynamics and Control*, *33*, 183–203.

Gates, D. J., Rickard, J. A., & Westcott, M. (1982). Exact cooperative solutions of a duopoly model without cooperation. *Journal of Mathematical Economics*, *9*, 27–35.

Gordon, H. (1954). The economic theory of a common property resource: The fishery. *Journal of Political Economy* , *62*, 124–142.

Grebogi, C., Ott, E., & Yorke, J. (1983). Crises, sudden changes in chaotic attractors and transient chaos. *Physica D*, *7*(1–3), 181–200.

Guckenheimer, J., & Holmes, P. (1983). *Nonlinear oscillations, dynamical systems and bifurcations of vector fields*. Berlin: Springer.

Gumowski, I., & Mira, C. (1980). *Dynamique chaotique*. Toulose: Cepadues Editions.

Gutierrez, C. (1995). A solution to the bidimensional global asymptotic stability conjecture. *Annales de l'Institut Henri Poincaré (C) Analyse Non Linéaire*, *12*(6), 627–671.

Hahn, F. (1962). The stability of the Cournot solution. *Journal of Economic Studies*, *29*, 329–331.

Hall, T., Szidarovszky, F., & Zhao, J. (2004). Some notes on a dynamic model of international fishing. *Pure Mathematics and Applications*, *15*(1), 45–54.

Halse, C., Homer, M., & di Bernardo, M. (2003). C-bifurcations and period-adding in one-dimensional piecewise-smooth maps. *Chaos, Solitons & Fractals*, *18*, 953–976.

Hanssens, D. M., Parsons, L. J., & Schultz, R. L. (1990). *Market response models: Econometric and time series analysis*. Dordrecht: Kluwer.

Hill, M., & Waterson, M. (1983). Labor-managed Cournot oligopoly and industry output. *Journal of Comparative Economics*, *7*, 43–51.

Hirsch, M., Pugh, C., & Shub, M. (1977). *Invariant manifolds. In Lecture Notes in Mathematics* (Vol. 583). Berlin: Springer.

Hommes, C. (1991). Chaotic dynamics in economic models: Some simple case studies. *Thesis University of Groningen*. Groningen: Wolters-Noordhoff.

Hommes, C. (1995). A reconsideration of Hick's non-linear trade cycle model. *Structural Change and Economic Dynamics*, *6*, 435–459.

Hommes, C., & Nusse, E. (1991). 'Period three to period two bifurcations for piecewise linear models. *Journal of Economics*, *54*(2), 157–169.

Hommes, C., Nusse, E., & Simonovits, A. (1995). Cycles and chaos in a socialist economy. *Journal of Economic Dynamic and Control*, *19*, 155–179.

Howrey, E., & Quandt, R. (1968). The dynamics of the number of firms in an industry. *Review of Economic Studies*, *35*, 349–353.

Howroyd, T. D., & Russell, A. M. (1984). Cournot oligopoly models with time delays. *Journal of Mathematical Economics*, *13*(2), 97–103.
Howroyd, T., & Rickard, J. (1981). Cournot oligopoly and output adjustment costs. *Economics Letters*, *7*(2), 113–117.
Invernizzi, S., & Medio, A. (1991). On lags and chaos in economic dynamic models. *Journal of Mathematical Economics*, *20*, 521–550.
Jackson, E. (1989). *Perspectives of nonlinear dynamics* (Vol. 1 & 2). Cambridge University Press, Cambridge, UK.
Karnani, A. (1985). Strategic implications of market share attraction models. *Management Science*, *31*, 536–547.
Kirman, A. (1975). Learning by firms about demand conditions. *In* R. H. Day & T. Groves (Eds.), *Adaptive economic models* ( pp. 137–156) . New York: Academic.
Kirman, A. (1983). On mistaken belief and resultant equilibria. *In* R. Frydman & E. S. Phelps (Eds.), *Individual forecasting and aggregate outcomes* (pp. 147–166). New York: Cambridge University Press.
Kirman, A. (1995). Learning in oligopoly: Theory, simulation, and experimental evidence. *In* A. Kirman & M. Salmon (Eds.), *Learning and rationality in economics* (pp. 127–178). Oxford, UK: Blackwell.
Kirman, A., & Salmon, M. (1995). *Learning and rationality in economics*. Oxford, UK: Blackwell.
Kopel, M. (1996). Simple and complex adjustment dynamics in Cournot duopoly models. *Chaos, Solitons & Fractals 7(12)*, 2031–2048.
Kopel, M. (2009). Oligopoly dynamics. *in* J. Barkley-Rosser Jr. (Ed.), *Handbook of research on complexity*. Northampton, MA: Edward Elgar. Chapter 6.
Kopel, M., & Szidarovszky, F. (2006). Resource dynamics under partial cooperation in an oligopoly. *Journal of Optimization Theory*, *128*(2), 393–410.
Kubıcek, M., & Marek, M. (1986). *Computational methods in bifurcation theory and dissipative structures*. Berlin: Springer.
Léonard, D., & Nishimura, K. (1999). Nonlinear dynamic in the Cournot model without full information. *Annals of Operations Resesarch*, *89*, 165–173.
Li, W., Rychlik, M., Szidarovszky, F., & Chiarella, C. (2003). On the attractivity of a class of homogeneous dynamic economic systems. *Nonlinear Analysis*, *52*, 1617–1636.
Li, W., & Szidarovszky, F. (1999*a*). An elementary result in the stability theory of time invariant nonlinear dynamical systems. *Applied Mathematics and Computation*, *102*, 35–49.
Li, W., & Szidarovszky, F. (1999*b*). The stability of Nash-Cournot equilibria in labor-managed oligopolies. *Southwest Journal of Pure and Applied Mathematics*, *10*, 1–22.
Lorenz, H. W. (1995). *Nonlinear dynamical economics and chaotic motion* (2nd ed.). Berlin: Springer.
Mailath, G. (1998). Do people play Nash equilibrium? Lessons from evolutionary game theory. *Journal of Economic Literature*, *36*, 1347–1374.
Maistrenko, Y., Maistrenko, V., & Chua, L. (1993). Cycles of chaotic intervals in a time-delayed chuas circuit. *International Journal of Bifurcation and Chaos*, *3*(6), 1557–1572.
Maistrenko, Y., Maistrenko, V., & Vikul, S. (1998). On period-adding sequences of attracting cycles in piecewise linear maps. *Chaos, Solitons & Fractals*, *9*(1), 67–75.
Maistrenko, Y., Maistrenko, V., Vikul, S., & Chua, L. (1995). Bifurcations of attracting cycles from time-delayed chuas circuit. *International Journal of Bifurcation and Chaos*, *5*(3), 653–671.
Marimon, R. (1997). Learning from learning in economics. *In* D. Kreps & K. Wallis (Eds.), *Advances in economics and econometrics* (pp. 278–315). Cambridge, UK: Cambridge University Press.
Matsumoto, A., & Serizawa, N. (2007), Strategic trade policy under isoelastic demand and asymmetric production costs. *The Annals of Regional Science*, *41*(3), 525–543.
McManus, M., & Quandt, R. (1961). Comments on the stability of the Cournot oligopoly model. *Review of Economic Studies*, *27*, 136–139.
Medio, A., & Lines, M. (2001). *Nonlinear dynamics: A primer*. Cambridge, UK: Cambridge University Press.

Milgrom, P., & Roberts, J. (1992). *Economics, organization and management.* Englewood Cliffs, NJ: Prentice-Hall.

Miller, R. K. (1972). Asymptotic stability and peturbations for linear Volterra integrodifferential systems. *In* K. Schmitt (ed.), *Delay and functional differential equations and their applications.* New York: Academic.

Mills, H. D. (1961). A study in promotional competition. *In* F. Bass, R. D. Buzzell, M. R. Greene, W. Lazer, E. A. Pessemier, D. L. Shawver, A. Shuchman, C. A. Theodore, & G. W. Wilson (Eds.), *Mathematical models and methods in marketing* (pp. 271–301). Homewood, IL: R.D. Irwin.

Mira, C. (1987). *Chaotic dynamics.* Singapore: World Scientific.

Mira, C., Fournier-Prunaret, D., Gardini, L., Kawakami, H., & Cathala, J. (1994). Basin bifurcations of two dimensional noninvertible maps: Fractalization of basins. *International Journal of Bifurcations and Chaos, 4,* 343–381.

Mira, C., Gardini, L., Barugola, A., & Cathala, J. (1996). *Chaotic dynamics in two-dimensional noninvertible maps.* Nonlinear Sciences, Series A. Singapore: World Scientific.

Mira, C., & Rauzy, C. (1995). Fractal aggregation of basin islands in two dimensional quadratic noninvertible maps. *International Journal of Bifurcations and Chaos, 5,* 991–1019.

Monahan, G. E. (1987). The structure of equilibria in market share attraction models. *Management Science, 33,* 228–243.

Munro, C., & Scott, A. D. (1985). The economics of fisheries management. *In* A. V. Kneese & J. Sweeney, (Eds.), *Handbook of natural resource and energy economics* (Vol. II, pp. 623–676). Amsterdam: North Holland.

Neary, H. M. (1984). Labor-managed Cournot oligopoly and industry output: A comment. *Journal of Comparative Economics, 8,* 332–327.

Negishi, T. (1961). Monopolistic competition and general equilibrium. *Review of Economic Studies, 27,* 136–139.

Nusse, H., & Yorke, J. (1992). Border-collision bifurcations including period two to period three for piecewise smooth systems. *Physica D, 57,* 39–57.

Nusse, H., & Yorke, J. (1995). Border-collision bifurcation for piecewise smooth one-dimensional maps. *International Journal of Bifurcation and Chaos, 5*(1), 189–207.

Okuguchi, K. (1964). The stability of the Cournot oligopoly solution: A further generalization. *Journal of Economic Studies, 31,* 143–146.

Okuguchi, K. (1970). Adaptive expectations in an oligopoly model. *Review of Economic Studies, 37,* 233–237.

Okuguchi, K. (1972). Quasi-competitiveness and Cournot oligopoly. *Review of Economic Studies, 40,* 145–148.

Okuguchi, K. (1976). *Expectations and stability in oligopoly models.* Berlin: Springer.

Okuguchi, K. (1993). Cournot oligopoly with product-maximizing and labor-managed firms. *Keio Economic Studies, 30,* 27–38.

Okuguchi, K. (1995). Decreasing returns and existence of Nash equilibrium in rent-seeking games. Mimeo, Dept. of Economics, Nanzan University, Nagoya, Japan.

Okuguchi, K. (1996). Existence of equilibrium for labor-managed Cournot oligopoly. *Nanzan Journal of Economic Studies, XI*(2), 111–119.

Okuguchi, K. (1998). Long-run fish stock and imperfetly competitive commercial fishing. *Keio Economic Studies, 35,* 9–17.

Okuguchi, K. (2003). Dynamic and comparative static analysis of imperfectly competitive international fishery. *Journal of Economics, 80*(3), 249–265.

Okuguchi, K., & Szidarovszky, F. (1999). *The theory of oligopoly with multi-product firms* (2nd ed.). Berlin: Springer.

Okuguchi, K., & Szidarovszky, F. (2003). Oligopoly with intertemporal demand interaction. *Journal of Economic Research, 8*(1), 55–61.

Ortega, J. M., & Rheinholdt, W. C. (1970). *Iterative solution of nonlinear equations with several variables.* New York: Academic.

Parthasarathy, T. (1983). *On global univalence theorems, Lecture Notes in Mathematics* (Vol. 977). New York: Springer.

Pérez-Castrillo, J., & Verdier, T. (1992). A general analysis of rent-seeking games. *Public Choice, 73*, 335–350.

Puhakka, M., & Wissink, J. (1995). Strategic complementarity, multiple equilibria and externalities in Cournot competition. Center for analytic economics working papers #95-18. New York: Cornell University.

Puu, T. (1991). Chaos in duopoly pricing. *Chaos, Solitons & Fractals, 1*(6), 573–581.

Puu, T. (2003). *Attractors, bifurcations, & chaos: Nonlinear phenomena in economics* (2nd ed.). Berlin: Springer.

Puu, T., Gardini, L., & Sushko, I. (2005). A Hicksian multiplier-accelerator model with floor determined by capital stock. *Journal of Economic Behavior and Organization, 56*, 331–348.

Puu, T., & Sushko, I. (2002). *Oligopoly dynamics, models and tools*. New York: Springer.

Puu, T., & Sushko, I. (2006). *Business cycle dynamics, models and tools*. New York: Springer.

Rand, D. (1978). Exotic phenomena in games and duopoly models. *Journal of Mathematical Economics, 5*, 173–184.

Reynolds, R., & Snapp, B. (1986). The competitive effects of partial equity interests and joint ventures. *International Journal of Industrial Organization, 4*, 141–153.

Ruffin, R. (1971). Cournot oligopoly and competitive behaviour. *Review of Economic Studies, 38*, 493–502.

Russell, D., DuBois, D. F., & Rose, H. A. (1986). Collapsing-caviton turbulence in one dimension. *Physical Review Letters, 56*(8), 838–841.

Sandal, L., & Steinshamn, S. I. (2004). Dynamic Cournot-competitive harvesting of a common pool resource. *Journal of Economic Dynamics and Control, 28*, 1781–1799.

Schaefer, M. B. (1957). Some consideration of population dynamics and economics in relation to the management of marine fisheries. *Journal of the Fisheries Research Board of Canada, 14*, 635–647.

Schmalensee, R. (1976). A model of promotional competition in oligopoly. *Review of Economic Studies, 43*, 493–507.

Schoonbeek, L. (1997). A dynamic Stackelberg model with production-adjustment costs. *Journal of Economics, 66*(3), 271–282.

Seade, J. (1980). On the effects of entry. *Econometrica, 48*, 479–489.

Shinkel, M. P., Tuinstra, J., & Vermuelen, D. (2002). Convergence of Bayesian learning to equilibrium in mis-specified models. *Journal of Mathematical Economics, 38*(4), 483–508.

Silvestre, J. (1977). A model of general equilibrium with monopolistic behavior. *Journal of Economic Theory , 16*, 425–442.

Sushko, I., Agliari, A., & Gardini, L. (2005). Bistability and border-collision bifurcations for a family of unimodal piecewise smooth maps. *Discrete and Continuous Dynamical Systems, Series B, 5*(3), 881–897.

Sushko, I., Agliari, A., & Gardini, L. (2006). Bifurcation structure of parameter plane for a family of unimodal piecewise smooth maps: Border-collision bifurcation curves. *Chaos, Solitons & Fractals, 29*(3), 756–770.

Sushko, I., Gardini, L., & Puu, T. (2003). The Hicksian floor-roof model for two regions linked by interregional trade. *Chaos, Solitons & Fractals, 18*, 593–612.

Szidarovszky, F. (1990). Multiproduct Cournot oligopolies with market saturation. *Pure Mathematics and Applications, 1*(1B), 3–15.

Szidarovszky, F. (1999). Adaptive expectations in discrete dynamic oligopolies with production adjustment costs. *Pure Mathmatics and Application, 10*(2), 133–139.

Szidarovszky, F. (2003). Learning in dynamic oligopolies. *In International Conference on Complex Oligopolies*, presented on May 18–21, Odense, Denmark.

Szidarovszky, F. (2004). Global stability analysis of a special learning process in dynamic oligopolies. *Journal of Economic and Social Research, 9*, 175–190.

Szidarovszky, F., & Bahill, T. (1998). *Linear systems theory* (2nd ed.). Boca Raton: CRC

Szidarovszky, F., Hu, Z., & Zhao, J. (2004). Dynamic oligopolies with market saturation. *In International Conference on Economic Dynamics*, presented on September 16–18, Urbino, Italy.

Szidarovszky, F., Hu, Z., & Zhao, J. (2006). Dynamic oligopolies with market saturation. *Chaos, Solitons and Fractals, 29*, 723–738.

Szidarovszky, F., Ilieva, V., & Okuguchi, K. (2002). Entry and merger in commercial fishing with multiple markets. *Journal of Economics, 76*, 247–259.

Szidarovszky, F., & Krawczyk, J. (2005). On stable learning in dynamic oligopolies. *Pure Mathematics and Applications, 4*, 45–54.

Szidarovszky, F., & Okuguchi, K. (1988). Perceived marginal costs in a oligopoly model. *In 3rd Congress of the European Economic Association*, presented on August 27–29, Bologna, Italy.

Szidarovszky, F., & Okuguchi, K. (1990). Dynamic oligopoly: Models with incomplete information. *Applied Mathematics and Computation, 38*(2), 161–177.

Szidarovszky, F., & Okuguchi, K. (1997). On the existence and uniqueness of pure Nash equilibrium in rent-seeking games. *Games and Economic Behavior, 18*, 135–140.

Szidarovszky, F., & Okuguchi, K. (1998). An oligopoly model of commercial fishing. *Seoul Journal of Economics, 11*, 321–330.

Szidarovszky, F., & Okuguchi, K. (2000). A dynamic model of international fishing. *Seoul Journal of Economics, 13*, 471–476.

Szidarovszky, F., Okuguchi, K., & Kopel, M. (2005). International fishery with several countries. *Pure Mathematics and Applications, 4*, 493–514.

Szidarovszky, F., Smith, V., & Rassenti, S. (2009). Cournot models: Dynamics, uncertainty and learning. CUBO, A Mathematical Journal, 11(2), 57–88.

Szidarovszky, F., & Yakowitz, S. (1977). A new proof of the existence and uniqueness of the Cournot equilibrium. *International Economic Review, 18*, 787–789.

Szidarovszky, F., & Yakowitz, S. (1978). *Principles and procedure of numerical analysis*. New York: Plenum.

Szidarovszky, F., & Yen, J. (1995). Dynamic Cournot oligopolies with production adjustment costs. *Journal of Mathematical Economics, 24*, 95–101.

Szidarovszky, F., & Zhao, J. (2004). Dynamic oligopolies with market saturation. *In International Conference of the Dynamic Games Society*, presented on December, 18–21, Tucson, Arizona.

Szidarovszky, F., & Zhao, J. (2006). Dynamics oligopolies with intertemporal demand interaction. *Computers and Mathematics with Applications, 52*(12), 1623–1626.

Theocharis, R. D. (1960). On the stability of the Cournot solution to the oligopoly problem. *Review of Economic Studies, 27(2)*, 133–134.

Tietenberg, T., & Folmer, H. (1998, 1999). *The international yearbook of environmental and resource economics 1998/1999: A survery of current issues*, Cheltenham, UK: Edward Elgar.

Tirole, J. (1988). *The theory of industrial organization*. Cambridge, MA: MIT.

Tuinstra, J. (2004). A price adjustment process in a model of monopolistic competition. *International Game Theory Review. 6*, 417–442.

Tullock, G. (1980). Efficient rent-seeking. *In* M. Buchanan, R. D. Tollison, & G. Tullock (Eds.), *Toward a theory of the rent-seeking society*. College Station: Texas A & M.

Van Huyck, J., & Battalio, R. (1998). *Coordination failure in market statistic games*. Laser-script from http://econlab10.tamu.edu/JVH_gtee/.

Van Huyck, J., Cook, J., & Battalio, R. (1984). Selection dynamics, asymptotic stability, and adaptive behavior. *Journal of Political Economy, 102*, 975–1005.

Van Huyck, J., Cook, J., & Battalio, R. (1997). Adaptive behavior and coordination failure. *Journal of Economic Behavior and Organization, 32*, 483–503.

Vives, X. (1999). *Oligopoly pricing : Old ideas and new tools*. Cambridge, MA: MIT.

Volterra, V. (1931). *Leçons sur la Théorie Mathématique de la Lutte pour la Vie*. Paris: Gauthiers-Villars.

Ward, B. (1958). The firm in Illyria: Market syndicalism. *American Economic Review, 48*, 566–589.

Witteloostuijn, A. V., & Lier, A. V. (1990). Chaotic patterns in Cournot competition. *Metroeconomica*, *41*, 161–185.

Xu, L., & Szidarovszky, F. (1999). The stability of dynamic rent-seeking games. *International Game Theory Review*, *1*(1), 87–102.

Zhanybai, T., Z., & Mosekilde, E. (2003). *Bifurcations and chaos in piecewise-smooth dynamical systems*. Singapore: World Scientific.

Zhao, J., & Szidarovszky, F. (2008). N-firm oligopolies with production adjustment costs: Best responses and equilibrium. *Journal of Economic Behavior and Organization*, *68*, 87–99.

Zhusubaliyev, T., Z., Soukhoterin, E., & Mosekilde, E. (2002). Border-collision bifurcations on a two-dimensional torus. *Chaos Solitons & Fractals*, *13*, 1889–1915.

Zhusubaliyev, Z., Soukhoterin, E., & Mosekilde, E. (2007). Quasiperiodicity and torus breakdown in a power electronic dc/dc converter. *Mathematics and Computers in Simulation* , *73*, 364–377.

# Index

LaVergne, TN USA
24 November 2009

165145LV00005B/45/P

9 783642 021053